W9-BUT-468

For All Mankind

ALSO BY HARRY HURT III

Texas Rich

FOR ALL MANKIND

HARRY HURT III

INTERVIEWS BY AL REINERT

A MORGAN ENTREKIN BOOK
THE ATLANTIC MONTHLY PRESS
NEW YORK

Published simultaneously in Canada
Printed in the United States of America

FIRST PAPERBACK EDITION

Library of Congress Cataloging-in-Publication Data
Hurt, Harry.
For all mankind.
"A Morgan Entrekin book."
1. Project Apollo. 2. Space flight to the moon.
I. Title.
TL789.8.U6A5419 1988 629.45'4 88-22324
ISBN 0-87113-170-6 (hc)
ISBN 0-87113-351-2 (pb)

The Atlantic Monthly Press
19 Union Square West
New York, NY 10003

FIRST PAPERBACK PRINTING

To the astronauts and cosmonauts who have given their lives in the human exploration of space . . .

Valentin Bonadarenko — March 23, 1961

Virgil I. Grissom
Edward H. White II
Roger B. Chaffee
Apollo 1
January 27, 1967

Vladimir Komarov
Soyuz 1
April 24, 1967

Georgy Dobrovolsky
Vladislav Volkov
Viktor Patsayev
Soyuz 11
June 30, 1971

Francis Scobee
Michael Smith
Judith Resnik
Ronald McNair
Ellison Onizuka
Gregory Jarvis
Christa McAuliffe
Challenger
January 28, 1986

TABLE OF CONTENTS

STAGE FOUR: 2001 AND BEYOND

PROLOGUE: FOR ALL MANKIND

We choose to go to the moon in this decade and do the other things, not because they are easy, but because they are hard, because there is new knowledge to be gained and new rights to be won, and they must be won for the progress of all mankind . . .

We shall send to the moon, more than 240,000 miles from the control center in Houston, a giant rocket more than 300 feet tall, made of new metal alloys, some of which have not yet been invented, capable of standing heat and stresses several times more than have ever been experienced, fitted together with a precision better than the finest watch, carrying all the equipment needed for propulsion, guidance, control, communications, food, and survival, on an untried mission to an unknown celestial body . . .

And therefore, as we set sail, we ask God's blessing on the most hazardous and dangerous and greatest adventure on which man has ever embarked . . .

PRESIDENT JOHN F. KENNEDY
September 12, 1962

Between December of 1968 and December of 1972, twenty-four men left the earth and flew to the moon. Twelve of those men landed on the moon and walked on the lunar surface. Man's first lunar landing occurred at 3:18 P.M. Houston time on the afternoon of July 20, 1969, when Apollo 11 astronauts Neil A. Armstrong and Edwin E. "Buzz" Aldrin, Jr., touched down in the Sea of Tranquility. At 9:56 that night, Armstrong hopped off the ladder of the lunar module *Eagle,* steadied his feet in the surrounding moon dust, and uttered the now famous sentence:

"That's one small step for man, one giant leap for mankind."

Back on Earth, where an estimated 600 million people (roughly one-fifth of the world's total population at the time) watched Armstrong's "giant leap" on live TV, there was a strange mixture of joy and disappointment, elation and outright disapproval. President Richard M. Nixon later proclaimed this to be "the greatest week in history since Creation." But in light of the ongoing war in Vietnam and erupting urban social problems, a highly vocal minority of poor, black, and disaffected young white middle-class Americans branded the expenditure of $40 billion to land two men on the moon a racist, sexist, and elitist waste of tax money.

After the Apollo 12 astronauts made man's second lunar landing four months later, even the most patriotic members of America's "silent majority" began to question the necessity of more moon shots. The mass media created the mistaken impression that Project Apollo was merely a repetitious public relations exercise, a technological boondoggle demanding little or no human heroics. That misperception was only temporarily challenged by the catastrophic explosions on Apollo 13 that nearly claimed the lives of the three astronauts on board. Once the Apollo 14 landing got the program back on track, the prevailing popular attitude toward subsequent lunar missions became, in the words of a widely reprinted photo caption, "So what?"

After the Apollo 17 mission in December of 1972, the National Aeronautics and Space Administration (NASA) terminated Project Apollo before accomplishing all of the program's ambitious lunar exploration objectives. Space agency officials had originally planned eight to twelve moon landing missions. Instead, there were only six, not counting the ill-fated Apollo 13, which circled the moon but could not land on the lunar surface.

Under pressure to give the American public a tangible return on their tax dollars, NASA embarked on a controversial new program based on the reusable space shuttle. Henceforth, the space agency paid only lip service to the noble theme etched on the plaque the Apollo 11 astronauts left on the moon: "We came in peace for all mankind." The first series of shuttle flights pioneered the commercialization and militarization of space, forsaking manned exploration of the solar system to concentrate on the pursuit of profits and the development of a Strategic Defense Initiative (SDI), also known as "Star Wars."

The tragic explosion of the space shuttle *Challenger* on January 28, 1986, has since prompted a long overdue reexamination of the goals and operations of the U.S. manned space program. The loss of the seven *Challenger* astronauts, who included a civilian grade school teacher, reminded all the world of the true dangers of what had come to be taken for granted as a routine exercise safer and more tedious than a cross-country automobile trip. Subsequent revelations of alleged malfeasance by NASA and the shuttle's private contractors have led critics to demand that future space ventures be modeled after the visionary—and successful—approach pioneered by Project Apollo.

Nevertheless, most of the world public still fails to recognize that Project Apollo was, just as President Kennedy had predicted it would be, "the most hazardous and dangerous and greatest adventure on which man has ever embarked." Future historians will probably regard the first lunar landings as the most significant accomplishments of the twentieth century. Although there have been numerous revolutionary breakthroughs in medicine, agriculture, manufacturing, communications, and transportation in previous eras, this is the first century in which human beings have left the earth and traveled to another planet. Neil Armstrong's "giant leap" ended man's cosmic isolation from the rest of the universe, and began a voyage into the unknown that will undoubtedly prove to be far more extraordinary than any science fiction fantasy.

Although live television coverage of the Apollo missions gave people back on Earth a rare opportunity to witness history in the making, the uniquely human drama of man's greatest adventure seldom if ever came across on the tube. The fault lay partly with the medium. The disjointed and abbreviated small screen news footage aired by the major networks, most of which was in black and white and blurred by transmission interference, simply could not communicate the personal sensations and wonder of exploring another planet for the first time. The messengers were also to blame. Most of the astronauts were by nature and training nonverbal men, dedicated overachievers who let their actions speak for themselves. Fixed in the public eye, they felt defensive and inarticulate, and so fell back on platitudes when asked to relate their exploits for the mass media.

For All Mankind is the true story of man's first moon landings, told through the actual words and deeds of the Apollo astronauts. It is not intended to be a definitive history of NASA or another exposé of the personal lives of the all-American boys with "the right stuff." This is the real stuff—the first complete and authoritative firsthand account of the "greatest adventure on which man has ever embarked."

Project Apollo is often compared to Christopher Columbus's daring voyage to the New World. But where Columbus and his crew kept only handwritten and deliberately exaggerated logs of their expeditions, man's first missions to the moon were the most extensively documented undertakings of all time. The NASA archives contain over 8 million feet of mission films, thousands of still photographs, and hundreds of reels of tape-recorded radio dialogue. In addition to drawing on these vast resources, the chapters that follow incorporate over ninety hours of exclusive taped interviews with the Apollo astronauts. Most of these interviews were conducted between 1980 and 1987, by which time the majority of Apollo alumni had long since retired from the space program and had finally begun to put their great adventure in clear perspective.

The narrative is organized in the form of an archetypal lunar mission. "Stage One: Go for the Moon" describes the three-day voyage from launch to lunar orbit, weaving in the history of man's conquest of the moon. "Stage Two: Men Walk on Moon" re-creates the drama of man's first lunar landing by Apollo 11 astronauts Armstrong and Aldrin, then details the untold triumphs and tribulations of the six subsequent lunar missions. "Stage Three: Return to Earth" begins with the homeward leg of the voyage from lunar lift-off to splashdown. It then reports on what has happened to the Apollo astronauts since their return to Earth, and describes how the tangible and intangible repercussions of the first lunar landings have changed their lives and the lives of everyone else on the planet. "Stage Four: 2001 and Beyond" discusses mankind's future in space, including current proposals to fly to Mars and/or return to the moon.

For All Mankind aims to answer the basic questions that all the news footage and documentaries of the Apollo lunar missions failed to answer: What was it like to be the first human beings to leave the earth and land on the moon? What did it mean—and what will it mean—for them and for all mankind?

STAGE 1:

GO FOR THE MOON

Astronomy compels the soul to look upward,
and leads us from this world to another.

PLATO
The Republic (342 B.C.)

1

THIS TIME IS FOR REAL

Cape Kennedy

Launch Day

Between the idea
And the reality
Between the motion
And the act
Falls the shadow.
T.S. ELIOT
The Hollow Men
(1925)

All twenty-four men who flew to the moon left Earth on the same day. This particular day happened to fall on nine different calendar dates between 1968 and 1972. But for each and every trio of Apollo astronauts, it always dawned in the same way. For this was the day when all the make-believe ended. This was the day when science would surpass science fiction. This was the day when they would attempt to make mankind's "impossible dream" come true.

This was Launch Day.

All the Apollo astronauts knew the officially scheduled dates of their Launch Day well in advance. But whenever that day actually arrived, the three prime crewmen in question could never quite believe it. At least not at first. For as Apollo 14's Ed Mitchell observes, up to the point of lift-off, all their training and preparations seemed "like a big game." Going to the moon was an undertaking so alien to human experience that the only way to prepare for it was through make-believe. And yet, the real thing was such an overwhelming experience that many astronauts consciously pretended it was just another simulation.

"The whole thing almost has a touch of unreality about it," recalls Apollo 15's Jim Irwin. "We realized that we were not practicing anymore, that this was it and everything was for real. But it seemed, in a way, to continue to be a simulation . . . You've rehearsed it and practiced it hundreds of times, and you really wonder, 'Is this another simulation or is this for real?' "

The wake-up calls came at nine different hours of the morning, afternoon, and night on all seven days of the week in all four seasons of the year. The timing had nothing to do with whether the sun happened to be rising or setting or shining on the other side of the planet. And it had nothing to do with whether the sky above the Cape happened to be clear and blue or cloudy and gray.

High winds, a heavy cloud cover, or a lightning storm could disrupt and indefinitely delay the launch. So could a sudden outbreak of solar flares. But the original countdown schedule for each and every Apollo mission was not determined by the sun or by the local weather forecast. Rather, and far more appropriately, it was dictated by the moon and the opening and closing of its invisible "launch windows."

A launch window was the brief period during each lunar month when the Apollo crewmen could lift off from the Cape on their three-day voyage to the moon and expect to arrive at their assigned destination when surface conditions were suitable for exploration. The specific date and duration of the launch window depended on the location of the landing site and the advent of the appropriate phase in the moon's orbital cycle.

"The way you timed your landing on the moon was usually keyed to the lighting at that particular spot," says Apollo 12 lunar module pilot Alan Bean. "You didn't want to land in the total darkness of the lunar night because it would be too cold and you couldn't see. But you didn't want to land in the direct sunlight of the lunar day because the surface temperature would be too hot. Ideally, you wanted to land when the line between day and night was just to the west of your landing site. You wanted to have the sun behind you, but it had to be at less than a ten-degree angle to the surface of the moon so that you could see the shadows thrown off by the craters and the boulders."

The Apollo 11 launch window happened to open early on the morning of July 16, 1969, when the waxing moon was nearly half full, because the astronauts were bound for a landing site on the lunar equator smack in the middle of the moon face visible from Earth. Those who landed farther west required fuller moons; those who landed farther east needed newer moons. Apollo 15, which aimed for the north central facial quadrant, launched early on the morning of July 26, 1971, during a crescent phase. Apollo 17 departed for a far northeastern landing site after midnight on December 7, 1972, when the moon was filling out her fourth quarter.

In most cases, a launch window opened for no more than a few hours at a time on no more than two consecutive days of a given lunar month, and did not reopen again for at least another twenty-eight days. And given the fact that postponing a mission for a single month could cost over $30 million, it was an appointment neither NASA nor the Apollo crewmen in question could afford to miss.

The three prime crewmen on each Apollo mission and their three backups always spent the night before Launch Day in the Manned Spacecraft Operations Building (MSOB—see "Glossary of Acronyms" at rear of book for explanation of all acronyms), strategically isolated a few miles from the Cape. Though the astronauts' dormitory on the second floor was considerably more spartan than the motels in nearby Cocoa Beach, it featured individually decorated private bedrooms, a gymnasium, a mess hall, and a sanctum of strictly enforced quiet accessible only to authorized personnel, a category that did not include family or friends.

During the preceding weeks, the astronauts gamely tried to revise or reverse their normal working and sleeping schedules to conform to the launch schedule. In theory, they hoped to synchronize their natural circadian rhythms with the arbitrary beat of the countdown clock so that they would wake up refreshed and ready to go when Donald K. "Deke" Slayton, the former astronaut who was director of flight crew operations, sounded the official prelaunch reveille.

In practice, the crew of Apollo 12 found it almost impossible to get in the proper circadian groove prior to embarking on man's second moon landing mission. They were slated to launch from the Cape shortly before noon EST on November 14, 1969, but due to the westerly location of their landing site, they would have to begin their descent to the lunar surface shortly after midnight on November 18. Accustomed to the early to bed and early to rise routine of professional test pilots, they planned to reorient themselves to the diametrically opposite cycles of the afternoon-to-graveyard shift by staying awake for over twenty consecutive hours after lift-off. According to Bean, the prospect of having to endure this unusually grueling first day schedule caused him anticipated prelaunch insomnia. "I remember waking up and then going back to sleep several times during the night," he reports.

Despite the unprecedented perils of having to attempt man's first lunar landing, the Apollo 11 astronauts enjoyed a much sweeter repose. That was partly because their launch window opened shortly after 8 A.M. at the start of a normal daylight shift; unlike the Apollo 12 trio, they would not have to readjust their circadian en route since their touchdown attempt was scheduled for mid-afternoon. All three crewmen were reportedly "sound asleep" by 10 P.M. EDT the night before launch, and continued to snooze comfortably until their wake-up call at 4:15 A.M. on the morning of July 16, 1969. The ostensibly imperturbable commander, Neil Armstrong, who faced greater performance pressure than any astronaut before or since, was the last of the three to roll out of bed.

But then again, anyone who attended Apollo 11's last preflight press conference could have predicted as much. For when asked to compare the launch to the rest of the mission in terms of relative risks and probability of success, Armstrong coolly replied: "It's one of the phases we have a very high confidence in. It's nothing new. It's the thing that's been done before, and done very well on a number of occasions, and we're quite sure this girl will go."

In spite of Apollo 13's near fatal accident on the way to the moon, most of the astronauts on later lunar missions say they slept even more soundly the night before Launch Day, and for similar reasons. They knew that all sorts of life-threatening hazards lay between them and the moon, but they also knew that every prior mission including unlucky 13 had managed to get off the launchpad and safely into outer space.

"I was surprised," admits Apollo 15's Jim Irwin. "You know, with the anticipation, you think that it might be a fitful sleep. But I realized that Deke Slayton would be awakening us early, and that we best get some rest because it was going to be an important day."

Once the three crewmen showered and shaved, they pulled on sport shirts and slacks, and marched down to the exercise room, where resident nurse Dee O'Hara weighed them in and took their temperatures. Although Nurse O'Hara was a seasoned veteran of both the Mercury and Gemini programs, she was still in her early thirties during the Apollo program, sprightly, and full of genuine good cheer at any time of the day or night.

"You always felt good going in to see her," Bean remembers, "because you knew that she was the kind of woman who would try to give you a break if she possibly could."

After Nurse O'Hara pronounced them fit for space flight, the crewmen would make a beeline to the mess hall. Deke Slayton was always there to greet them along with several high-ranking NASA bureaucrats, and so was a resident artist who sat unobtrusively off to one side sketching their official portraits.

For the past several weeks, the astronauts had been on a "low residue" diet that featured steak at almost every meal to reduce their bodily wastes in space. ("I'm sick of steak," Apollo 11's Armstrong complained to his wife a few days before lift-off.) On Launch Day, they sat down to yet another steak, eggs, toast, juice, and coffee.

Like every other item on the checklist, launch breakfasts were scheduled to be completed within a strict time limit, typically no more than twenty-five minutes. Apollo 11's Mike Collins claims that the earliest launch breakfasts "had an air of studied casualness about them" with the astronauts acting as if they were "bored at the prospect of another empty day." Even so, the unique nature of man's first lunar landing mission prompted the space agency administrator Dr. Thomas O. Paine to give the Apollo 11 crewmen some unique last-minute operating orders.

"[He] told us that concern for our safety must govern all our actions, and that if anything looked wrong we were to abort the mission," Apollo 11 lunar module pilot Buzz Aldrin reports in his post-mission memoir. "He then made a most surprising and unprecedented statement." Paine told the astronauts they did not have to take a do-or-die approach to the landing for fear they would never get another chance if they blew it. If Armstrong and Aldrin had to abort the landing for any reason, they would be allowed to bypass the regular crew rotation system and get first crack at trying again on the next Apollo mission. Paine gave them his word. According to Aldrin, "What he said and how he said it was very reassuring."

In light of Apollo 11's success, the astronauts on the next six lunar landing missions were not offered such last-minute reassurance, nor did they seem to need it. The Apollo 12 crewmen ate their launch breakfast in the company of a stuffed gorilla sent over by a prankster friend of Pete Conrad's. Apollo 16's Charley Duke was almost overly relaxed at his launch breakfast. "I had a hard time keeping my eyes open," he confesses.

Upon completing their prelaunch meal, the three crewmen returned to their dorm rooms to brush their teeth and pick up their PPKs, the "personal preference kits" the trio were allowed to take with them to the moon. Slightly smaller than the average dop kit, each PPK could hold up to eight ounces of material, and so long as the contents weren't flammable, alcoholic, or drug related, NASA officials would routinely approve them.

Since the majority of astronauts were married, Protestant, active military officers, and staunchly patriotic, most PPKs included the same types of items: innocuous good luck charms, family heirlooms, commemorative medallions, flags, armed service emblems, and religious icons.

Apollo 16's Charley Duke lists a typical inventory: "I had some special jewelry that was designed for our flight that I carried to the moon for my wife, my mother, and my mother-in-law; some coins that were minted by a friend of the Apollo 16 mission patch emblem, two medallions minted for the twenty-fifth anniversary of the air force, and an air force flag. I also had some miniature flags of the U.S. and other countries, which were nice because they were so light you could carry about a hundred of them, and they only took up about an ounce of weight."

Several personal preference kits, however, contained various contraband and other controversial items. Despite the prohibition against alcohol, Apollo 11's Buzz Aldrin packed his PPK with a tiny gold chalice and a thimble of red wine. He later reported that NASA officials approved this sacred cargo, but because of atheist Madalyn Murray O'Hair's suit against Apollo 8 for reading the Book of Genesis, he was ordered not to reveal their existence or his intended use for them to the news media.

Aldrin and most of the other astronauts on the early Apollo missions also stuffed their PPKs with caches of coins, first-edition postage stamp covers, and other collectors' items potentially worth millions of dollars. According to Aldrin, the Apollo 11 crewmen also left behind several hundred signed first-edition stamp covers that would "be canceled on the day of lift-off and kept as a sort of insurance policy for our families in the event of a disaster."

Following the Apollo 15 flight, media exposés of these not-so-innocent practices would eventually scandalize the entire astronaut corps either directly or by association. But in the early days, both NASA officials and the astronauts were still blissfully naive. According to Aldrin, "It was of such small import that NASA's official compilation of what I took was not even deemed worthy of being typed up. The confirming list they gave me back was handwritten and complete."

After zipping up their PPKs, the three crewmen usually paused for a few fond farewell glances at photos of families and friends. Then they left their dorm rooms for the last time, and trekked down the hall to the suit room. It was now a little over an hour since their wake-up call, and the threesome was often still a bit bleary, not quite fully comprehending what was already happening. But upon crossing the threshold of the suit room, they were reawakened by a startling scene one Apollo veteran compares to "something right out of Flash Gordon."

White walled, windowless, and surgically sterile, the suit room was an otherwise off-limits area about the size of a large hotel suite that looked like a cross between an intensive care unit, a locker room, a costume parlor, and the laboratory of a mad scientist with a fetish for reclining leather couches and oxygen hoses. It was inhabited by a white-coated army of NASA "suit techs," the valets of the most elaborate and expensive wardrobe ever made. And it was here, in this room that resembled a set for a science fiction movie, that most of the astronauts were jolted by their first Launch Day reality shocks.

"It's like the big game is about to start," recalls Charley Duke. "You get sweaty palms, butterflies in your stomach, your heart starts pounding . . . I remember saying to myself, 'Gee, here we are in the suit room, and pretty soon we're gonna be getting down close to launch. If all goes well, it looks like we really are gonna get airborne today.' You start thinking about the little things, things like, 'Boy, I sure hope the pressure suit checks out okay. I sure hope I got it fitted right.' I'd sat in it and worked in it, and we thought we had it right, but it's a trial-error thing because you can shorten or lengthen the arms and legs only so much. I had a backup suit, of course, but I didn't want all that worry of having to take off the first suit and try to put on the other one in time for launch."

Before the crewmen could try on their pressure suits, they had to strip down and "get instrumented." Led by supervisor and majordomo Joe Schmitt, the suit techs shaved the astronauts' chests, then plastered each of them with four silver chloride electrodes wired to a biomedical instruments belt that monitored the crewman's heart rate and respiration. Compact power converters inside each belt would transmit signals to an electrocardiograph and a pneumograph inside the spacecraft, thereby maintaining an up-to-the-second medical record without inhibiting the astronaut's freedom of movement.

Each crewman then massaged his buttocks with a special salve, and slipped on a "fecal containment bag," a large plastic diaper with a hole in front for the penis. This precaution was taken not for fear that the astronauts might defecate in their suits during the launch, but in case something went seriously wrong after launch.

"Had we, say, a major emergency after leaving the earth's orbit or some problem docking in space, the cabin could possibly lose pressurization," explains Buzz Aldrin. "It would be some five or six days before we were back floating in the Pacific and, in spite of our low residue diets, a bowel movement while in the pressurized space suits was likely. The fecal containment garment was designed to do what its name implies—also to keep the odor from making the culprit and his buddies sick."

The next items to go on were a jock strap with a penis hole, a pair of wrist- and ankle-length long underwear, and a condom. The condom was then clipped to a valve on the back side of a "urine collection device" that resembled a hot water bottle. It was almost inevitable that the astronauts would use these liquid waste receptacles because even if the launch was successful at least five or six hours would elapse before they would be able to remove their pressure suits inside the spacecraft. It was also inevitable that the astronauts found macho humor in the prophylactic part. As Aldrin notes: "The rubbers must fit snugly in order to work, consequently they were a subject of much joking. Our legs weren't the only things with a tendency to atrophy in space."

Now the crewmen were ready to don their bulky white pressure suits, a task that required the assistance of the suit techs at every step. These custom-tailored space togs cost about $100,000 apiece, and came in two basic styles. The command module pilot, who usually didn't venture outside the spacecraft, wore the thirty-five-pound "intravehicular" model. The commander and the lunar module pilot wore the heavier "extravehicular" model that weighed about fifty pounds for extra protection on their upcoming moon walks.

Each suit had over five hundred separate parts and required four days to strip down and reassemble. The theory behind this complex design was borrowed from high-altitude mountain climbers who wore several layers of undergarments beneath their snow parkas. Similarly, the pressure suits provided the astronauts with three distinct layers of inner and outer protective clothing in addition to their long johns.

Tougher than a fire hose and similar in texture, the white outside thermal covering was itself composed of two sublayers of Beta cloth, an extraordinary nonflammable synthetic fiber developed for the Apollo program by ILC Industries Inc. of Dover, Delaware. There were extra Beta cloth patches at the shoulders, elbows, and knees to guard against any rips or punctures that might be incurred in the event of an accidental fall. The astronauts' gloves, boots, and PPKs were covered with the same material.

Before attempting to walk on the lunar surface, the commander and the lunar module pilot would add yet another protective coat, the Integrated Thermal Micrometeoroid Garment, which zipped over the outer shell of the pressure suit. These moon-walking overalls were made of two inner layers of nylon, seven intermediate layers of Kapton-laminated Beta cloth, and an exterior armor of Beta cloth coated with Teflon, the synthetic paste later adapted to the commercial manufacture of non-char cookware.

The lining of the pressure suit was woven from five-ounce Nomex, a silky smooth fabric that would not irritate the skin of the wearer. The middle layer of the suit between the Nomex inner lining and the thermal shell—the "pressure garment assembly"—was the most intricate and the most important.

By the time the suit arrived at the Cape, the pressure garment had already been stitched in place, but when dissected, it appeared to be, as Duke put it, "real Flash Gordon stuff." Colored an eerie blue and shaped like a human torso, it consisted of a neoprene bladder laced with a labyrinthine network of hoses, rings, valves, and pulleys. When inflated, the bladder sandwiched the astronauts' bodies with about one-quarter G of pressure. According to NASA doctors, this pressure would keep the crewmen's blood from boiling and prevent their hearts, lungs, and arteries from popping in the vacuum of space.

Like the pressure suits, the helmets also came in two styles. All three crewmen wore the lightweight "intravehicular" models at launch. Ground from a transparent polycarbonate called Lexan, they were shaped like giant fishbowls, affording maximum visibility. The commander and the lunar module pilot carried two extra visors to convert their helmets into the "extravehicular" mode needed for their moon walks. One visor was a lightly tinted glare reduction lens. The second was coated with a shiny gold film to reflect solar radiation, and worked like a two-way mirror: The astronauts could see out, but whoever was on the outside could not see their faces.

Putting on the helmet also required assistance. First, a pair of suit technicians guided each crewman to one of the reclining couches. A second pair of suit techs helped the astronaut wiggle into a black and white "Snoopy" skullcap mounted with his voice communication gear, while the first pair plugged his pressure suit with hoses connected to the couch-side air-supply consoles. Finally, two more suit techs appeared with the helmet itself, lowered it over the astronaut's head, then screwed it into a thin metal ring at the collar of the suit.

Though some pressure suits fit better than others, the Apollo astronauts unanimously complain that wearing all this armor was never comfortable, and was often downright debilitating, in Earth gravity. The sheer weight was a problem whenever they had to walk, but the relentless grip of the pressure garment was a constant annoyance. Inflated like plastic party dolls and almost as inflexible as concrete sculptures, the astronauts could not perform even the simplest physical movements like raising their arms without going through a long and exhausting series of mechanistic contortions. Nor could they get out of their suits for a few minutes to take a breather.

"I guess I always felt a twinge of claustrophobia when they locked me into the suit," admits Apollo 15's Irwin. "We referred to it as the 'moon cocoon,' and that's what it was. Inside was everything we needed for life: a little bit of water, a lot of oxygen, some protection from micrometeorites, a lot of thermal protection from the cold and heat . . . It became our second skin, and we were completely enveloped in it, completely dependent on this new technology . . . I realized, 'I'll suffocate if for some reason the supply of air should be cut off.' "

Getting "locked into the suit" was also a symbolic turning point for many of the Apollo astronauts: a clear and somewhat sobering sign that the first phase of the Launch Day checklist was about to end and that the second phase—boarding the spacecraft and actually preparing for liftoff—was about to begin.

"When you go ahead and buckle yourself into your helmet, that's kind of a note of finality," recalls Apollo 14's Stu Roosa. "It's a signal that 'Yeah, we're really getting ready to do it.' "

But before the three crewmen could proceed to the spacecraft, they had to lie back and wait for at least another half hour. Their air-supply consoles were filling their suits with 100 percent oxygen, purging their systems of bubble-producing nitrogen. Prior to embarking for the launchpad, the crewmen would switch to a less flammable mixture of sixty percent oxygen and forty percent nitrogen to reduce chances of a fire like the one that claimed the lives of the Apollo 1 crew, then go back to pure oxygen when they were safely in space. In the meantime, the astronauts had to complete their prebreathing exercises.

"You're like a person waiting in the wings to go on stage," recalls Al Bean. "You have less time to daydream because you're trying to remember what you've got to do when the curtain rises. I had the feeling that I really had to try to concentrate on the things I had to do on the moon, on the way to the moon, and all that."

Adds fellow Apollo 12 crewman Dick Gordon: "You're in your work clothes, and you're ready to go to work. It doesn't take long to get dressed, maybe twenty minutes or so, but you still have to stay there and do everything in conjunction with the booster schedule and the spacecraft schedule. It's kind of a catch-up time. If there's a problem with the booster, we're in the suit room and not out on the launchpad."

The astronauts on later missions were considerably more relaxed. Apollo 15's Irwin, who launched over a year and a half after Apollo 12's Gordon and Bean, is an excellent case in point. Rather than continuing to fret over the "claustrophobia" of being locked into his pressure suit, he asked the suit techs to drape a towel over his helmet to block out the overhead lights. Then, he says, "I closed my eyes and took a little nap."

If the countdown continued without a hitch, the astronauts usually got approval to depart for the launchpad about two and a half hours after their wake-up call. This usually happened to be about two and a half hours before they were scheduled to lift off. Suit room supervisor Joe Schmitt ordered his troops to cover their boots with yellow galoshes, and shift their air-supply hoses to suitcase-size portable units. Then, slowly and methodically, the suit techs would lift the astronauts up from their reclining couches and onto their feet so they could begin the short walk to the tranfer van.

As the three crewmen emerged from the suit room, they were jolted by an even more disorienting change of scene. The hallways along their path out of the MSO building were always lined with NASA support personnel and specially invited friends and family members. But although the astronauts could see the familiar faces all around them, the only sounds that penetrated the artificial quietude of their "moon cocoons" were the hissing of their portable supply units and the rhythmic squish-squish-squishing of their yellow galoshes.

"My wife and daughter were there," recalls Apollo 17's Gene Cernan, "but I couldn't talk to them. I could only look at them, and sort of grab their hands, hold them, hug them, and throw them a kiss. They're standing like two feet from you, but because you're in that suit, isolated from noise and everything else, they might as well be miles and miles away."

Cernan adds that the frustration of not being able to communicate with his wife and child made him realize that they and the families of his fellow crewmen had to bear a burden heavier than any pressure suit. "From then on, they've got their ordeal just as we've got our ordeal. Ours is fun, but theirs is sort of wait and see and keep you fingers crossed because they can't do anything else."

Fortunately, the Apollo astronauts could look forward to some unofficial preflight rituals to help them regain their bearings. The first was saluting Cape security chief Charlie Buckley, who always stationed himself on the ramp between the door of the MSOB and the door of the transfer van. The second was making sure that Buckley did not confiscate the going-away present each was carrying for Guenter Wendt, the "führer of the launchpad," whose presence at lift-off, according to astronaut superstition, was a necessary prerequisite for success.

Most of the gifts the astronauts got for Wendt referred to an inside joke. Having listened to him brag ad nauseam about his prowess as a fisherman, Apollo 11's Mike Collins nailed a thumb-size minnow to a wooden plaque engraved "GUENTER'S TROPHY TROUT." Neil Armstrong printed up a "space taxi" ticket "good between any two planets." Fellow crewman Buzz Aldrin carried a personally inscribed Bible. Apollo 14 commander Alan Shepard brought Wendt a World War II vintage German army helmet stenciled "Col. Klink," a bumbling character in the TV sitcom *Hogan's Heroes.* Like Collins, most of the astronauts smuggled their booty past security in brown paper sacks, or, like Armstrong, in the pockets of their pressure suits.

Instinctively wary of the paparazzi swarming around outside the MSOB in hopes of one last "photo opportunity," the three crewmen would cross the gangplank to the transfer van as fast as their pressure-suited legs could carry them. Then Schmitt and the suit techs helped them pile into their assigned seats.

The distance from the MSOB to the launchpads on Merritt Island was only about eight miles via U.S. 1, the main coastal highway, and the route had already been cleared by local law enforcement authorities. But according to the astronauts, many of whom were speed demons on land as well as in the air, the drive always seemed to take longer than it should.

"We drove very slowly," says Jim Irwin, "and I still don't completely understand why. It was almost as if we were afraid that the van was going to break down. But just in case it might, there was a backup van that followed us, just as there is a backup for everything we do in space."

Even on early morning launches when it was still dark at the Cape, all the local byways, back streets, bridges, bays, and navigable canals were clogged with car, camper, bus, truck, van, and boatloads of anxious spectators. Apollo 11 naturally drew the largest turnout, an estimated one million strong. Henceforth, the Launch Day crowds steadily dwindled, hitting a low of a few hundred thousand for Apollo 16. Apollo 17, the last lunar mission, prompted a resurgence of nostalgia for manned space exploration, attracting well over half a million people.

In addition to hordes of ordinary citizens, the Apollo launches were attended by throngs of media and official "guests"—celebrities, politicians, foreign dignitaries, former astronauts, families, and selected personal friends of the three crewmen—who viewed the lift-off from a VIP grandstand at the Cape. As the countdown proceeded, these VIPs were treated to a running commentary by the space agency public affairs officer (PAO) broadcast via loudspeaker.

The grandstand crowd at the historic Apollo 11 launch was estimated at over 20,000. This tally included some 3,500 reporters and photographers from all over the nation and the world. Japan, which sent a team of 118 correspondents, boasted the largest foreign contingent. The domestic media army of over 2,500 strong was led by America's "most trusted man," CBS television news anchor Walter Cronkite, who had made documenting the space program his personal specialty.

Also on the scene was one of America's most famous and controversial authors, novelist Norman Mailer, who had recently run for mayor of New York and was now writing a personal history of Apollo 11. An outspoken critic of the war in Vietnam and the so-called Establishment, Mailer did not share Cronkite's patriotic enthusiasm for the space program. On Launch Day, he could not decide whether Project Apollo was "the noblest expression of the twentieth century or the quintessential statement of our fundamental insanity."

President Richard M. Nixon, who had been denied the privilege of a prelaunch dinner with the Apollo 11 crewmen by overcautious NASA doctors, did not witness the launch in person, but his predecessor, former President Lyndon B. Johnson, would not miss it for the world. Upon his arrival at the Cape, LBJ declared, "I have ridden on every flight, and I doubt that any human being could be as concerned or troubled until splashdown as I am or have been."

Joining LBJ in the grandstand were 230 incumbent congressmen and 34 senators from both sides of the aisle, 275 leading U.S. businessmen and industrialists, Leon Schacter of the Amalgamated Meat Cutters and Butcher Workers union, Cardinal Cooke, Prince Napoleon of Paris, some four hundred foreign ministers and diplomatic attachés, and comedian Jack Benny. Vice-President Spiro T. Agnew represented the White House, and GOP patriarch and former presidential candidate Barry Goldwater showed up in a red golf shirt and slacks.

The NASA contingent personified the history of manned space flight. Reigning administrator Dr. Thomas O. Paine was flanked by his predecessor, former space agency boss James E. Webb, and nearly all of NASA's original geniuses: Dr. Robert R. Gilruth, Max Faget, Abe Silverstein, George Low, and many others, including the venerable Wernher von Braun. There was also a distinguished special guest standing beside Von Braun—the octogenarian Dr. Hermann Oberth, the founding father of German rocketry.

Black civil rights activist Reverend Ralph B. Abernathy was also in the grandstand area leading a group of "poor people" who planned to march in protest of spending billions of tax dollars to land a man on the moon instead of on sorely needed social programs. The day before, NASA administrator Dr. Thomas Paine had assured Abernathy, "If it were possible for us not to push the button tomorrow and solve the problems with which you are concerned, we would not push the button."

Abernathy would later become so "awed" by the Apollo 11 launch spectacle that, for a moment, he reportedly forgot what he had come to protest. "This is really holy ground," he conceded when he finally regained his voice, adding, "and it will be more holy once we feed the hungry, care for the sick, and provide for those who do not have houses."

The Apollo 11 crewmen, who were insulated from the outside world by their pressure suits and the body panels of the transfer van, could manage only a few fleeting glimpses of the Launch Day crowd as they drove to the pad. That was just as well, as far as they were concerned, for Armstrong, Aldrin, and Collins were simply too preoccupied to indulge in people watching. So were the crewmen on later Apollo missions. Instead of trying to peer through the windows of the transfer van, many of the astronauts used this last peaceful interlude to contemplate the tumultuous events to come.

"As we drove out," recalls Irwin, "there was plenty of time to reflect on your life . . . Where are you going today? Or where are you hoping to go to today? Did you have your life in order? By this time it was light, the sun was up, and I could see that it was going to be a beautiful morning, no clouds that might interfere with our launch. I could imagine that birds were singing on the outside, but on the inside of the van it was quiet. We weren't talking. We looked at each other a little bit and smiled, that smug look. We knew where we were going, but I think that we were quiet because there were so many thoughts racing through our minds as we drove out to that rendezvous with destiny."

When the transfer van turned off the highway and started down the access road to the launchpad, the three crewmen were jarred by another change of scene. Now, for the first time since Launch Day began, they could see the needle-nosed profile of the Saturn 5 rocket that was supposed to hurtle them into space.

On early morning launches, the rising sun would splash the Merritt Island sky with sensational bursts of pink, orange, gold, and indigo, making the oyster-white shell of the Saturn 5 sparkle like a giant pearl bullet next to the fiery amber scaffolding of the launch tower. On night launches, the pad was illuminated by a battery of lamp pods even brighter than the noonday sun, and thanks to the coincidental timing of the lunar launch window, the booster's gleaming silhouette was often crowned with the magnificent halo of its ultimate destination, the moon.

But as Apollo 11's Mike Collins observes, no matter what time of day or night it was, the smoke-snorting Saturn 5 always appeared to be "truly a monster." This was not merely an optical illusion, but an apparition of the true nature of this terrifying technological beast.

"There's frost on the sides of the cold cryogenic tanks," reports Apollo 16's Ken Mattingly, "and there's little vapors coming out . . . And when you look at it, it doesn't look like just an inanimate hunk of machinery . . . You have this feeling that it's alive!"

Although the three crewmen knew better, they also knew enough about the billion-dollar booster's actual dimensions and demonstrated performance to be awestruck. Including the attached Apollo spacecraft, the Saturn 5 stack rose almost forty stories (a total of 363 feet) from her truncated bottom to her needle-tipped top. Higher than the highest redwood tree, taller than a football field standing upright on one end zone, she weighed almost six and a half million pounds fully loaded (6,487,354 pounds, to be exact), which made her over three thousand times heavier than the proverbial ton of bricks.

In order to get from her birthing area in the Vertical Assembly Building (VAB), the world's largest building, to the launchpads on Merritt Island, the rocket had to be towed in vertical position by the world's largest portable structure—the "crawler-transporter," a steel-treaded tractor-trailer 131 feet long and 114 feet wide driven by a 3,000 h.p. diesel engine, which itself weighed close to six million pounds without any cargo. Capable of a top speed of only one m.p.h., it took the crawler-transporter at least six to eight hours to make this short haul of just 3.5 total miles.

The Saturn 5 boasted the biggest bang—both for the buck and for absolute upward boost—of any rocket known to man. Her five F-1 jet engines were capable of delivering over 7.7 million pounds of thrust, 1.2 million pounds more than the booster's total weight, and over ten times the thrust of the Gemini program's Titan 2. The explosive force that erupts at lift-off is equivalent to TNT.

On Launch Day, the process of fueling the Saturn 5 began about nine hours before lift-off, and took over four and a half hours to complete. Huge reptilian hoses filled the booster's tanks with over six million pounds of cryogenic (low temperature) propellants, mostly liquid oxygen (LOX) cooled to minus 297 degrees Fahrenheit and liquid hydrogen cooled to minus 423 degrees Fahrenheit, along with liquid nitrogen, liquid helium, and kerosene, components of a hypergolic fuel mix that would self-ignite when the chemicals made physical contact in the combustion sequence.

After the initial spark from a ground cable on the launchpad, most of the billions and billions of internal combustion explosions in the Saturn 5's belly would be set off by a chain reaction of inflammable gas colliding with inflammable gas. The on-board electrical package needed to keep the first stage going and help light stage two consisted of just eleven batteries with a total weight of less than 1,000 pounds and a combined juice of only 364 volts. There were no alternators, generators, spark plugs, pistons, or crankshafts. None were needed, for the booster could propel herself even more simply than the average family car—she just burned up her gases and blew the exhaust out her rear nozzle.

Yet the Saturn 5's awesome bang and enormous bulk belied the amazing grace and delicacy of her design. Stylishly streamlined from nose cone to exhaust nozzle, she measured no more than thirty-three feet around at the broadest part of her lower beams. The vast majority of her 6.5-million-pound body weight was fuel, five million pounds of which would burn off within the first three minutes after lift-off. The three sections of the Apollo spacecraft (the command, service, and lunar modules), with a combined length of eighty-two feet, accounted for about 100,000 pounds. But the three stages of the booster, with a combined length of 281 feet, weighed just a little over 400,000 pounds, one-fifteenth the weight of her fuel supply.

One of the secrets to the Saturn 5's monstrous power was her amazingly thin skin. Composed of an aluminum alloy jointly developed by Reynolds and Alcoa, the booster shell was never more than a quarter of an inch thick at any point along the fuselage; some of the lower sections, which were basically fuel tanks, could be less than 80/1,000ths of an inch. This tissuous construction, yet another Von Braun inspiration, was what really made it possible for her to fly. For without the dramatic weight reductions it afforded, the giant booster would simply have been too heavy to get off the ground.

Even so, the Apollo astronauts confess that knowing about the fragility of the Saturn 5 could be rather unsettling. Though buttressed with steel bulkheads and corrugated for extra support, the booster shell was also easy to puncture. "If you stood right next to it and jammed a pencil into the side, you could get the pencil to go right through the booster shell," reports Al Bean. "I don't know of anyone who ever tried that, I just know that it was possible . . . It might take you a while if you only used your hands, but I'm sure you could do it pretty easily if you used a hammer."

Less than twenty minutes after departing from the MSOB, the transfer van would reach the launchpad, and pull up at the foot of the giant booster. The doors of the van would swing open, and the three crewmen would hop out onto the tarmac with the help of the suit techs. Many of the astronauts say that was where they felt their first profound prelaunch reality shocks. Apollo 16's Ken Mattingly explains why:

"Normally when you're out on the pad, there's a lot of noise. There's machines running, elevators going up and down, people clanging and banging and talking . . . But on the day of the launch, there was an eerie silence, and the pad was deserted except for a couple of support people . . . That was the first thing that was really different, the kind of thing that brings home for the first time that today is not the game we've been playing in practice. This is reality."

At the same time, many of the astronauts report that arriving at the launchpad also prompted the raising of an internal red flag, a defense against the possibility that their very real lunar mission might yet fail to get off the ground. As Mattingly remembers, "Right at the last minute, there was a psychological block in there that said, 'Don't count on this so heavily. It might not happen.' "

The Apollo 17 crewmen were actually prepared for a last-minute disappointment, due to the fact that they arrived at the pad during a nasty electrical storm. "I was absolutely sure we were not going that night," recalls mission commander Gene Cernan. "I thought, 'We'll just sit here until they tell us to get out, and we'll try again tomorrow.' I wasn't being a pessimist. I just knew there was no way they could get all the problems solved or be satisfied they were solved. Or there'd be some witch hunt that would find out that everything was okay and why that wasn't okay, so we can't go through with it even though it is okay. The only thing I could think of was, 'Gee whiz, if we don't launch today all my friends are going to go home, and won't stay around to watch this spectacular event.' "

For the Apollo 15 crewmen, who also anticipated a launch delay due to technical difficulties with the booster systems, the sight of the Saturn 5 towering over the desolate launchpad conjured much darker thoughts. "We were reminded that this was, in a way, a monument to the genius of man, to the technology that we'd achieved at this point in history," recalls lunar module pilot Jim Irwin. "But as we viewed it, we viewed it in a little more personal terms. And we just wondered, 'Will it work? Will it really take us to the moon and bring us back?' "

Irwin hastens to add that the Apollo 15 astronauts were bolstered by an abiding "faith" born of hard-earned knowledge and firsthand experience—"faith in people and in hardware, faith in ourselves." But he also admits the realization that he and his crew mates were going to find out whether or not the Saturn 5 worked by putting their own lives on the line. "We walked a little slower that morning," he recalls. "We were looking around a lot. We didn't want to miss anything . . . Because for all we knew, it might be our last time to see things on Earth."

Still carrying their portable air-supply units, the three crewmen would be ushered into the steel mesh elevator at the base of the launch tower scaffolding by Joe Schmitt and his suit techs. Then one of the few remaining pad techs, yet another anonymous but indispensable man in a white plastic hard hat and white coveralls, would slam the elevator door with a resounding clang, and send the astronauts skyward with a snappy salute and a Florida drawl: "Good luck, gentlemen, have a good flight!"

2

MAN AND THE MOON

Science Fiction vs. Scientific Fact

2000 B.C.–20th Century

The Yankees wish to take possession of this new continent of the sky, and plant upon the summit of its highest elevation the star spangled banner of the United States of America.

JULES VERNE
From the Earth to the Moon (1865)

The first man who claimed to have landed on the moon was not an Apollo astronaut but a self-styled literary argonaut named Lucian of Samosata. Born on the banks of the Euphrates River about 120 A.D., Lucian was an impoverished Syrian who somehow managed to educate himself in Greece for a career as a traveling lecturer and satirist. He recounted the details of his fantastical lunar voyage in *A True History* published in 165 A.D.

The odyssey begins as Lucian and his crew sail westward from the Straits of Gibraltar in hopes of discovering "what formed the farther border of the Atlantic Ocean and what peoples lived there." But as their ship crosses the ocean, it is swept up by a ferocious storm and blown to "an island in the sky" that turns out to be the moon.

The resident monarch of the moon is King Endymion, a Greek mythological figure allegedly kidnapped from Earth by Selene, the jealous goddess of the moon. Endymion's subjects, the so-called moonmen, are bizarre extraterrestrial creatures in humanoid form. They have "removable eyes" and ears made of wood or leaves. They drip bitter honey from their noses, sweat milk from their pores, and dine exclusively on frogs and dew compressed from thin air. And as Lucian archly notes, their society is entirely homosexual for there are no women on the moon.

Lucian and his crew join the moonmen in a cosmic war against King Phaethon and the sunmen over the right to colonize the morning star. The opposing armies, each numbering over 120,000 strong, ride into battle on three-headed buzzards, winged acorns, giant insects, and "saladbirds" with feathers made of lettuce leaves. Just as the moonmen appear to have victory in hand, a horde of "cloud-centaurs" led by Sagittarius, the archer of the Zodiac, erects a barricade that blocks out the light of the sun and leaves the moon in perpetual eclipse. Endymion and the moonmen are forced to pay tribute to the victors. In return, Phaethon and the sunmen agree to lift the cloud barricade, and allow joint colonization of the morning star.

Despite their defeat, the moonmen insist on entertaining Lucian and his crew for several days before allowing them to return to Earth. In the course of these revels, Lucian stumbles on yet another marvel in Endymion's royal palace—"an enormous mirror suspended over a rather shallow well." According to his report, "If you stand in the well, you hear everything said on Earth; if you look at the mirror, you see each city and nation as clearly as if you were standing over it. When I took a look, I saw my own homeland and my house and my family; I can't say for sure whether they saw me."

We can be certain that Lucian of Samosata did not actually go to the moon, for as he blithely admits in the preface to *A True History,* "the one and only truth you will hear from me is that I am lying." By intention and profession, Lucian was the Mark Twain of his day, not the Neil Armstrong. He deliberately told lies in order to tell the truth about history and Greek society. His tall tale was meant to be a parody of the preposterous accounts of one-eyed giants, hydra-headed monsters, and apocalyptic sea battles that venerated scholars like Homer, Herodotus, and Thucydides had tried to pass off as true stories.

But thanks to *A True History,* Lucian of Samosata can rightfully claim an important place in the history of man's conquest of the moon. What became twentieth-century science began as science fiction. Fact grew out of fantasy, not the other way around. In order to actually get to the moon, man first had to dream about getting there. If Lucian was not the first human being to fantasize about going to the moon, he was the first to commit his fantasy to writing. By so doing, he became the ancient ancestor of modern science fiction. Although the title was admittedly false advertising, *A True History* marked a "giant leap" in the evolution of consciousness, for it was the first time on record that man had depicted the moon as a place he might visit.

Before Lucian, virtually all of mankind had either worshiped the moon or revered her as the exclusive province of the gods. In most primitive cultures, the sun was portrayed as the destructive Earth-scorching force, while the moon was seen as a benign deity widely associated with hunting, harvesting, feminine fertility, sexuality, love, mystery, magic, and the occult. Likewise, the moon is recognized in some of the earliest human etchings as the ruler of the sea and the tides. The ancient Greek mythologists saw the moon both as a goddess personified by the vixen Selene, and as the celestial sister of Mt. Olympus, inhabited by several different goddesses. One of the female deities who ruled the moon was Artemis, whose dual personality governed both fertility and deadly plagues. Artemis was the sister of Apollo, the god of music, prophecy, and poetry, whose fiery chariot pulled the sun across the sky.

For over 1,400 years after Lucian, man continued to cling to ancient moon cults and fabricate new ones, as the proliferation of popular fantasy far outpaced the spread of scientific fact. The Chinese believed that the moon had magical powers over sexuality, and that it was inhabited by a giant rabbit and a beautiful princess, a legend that would surface many centuries later at a crucial juncture in man's first real lunar landing mission. Various tribes and clans from the Zulus of Africa to the Tibetans and the Amerinds also claimed to see some form of hare on the lunar surface. The natives of North America saw a coyote in the moon. The Europeans of the Middle Ages sighted the famous "Man in the Moon," a figure William Shakespeare described in *A Midsummer Night's Dream* as the "man, with lanthorn, dog, and bush of thorn." In addition to concocting early legends about werewolves and vampires brought to life by the full moon, they also started the folktale that "the moone is made of a greene cheese."

The Renaissance wizard Galileo Galilei was the first man to see the moon more or less as it really is. The moon in Galileo's telescope was not a chunk of "greene cheese" or, as some Medievalists suggested, a "perfect" wafer smooth and thin as glass. It was an "imperfect" orb bristling with jagged mountain peaks not unlike those on Earth. The moon also had large flatland basins that Galileo termed *"maria,"* or seas, even though they did not actually hold water. Four of the larger *maria—Frigoris* (cold), *Imbrium* (rain), *Serenitatus* (serenity), and *Nubium* (clouds)—corresponded with the brow, eyes, and mouth of the "Man in the Moon." Though Galileo did not observe signs of life on the moon, he did conclude that the place was potentially habitable, and said as much in *Siderius Nuncius* ("Messenger from the Skies"). But the book so defied the orthodoxy that it was immediately banned, and its author spent the remainder of his days going blind while under house arrest.

In 1634, the German mathematician Johannes Kepler, the father of modern astrophysics, became the father of modern science fiction with the publication of his astounding *Somnium* ("Dream"). Like *A True History,* Kepler's novel recounts a fictional voyage to the moon, but what distinguishes it from previous space fantasies is its basis in scientific fact. Though the astronauts are swept up from the sea by the unlikely intervention of "spirits," their trans-lunar trajectory conforms with Kepler's laws of planetary motion, the first of which posits that celestial bodies do not orbit in perfect circles but in ellipses. Likewise, what the astronauts see when they reach the moon is described in accordance with Kepler's telescopic observations. Kepler did not seriously propose that man attempt to go to the moon. But in the course of explicating his theories, he showed how such a voyage was theoretically possible.

Half a century after Kepler's death, Sir Isaac Newton (1642–1727) formulated his now famous theory of gravity and his laws of motion. Two of Newton's principles—that objects in motion tend to stay in motion until acted upon by an outside force (like gravity) and that every action produces an equal and opposite reaction—would form the basis for the modern theory of jet propulsion that enabled the Saturn 5 rocket to launch the Apollo astronauts to the moon.

Man continued, however, to regard the idea of going to the moon as totally preposterous until the birth of Jules Verne, the greatest science fiction writer of all time. Born in Nantes, France, in 1828, Verne gave up a secure career in law to become a man of letters. In 1865, the year the U.S. Civil War ended, he published *From the Earth to the Moon,* one of the most uncannily prophetic novels in modern literature. The only major scientific flaw in the story is the means Verne uses to get his astronauts off the ground. Instead of boarding a rocket, the crewmen climb inside a conical-shaped metal projectile that is shot out of a giant cannon. But Verne's estimate of the speed necessary to escape the earth's gravity (12,000 yards per second) turned out to be a remarkably close approximation of the actual "escape velocity" needed by the Apollo astronauts.

Similarly, the narrative details of Verne's novel bear a startlingly close resemblance to the true story of man's first lunar mission over 100 years later. The sponsors of the moon voyage in Verne's book are members of an American gun club looking for some way to use their great artillery now that the Civil War is over. Although the sponsors eventually dedicate their lunar expedition to all humanity, Verne notes that their true motive for sending astronauts to the moon is "to take possession of this new continent in the sky, and plant upon the summit of its highest elevation the star spangled banner of the United States of America."

Verne's spacecraft, which is named *Columbiad,* is finally launched from the southern tip of Florida after the state of Texas wages a valiant though ultimately unsuccessful campaign to be chosen as the launch site. The three astronauts on board *Columbiad* do not actually land on the lunar surface, but they orbit close enough to conclude that the moon could abide a human colony and/or military base. After splashing down in the Atlantic Ocean, the crewmen are hailed as conquering heroes, and dispatched on a cross-country victory tour with the president of the United States. Then the sponsors of the voyage form a company to exploit the business potential of future space travel.

Jules Verne had no way of knowing just how faithfully life in the twentieth century would imitate his nineteenth century art, but he managed to get some inkling before he died. Over the next four decades, as Verne cranked out such classics as *Journey to the Center of the Earth* and *Twenty Thousand Leagues Under the Sea,* he witnessed the invention of the telephone, the internal combustion engine, and the electric light. Contemporary science still lagged behind the science fiction of his literary protégé H.G. Wells, who published *War of the Worlds* (1897) and *The First Men in the Moon* (1901), a romantic novel in which the hero encounters lunar creatures far more horrific than those described by Lucian of Samosata in *A True History.* Even so, fact was starting to catch up with fantasy. Verne's death in 1905 coincided with the true dawn of the modern space age, for that was the year an obscure Russian visionary provided the theoretical missing link that would enable man to transform his dream of going to the moon into reality.

His name was Konstantin Eduardovich Tsiolkovsky (1857–1935), and he was a partially deaf Russian schoolteacher who starved himself to buy books so he could develop his outlandish theories of math and physics. His unprecedented contributions to the advancement of rocketry science were summarized in his 1903 science fiction novel *Outside the Earth.*

"For a long time I thought of the rocket as everybody else did—just as a means of diversion and petty everyday uses," Tsiolkovsky wrote later. "I do not remember exactly what prompted me to make calculations of its motions. Probably the first seeds of the idea were sown by that great author of fantasy Jules Verne—he guided my thought along certain channels, then came a desire, and after that, the work of the mind."

Like Kepler's *Somnium, Outside the Earth* recounted a fictional space voyage, artfully weaving in serious mathematical and physics theories along the way. But Tsiolkovsky made a giant leap beyond Kepler and well over the heads of most of his contemporaries by providing the first realistic blueprint for a rocket capable of getting a manned spacecraft off the ground.

Tsiolkovsky was the first man to discover the fundamental laws of jet propulsion. The basic design of his imaginary rocket was virtually identical to that of the Saturn 5 booster that would launch the Apollo astronauts to the moon. So was the basic operating principle, which was itself based on the Newtonian law that every action (in this case, the rocket's discharge of thrust through the exhaust nozzle) produces an equal and opposite reaction (the upward lift of the vehicle). Along with divining the fundamentals, Tsiolkovsky added some key practical refinements. His booster was equipped with engines that burned liquid fuel, which heated into gas vapor more rapidly and with less energy than solid fuel. The booster was also divided into three stages; in order to offset the added weight of the liquid fuel, each of the first two stages was designed to drop off when its fuel tank was emptied, thereby enabling the engines to deliver more thrust per pound.

Though Tsiolkovsky would later be heralded as the "father of Russian rocketry," he did not intend his fictional creation to become the model for awesome military weapons capable of destroying whole cities. He envisioned it solely as a vehicle for manned space exploration, an instrument of peace dedicated to the advancement of human knowledge. As he put it, "not only the earth, but the entire universe is the heritage of mankind."

American physicist Robert Goddard, a long-legged professor at Clark University, grabbed the world leadership in rocketry before the Russians ever took Tsiolkovsky seriously. In 1926, Goddard launched his first successful test rocket from his Aunt Effie's farm in Eden Valley, New Mexico. Powered by liquid oxygen–fueled engines, the rocket reached an altitude of forty-one feet in two and a half seconds, then crashed to the ground. Goddard went on to test several more advanced models, with gyroscopes to stabilize their trajectories and parachutes that enabled them to land intact.

Meanwhile, various independent scientists in England tried to pick up where Goddard had left off by exchanging ideas through the newly formed British Interplanetary Society. Their German counterparts organized the Verein für Raumschiffahrt ("Society for Space Travel") under the leadership of Dr. Hermann Oberth, the author of *The Rocket into Interplanetary Space,* a booklet urging that the work of Tsiolkovsky and Goddard be used to develop a manned spacecraft and a manned space station in the earth's orbit. Oberth's booklet became a surprise international best-seller, sparking a dramatic new wave of interest in rocketry by serious scientists and, for the first time, by a large segment of the general public.

Oberth also carried on the noble tradition of Jules Verne by assisting in the birth of a whole new genre of science fiction films. In 1926, he acted as the scientific advisor to director Fritz Lang on *Die Frau Im Mond (The Woman in the Moon),* the saga of a make-believe lunar voyage that would win critical acclaim as "the first real space travel movie." He later served in the same capacity on *Destination Moon,* an astonishingly prophetic 1950 science fiction epic based on a script by Robert Heinlein. In addition to depicting a realistic scenario for a lunar landing mission, *Destination Moon* climaxed with the hero's dramatic declaration that he had come to "take possession" of the Earth's sister planet "for all mankind."

But it is quite likely that twentieth-century man would never have actually attempted to land on the moon had it not been for the ignominious rise and fall of Adolf Hitler. In 1927, the Nazis recruited Oberth and several other leading members of the Society for Space Travel to staff a new rocketry research center at Peenemünde, a remote village on the Baltic Sea. Hitler had no interest in using rockets to conquer outer space. Instead, he ordered the scientists at Peenemünde to build Vengeance Weapon No. 1 (V-1) and Vengeance Weapon No. 2 (V-2), the first guided missiles. Peenemünde's most prominent dissenter, however, the outspoken young Wernher Von Braun, kept complaining that he was not interested in making weapons of war, only in developing rockets for use in manned space exploration. In 1944, Von Braun was arrested on orders from Heinrich Himmler and thrown into an SS prison for two weeks. He might have stayed there for the rest of the war if it were not for the personal intervention of Peenemünde's commanding general, who insisted that the V-2 project could not go on without him.

In 1945, the U.S. began the nuclear age by dropping the first and second atomic bombs on Hiroshima and Nagasaki. That was also the year the Russian army captured Peenemünde and about 2,000 Nazi rocketry experts along with a huge cache of drawings and data files. Fortunately for the U.S., Von Braun managed to move himself and 130 other top German scientists from Peenemünde to a mountainous site in the interior so they could be captured by the Americans. Though neither side issued an official announcement, this bilateral division of the Peenemünde braintrust marked the true beginning of the space race.

Unfortunately for the U.S., the Truman administration failed to use the superior brain power of its ex-Nazi rocketeers to full advantage. Von Braun and several other German scientists were hustled off to Fort Bliss, Texas, to work on Project Bumper, which consisted mainly of test firing captured V-2s in hopes of developing a similar rocket for the army. Much to their chagrin, the Project Bumper team discovered that the long-neglected designs Robert Goddard had conceived back in the 1920s were actually superior to the German rocket models. But the message implicit in that historical irony was apparently lost on U.S. leaders. Between 1945 and 1951 the United States had "no ballistic missile program worth mentioning," according to Von Braun, because there was no obvious need for one and no taxpayer support. The nation's only major triumph in aerospace engineering occurred on October 14, 1947, when ace test pilot Chuck Yeager broke the sound barrier in an experimental rocket plane called the X-1.

3

IT FEELS JUST LIKE IT SOUNDS

Merritt Island and Above

Lift-off and Launch

The ship of state—the gods once more,
After much rocking on a stormy surge,
Set her on an even keel.

SOPHOCLES
Antigone (410 B.C.)

All twenty-four men who flew to the moon left Earth at precisely the same moment. Their official times of departure were at nine different hours on nine different Launch Days. But their common moment of ascension always came when the launch tower elevator left the ground. "This ride up the elevator, this first vertical nudge," observes Apollo 11's Mike Collins, "has marked the true beginning . . . for we cannot touch the ground any longer."

Unless the launch was canceled at the last minute, the three crewmen would not ride back down the elevator. Nor would they even be able to once the pad crew sealed the hatch of their spacecraft, and retracted the service bridges in preparation for launch.

If a catastrophic accident occurred before lift-off, the astronauts had only one way to get back on the ground without waiting for the elevator—an emergency escape gondola hooked to a slide wire at the top of the launch tower. Installed in the wake of the Apollo 1 fire, the gondola was supposed to whisk the escapees to a landing area about four hundred feet from the base of the Saturn 5 booster in less than six seconds. But as all potential escapees were well aware, the gondola had never been tested under fire.

The best means of emergency escape for the three crewmen was also the one of last resort: a full-scale abort. This they could do at any time before, during, or immediately after lift-off via the launch escape tower (LET). Not to be confused with the four-hundred-foot-high launch tower scaffolding, the LET, also and more appropriately known as the "escape booster," was a mini-rocket affixed to the nose of the command module that could blow the astronauts clear of the Saturn 5 and parachute them into the Atlantic.

But assuming the countdown continued, the astronauts knew they would not be able to set foot on their native planet again that day or for several more to come. The crewmen on the earliest and briefest lunar missions (including Apollo 8 and 10, which orbited the moon but did not land) could expect to be gone for at least a week. The crewmen on later missions like Apollo 16 and 17 realized that if all went well they would not be back for up to twelve days. At the same time, all twenty-four men who flew to the moon were aware of a fact none would ever dare mention as they rode up the launch tower elevator—that if all did not go well, they might never set foot on Earth again.

Thirty seconds later, the elevator jerked to a stop, and the crewmen found themselves on a narrow orange gangplank 320 feet above the marshes of Merritt Island. Apollo 11's Mike Collins later admitted that the breathtaking view inspired him to indulge in "one last little bit of schizophrenia" before boarding the spacecraft: "You can face due south toward Miami and close your left eye, and all you see with your right eye is this gigantic pile of machinery . . . Then if you open your left eye and close your right eye, you see the Florida that Ponce de Leon saw, the primeval beach undisturbed. You see not a soul, only the earth and the sky. That's kind of an interesting juxtaposition when you're about to go to the moon. You have the feeling of this pile of junk you see out of your right eye being transferred into the virgin reaches of the sky that you see out of your left eye. And that makes you wonder whether the two are going to coexist happily."

Few other astronauts shared Collins's bifocal vision or his gift for poetic description. But the view from the top of the launch tower prompted all of them to take one last long look at the big picture. Rather than contemplating glimpses of the primeval coastline, many couldn't help but dwell on the sight of the Saturn 5.

"Your attention is really on the rocket," recalls Apollo 14's Stu Roosa. "Charley Duke [of Apollo 16] and I worked in launch operations, and we knew the booster systems better than anyone else in the astronaut office because that was our job. But this doggone Saturn 5 launching was like giving birth. You start with an inanimate hunk of metal, and over the long period of the countdown, you breathe life into this thing. And by the time you release those hold-down arms, it's got a mind. You've created almost a live object . . . So when you're standing out there on the gangplank, and this thing's venting, and there's six hundred pounds of ice on the sides, you know it's just a few hours away from really coming to life."

These eleventh-hour existential musings were invariably interrupted by a wiry martinet with black-rimmed glasses and a German accent. He was Guenter Wendt, the irrascible "launchpad führer," and he would grin with childlike anticipation as he ordered the three crewmen to hand over the mysterious brown paper bags they had smuggled out of the suit room.

Unlike Von Braun, Wendt was not an alumnus of the Peenemünde group. He was a former flight engineer who completed his hitch in the Luftwaffe about a year before V-E Day and followed his divorced father to St. Louis, where he drove a delivery truck until his immigration problems were resolved. He later signed on with McDonnell Douglas, and became the pad leader for Project Gemini, cracking the whip on both his NASA minions and outside contractors like his own employer. As Apollo 12's Pete Conrad once observed, "It's easy to get along with Guenter—all you have to do is agree with him."

Yet, all twenty-four moon voyagers appreciated Wendt's authoritarian insistence on perfection more than their token going-away presents could ever express. Many believed that the Apollo 1 fire, which occurred shortly after North American took over as prime contractor, would never have been allowed to happen if Wendt had been in command. Chief astronaut Deke Slayton evidently felt the same way, and pulled the necessary strings to rehire him to supervise every subsequent launch starting with Apollo 7.

Happily, the launchpad führer's obsession with making his rockets run on time did not prevent him from popping a few surprises of his own. On the hectic morning of the Apollo 11 launch, for example, the three astronauts awoke to find the walls of the suit room papered with signs that read, "The Key" and "To the Moon" and "Is Located." When the crewmen arrived at the top of the launchpad a few hours later, Wendt presented them with a four-foot-long foam rubber "key to the moon" covered with nonflammable silver masking tape. His only lament, as he recalled afterward, was that the key was so big "they couldn't take it along."

When the officially scheduled "thirty seconds of recreation period" on the launch tower was over, Wendt would lead the three crewmen across the gangplank and into the tentlike "white room" shielding the hatch of the Apollo spacecraft. Airtight and surgically sterile, the white room was the last checkpoint the astronauts had to pass before entering the command module. As soon as the pad techs in the white room finished inspecting their pressure suits, helmets, and oxygen hoses one more time, the astronauts commenced the gymnastic boarding ritual former gymnast Alan Bean aptly refers to as "saddling up."

The protocol of the boarding procedure was determined by the seating assignments. The mission commander (CDR) always sat on the left next to the abort handle and therefore boarded first. The lunar module pilot (LMP) usually occupied the right side couch and got in second. The command module pilot (CMP), who would not land on the moon but would be responsible for piloting the mothership into and out of lunar orbit, usually occupied the middle couch beneath the main control panel and boarded last.

Because of the Saturn 5's vertical posture on the launchpad, all three crewmen had to perform some apelike acrobatics to squeeze their big white bloated "moon cocoons" inside the spacecraft. One by one, they kicked off their yellow Earth-walking galoshes, grabbed the short metal bar above the hatch, and plunged in feet first, swinging their legs and hips as far upward as they possibly could in one spasmodic motion so that the soles of their boots, like the padded leather couches in which they would sit, were facing toward outer space. As Apollo 15's Jim Irwin describes the maneuver:

"There are several men on the outside and two more on the inside who guide our bodies as we slide into the couches, and we welcome their assistance because we're so bulky with those big pressure suits on . . . Once they get us into proper position, they attach the hardware to the suit—the electrical umbilical that will bring power in and take out biomedical perimeters and voice communication, the umbilicals that bring in oxygen and circulate the air we exhale . . . Then they attach the seat belts with the shoulder harnesses, and they pull them down very tightly because they want us absolutely anchored to our couches for the acceleration of lift-off."

Although the couches had been custom molded to the dimensions of their occupants, the crewmen always found their cushioned recliners insufferably snug on the launchpad. Their discomfort was compounded by the fact that they were unable to sit or lie back in conventional fashion.

Instead, they literally had to hang upside down from the rafters with their feet locked in titanium clamps bolted to a crossbeam directly above their heads.

To make matters worse, the mission commander and the lunar module pilot could not look out their windows because the launch escape tower was still shrouding the sides of the spacecraft. The command module pilot in the middle couch was the only one able to get a glimpse of the outside world, and to do so, he had to squint through a tiny steel-collared portal located up near the nose cone. But the CMP's pinhole view was often quite stunning.

"I had the only window at this point," recalls Apollo 16's Ken Mattingly, "and doggone if the moon wasn't visible in the daylight right straight out the top of the spacecraft . . . I couldn't help but think to myself, 'I know they're doing their job right because the moon's right there in front of us, and that's where we're pointed, and they're just gonna launch us straight to that thing . . .' "

For the next half hour or so, the three crewmen would try to dangle motionlessly as one of the astronauts assigned to their backup crew policed the cabin of the command module. This faithful understudy had already been out on the launchpad for several hours running through a checklist 447 items long. He had dozens of instrument lights, control knobs, and switch positions to identify, verify, rectify, and certify, and he would go about his tasks as methodically as if he himself were about to be launched into space that day.

Finally, the backup would pat each of the prime crewmen on the shoulder, wave farewell, and scramble back out to the gangplank. Then he would lean against the hatch cover, and signal Wendt and the pad techs to help him push it shut.

What followed was by far the most primitive of all Launch Day rituals. Opening the hatch from the inside required no more than fifteen seconds thanks to improvements made after the Apollo 1 fire. But partly because of the added weight of the escape gadgetry, closing the hatch from the outside was a physical ordeal that required the collective grunting and groaning of at least three strong men for a good minute and a half.

When the hatch cover finally dropped in place, the insulation sealed with a pneumatic smack that should have been almost inaudible to the three pressure-suited crewmen inside the spacecraft. But perhaps because they knew it was coming, the astronauts claim that the smack resounded with a loud and clear ring of finality.

"When they closed that hatch it went 'clang' like a dungeon door," remembers Apollo 15's Jim Irwin. "You realized that this was a moment of reality, that this was no longer simulation, that they weren't about to open that door until twelve days had passed and you had returned from the moon. It was too late now to knock on the door and say, 'Fellows, I changed my mind. I'd like to get out.' "

As the astronauts proceeded to run through yet another exhaustive prelaunch checklist, they felt a familiar but always unnerving claustrophobia. Their spacecraft consisted of three main sections: the command module (CM), the service module (SM), and the lunar excursion module (LEM or LM). But the service module, which measured twenty-four feet seven inches in length and twelve feet ten inches in diameter, was an uninhabitable tube full of engines and fuel tanks. The lunar module, a two-passenger vehicle twenty-two feet eleven inches long and fourteen feet one inch in diameter, was inaccessible on the launch pad.

For most of the voyage to and from the moon, the three crewmen had to remain in the cramped confines of the command module. The cone of the CM was only twelve feet high and twelve feet ten inches wide, and the usable volume of space inside the equipment-crammed cabin was considerably less than that. Most of the cockpit area was taken up by the three crew couches, each covered with fireproof Armalon fiberglass stained battleship gray. Below the couches was a one-foot-wide crawl space that could double as a sleeping berth. Below that was the lower equipment bay, an area about the size of a small shower, which contained the sextant and the navigational platform, the water spout, and the food lockers. But prior to lift-off the astronauts had to stay strapped in their couches three abreast.

"It was like you had squeezed three guys in the front seat of a Volkswagen Beetle," recalls Apollo 16's Charley Duke. "We were sitting shoulder to shoulder, touching each other. If I moved, Ken Mattingly [the command module pilot in the center seat] could feel it. And if he moved, John Young [the commander in the left seat] could feel it. Once you got into Earth orbit, you could take off your suit, and go down to the lower equipment bay, and you could stand upright or float upright in there. So when you were in space, in weightlessness, you felt like there was plenty of room. But when you were on the ground, you felt like there wasn't any room at all."

The walls of the command module cabin were painted battleship gray to match the couches, and featured a mind-boggling array of over 600 gauges, dials, control switches, and circuit breakers interconnected by over fifteen miles of electrical wiring. The main control panel located above the CMP's center couch itself consisted of 300 separate switches, forty-eight warning lights, two artificial horizon monitors, and a computer keyboard and display. (A second computer was located in the lower equipment bay.) On the right armrest of the commander's left-side couch was an Attitude Controller Assembly which enabled him to guide the spacecraft's pitch, roll, and yaw manuevers. On the left armrest was the Thrust-Translator Controller, which allowed him to control the spacecraft's acceleration along three directional axes, and a T-shaped abort handle.

The wall, ceiling, and floor spaces in between the instrument panels were punctuated with hundreds of tiny black strips and squares of Velcro. These prickly fibered "male" patches were designed to mate with the soft fuzzy "female" patches glued to the hand tools, pens, scratch pads, and eating utensils stashed in the storage compartments in the lower equipment bay. Commonly used as apparel and fabric fasteners on Earth, the strategically located Velcro adhesives would provide the astronauts with convenient places to "put things down" in the weightlessness of outer space.

On Launch Day, the stark technological decor of the command module was usually littered with scraps of handwritten memos from the crewmen to themselves. Most of these notes contained cryptic phrases like "BOIL > 50" and "S-BAND AUX TO TAPE 90 SEC PRIOR TO DUMP." These uniquely human mnemonic messages helped the astronauts remember such things as last-minute revisions in the guidelines for monitoring cabin radiator temperature or when to turn on their onboard tape recorders to document key phases in the flight plan.

Thanks to some ingenious NASA engineering, the Apollo spacecraft used the same basic fuel substances both for propulsion and for the sustenance of the three crewmen. The service module, which would do most of the propulsion work after the main booster dropped off, took 41,000 pounds of LOX and various hypergolic liquids. The lunar module held 23,245 pounds. The command module, which would not detach from the SM until shortly before reentering Earth atmosphere, carried only three hundred pounds. But the fuel cells of the CSM combination were equipped to convert unburned hydrogen and oxydizer residuals into good old H_2O, which the astronauts could then use for drinking and food-mixing water.

The first order of business after the hatch was sealed called for the commander to find out if all spacecraft systems were operating properly. He would begin by making sure that neither the crewmen nor their backup had accidentally bumped any switches or buttons during the bustle of the boarding procedure, dutifully radioing a detailed readout of his inspection to the ground controllers. Next, he checked the emergency detection system, and confirmed that it was activated. He then tested the rotating gimbals on the service module engine to verify that they would turn in response to the hand lever beside his couch.

Finally, as a precaution against a repeat of the Apollo 1 tragedy, the three crewmen purged the cabin of the pure oxygen they had been breathing since they left the suit room, and switched to a less flammable mixture of 60 percent oxygen and 40 percent nitrogen. Later, if and when they managed to launch safely into outer space, they would purge out the intestine-bubbling nitrogen, and switch back to 100 percent oxygen. Along with eliminating the need for extra nitrogen tanks and mixing mechanisms, breathing nothing but pure oxygen would help them avoid painful intestinal cramps and, perhaps even more important over the long haul in their closely confined spacecraft, help reduce flatulence and cabin odors.

By the time the crewmen completed their second prebreathing exercises of the day, the countdown to lift-off was approaching the one-hour mark—T minus sixty minutes, and counting—and they suddenly found themselves hanging by their boot clamps with nothing left to do until well after launch.

True to character and training, many Apollo astronauts claim they tried to keep busy by immersing themselves in the checklists for the next stages of the mission. One practiced high-speed instrument readouts. Another studied photos of his intended lunar landing site shot by unmanned reconnaissance probes. A third reviewed his role in the separation, transposition, and docking maneuvers the astronauts would have to perform later that day after thrusting out of Earth orbit toward the moon.

The Apollo 11 crewmen spent much of their wait on the launchpad worrying that an extra pouch on the left leg of commander Neil Armstrong's pressure suit might snag on the abort handle. Apollo 12's Dick Gordon was convinced that his launch would be postponed until the following day because of lightning, and drifted off to sleep. But most of the other astronauts say all they did was wait and wait and wait.

"After you've done everything you can do in the cockpit," recalls Apollo 16's Ken Mattingly, "there's a long period of time in there when you've got nothing left to say and you don't know any new jokes to tell . . . Typically, people who are going out to play in a football game or compete in some sport can do something to occupy themselves until it's time for them to actually participate. They're psyched up for only one thing, but they can't release that energy just yet, so they make small talk or play silly games just to pass the time. But there weren't many of those things you could think up to do in the spacecraft . . . so you just wait."

The crewmen of Apollo 15, the first "moon rover" mission, had to endure an extended wait on the launchpad due to booster problems, but in contrast to Apollo 12's Gordon, who slept away his stay, they anguished over each nerve-racking minute of their hold. "We had to lie there several hours waiting as they continued to test the rocket," remembers Apollo 15 lunar module pilot Jim Irwin, "and I must admit that during that waiting period, there were times when I was a little tense, times when my heart beat a little faster . . . Those were anxious hours of anticipation, of expectancy . . . And yet the time went by so slowly . . ."

The wait was made more maddening for each Apollo crew because from this point forward their fate was almost entirely beyond their control—at least until the spacecraft inserted itself into Earth orbit. For the remainder of the countdown, all the major decisions—except the commander's prerogative to throw the emergency abort handle—were up to the men on the ground and their high-speed computers.

The official nerve center of every Apollo flight was located over a thousand miles from the launchpad in the Mission Operations Control Room (MOCR) on the second floor of the Mission Control Center (MCC) building at the Manned Spacecraft Center (MSC) in Houston. This amphitheater full of acronym-babbling technocrats had already become a familiar scene to millions of American television viewers during Project Gemini. But here again, appearances did not always correspond with reality. Apollo 16's Charley Duke describes Houston from the insider's point of view:

"When you look at the control room on TV it looks huge, but it's actually a very small facility, maybe sixty feet by thirty feet. There were three main tracking screens on the wall with a lot of clocks and telemetry information display lights, then there was a big wide area in front of the screens that went back in four tiered rows like a movie theater except they were full of computer consoles. But there were only about four main stations on each row, and you never had more than thirty-five or forty people in the room at any one time . . .

"The first row was nicknamed 'the trench,' and that's where the systems monitoring teams sat, and you had a separate station for each system—one for the lunar module systems, one for guidance and navigation, one for propulsion, one for control. In the next row, they had an area for monitoring the Saturn 5 booster, a console for the command module systems, a place for the flight surgeons, and a station for the CAPCOM who is in radio contact with the flight crew.

"The third row back was for the flight director, the mission planning guys, and the public affairs officer who announced what was happening over the intercom to the press room. Behind that, there was a row for the bosses, the senior managers of NASA. And then behind all that, there was a glassed-in area called the viewing room, which was for the VIPs, people who had been very interested in the space program but weren't actually involved in the day-to-day operations."

In addition to the VIPs, there was also an unseen presence at Mission Control: the computer. Or to be more precise, the computers, for there were actually five of them. They were housed in the Real Time Computer Complex (RTCC), then touted as "one of the world's largest computer centers," which took up virtually the entire first floor of the building. Wired directly to the consoles and display panels in the control room, each computer was an IBM 360 Model 75J mainframe capable of processing 1.3 million bits of information per second. The prime contract for IBM's installation, programming, and maintenance of these high-speed electronic thinking machines totalled $186 million.

Of course, as the Apollo astronauts and their guardian angels at Mission Control were well aware, the RTCC computers were not infallible. Not by a long shot. They, too, made mistakes and miscalculations large and small which could only be caught, corrected, and properly compensated for by their human keepers. Contrary to widespread public misconception, the men up in the control room were the true masterminds of every lunar mission. They couldn't add, subtract, or multiply figures as fast as the computers, but they could come up with alternative ways to succeed if and when the computers failed. They were also the ultimate decision makers. They could overrule an automatic GO or NO GO signal. And, if worse came to worst, they could manually trigger the spacecraft's abort mechanisms at any time.

The ringmasters of the control room show were the flight directors, the on-site commandants respectfully addressed as "Flight." Supervised by former Gemini controller Christopher Columbus Kraft, their roster included the reliable Glynn Lunney and future MSC chief Gerald T. Griffin.

But the most famous and striking Apollo flight director, the one who somehow wound up being the man in charge in Houston during the most critical hours of almost every mission, was Eugene Kranz.

Always distinguished by his starched white vest, Gene Kranz was a rock-jawed former fighter pilot with a blond crew cut and a very cool head. Kranz was the flight director on duty when a series of computer alarms nearly forced the Apollo 11 astronauts to abort man's first attempt to land on the moon. He was also at the helm in Houston when Apollo 13's oxygen tanks exploded, and led the contingency planning team that figured out how to get the astronauts home alive. Like all his fellow astronauts, Apollo 16's Duke gushes with praise for Kranz:

"Gene Kranz was like a drill sergeant in the marines, barking orders at everybody in sight. But he also had the ability to take in just about everything that was going on all around the room simultaneously, make sense of it, and know what to do next. He always wore that white vest because he had worn it during one of the Gemini missions that got in trouble and then pulled through okay, and it was supposed to be good luck. But his success as an Apollo flight director had a lot more to do with hard work than luck."

Every bit as dedicated as the astronauts, Kranz's troops were by definition a much more diverse and odd-looking lot. Many were either former military pilots like their boss or former airfield controllers who wore short-sleeved white shirts and skinny black ties. But the ranks also included peach-faced engineering prodigies fresh out of colleges like MIT and Cal Tech, balding computer specialists with Coke-bottle glasses, and paisley-shirted systems analysts with sideburns, goatees, and ongoing mid-life crises.

Unlike the computers, the men in the trench could not and did not try to work nonstop from Launch Day to splashdown. Hoping to maximize manpower and alertness, they maintained Houston's round-the-clock vigil by dividing into three separate but equal teams rotating every eight hours. The White Team handled the day shift. The Red Team took over in late afternoon or early evening. The Black Team kept the night watch.

And yet, as even Kranz acknowledged, Mission Control, for all its wizardry, was a windowless world of make-believe. Except for remote TV footage of the Saturn 5 available before and shortly after launch, the men in the trench could not see the mission in action, only a simulated representation of the real thing. The spacecraft on their main tracking chart was a little green light ("a bug with an oil can up its butt," in NASA vernacular). As the mission progressed, the green bug would crawl across the tracking chart at so many thousand-miles-per-inch. But that reassuring little light show was only a computer-simulated guestimation of the actual flight based on telemetry data.

The Apollo astronauts knew the inner workings of Mission Control from firsthand experience, for at least one of them had to be on the premises during every phase of every lunar mission. Although everyone in the control room could monitor what was happening, only one individual—the CAPCOM—was authorized to speak directly to the astronauts aboard the spacecraft. And the CAPCOM was always an astronaut who hoped to fly to the moon himself one day and/or already had, for NASA officials believed that a fellow astronaut would have a natural rapport with the three Apollo crewmen, an insight into how they thought and felt, which might make the difference between life and death in a crisis.

The Apollo astronauts also knew that Launch Day was the one day when Houston did not call all the shots, not even the most important shot of all. The men in the trench would not take charge of the mission until after the Saturn 5 cleared the top of the launch tower scaffolding. All major events prior to that time were managed and monitored by another team of technicians much closer to the action—the crew of Launch Control at the Cape.

Back in the early days of the manned space program, the Launch Control Center (LCC) was huddled in a concrete bunker about a mile from the launchpads. By the advent of Project Apollo, it had moved to a more spacious new facility in the VAB complex. The LCC had roughly six times more equipment and personnel per shift than Mission Control because it was in charge of all launchpad activities as well as the actual lift-off and the first ten seconds of flight.

Like the main arena in Houston, the firing room at Launch Control was full of computer consoles and wall-size tracking screens, but it was not blind to the outside world. There were no less than sixty TV screens showing the Saturn 5 and its passenger ship from sixty different angles. In the viewing room behind the trench area, there was also an enormous picture window affording a panoramic view of the Merritt Island launch complexes three and a half miles away.

Although the launch controllers were kindred spirits of the men at Mission Control by training and expertise, they were also proudly independent by nature and job description. They were the NASA equivalent of blue-collar technicians, the men who literally had to get the mission off the ground by doing all the "dirty work" associated with the care and feeding of the Apollo/Saturn 5. Appropriately, the director of launch operations at the Kennedy Space Center was a powerfully proportioned former West Point football player named Rocco A. Petrone, affectionately known as "Rock."

"This is Apollo-Saturn Launch Control. We've passed the eleven-minute mark. Now T minus ten minutes fifty-four seconds in our count-down . . ."

The voice of the public affairs officer in the firing room boomed out of the loudspeakers at the VIP grandstands and into millions of homes around the world via network TV. As the countdown proceeded, the PAO provided a play-by-play commentary worthy of a professional sports-caster. But the three astronauts out on the launchpad heard another voice.

The commentator who kept the Apollo crewmen apprised of the countdown was nicknamed "Stoney," and he was the LCC counterpart of the CAPCOM in Houston. His unlikely radio handle was a legacy from the days when the LCC was buried in a concrete bunker. Like CAPCOM, Stoney was always an astronaut. He was also the man who would, if all went well, call out the magic words his fellow astronauts had been waiting so long to hear: "Three . . . Two . . . One . . . Zero!"

Yet, even the men in the LCC firing room would not call all the shots, for they were ultimately more computer-dependent than Mission Control. At T minus three minutes, the countdown would go on automatic sequence all the way to ignition and lift-off. If something went wrong, the launch controllers could still activate the abort mechanisms at any time. But if the Saturn 5 was to launch successfully, the LCC master computer had to execute its multimillion commands per second programs without fail.

At T minus five minutes and counting, Rocco Petrone's troops would begin their final preautomatic GO/NO GO check. Console by console, the men responsible for each major system and subsystem answered the roll call.

Guidance?

"GO!"

Communications?

"GO!"

Environmental control?

"GO!"

Abort systems?

"GO!"

Computer systems? Engine systems? First stage? Second stage? Third stage? Lunar Module? Service module? Command module? Instrument Unit? Launch Escape Tower?

"GO! . . . GO! . . . GO!" echoed throughout the firing room.

Last but by no means least came the three Apollo astronauts waiting atop the Saturn 5. The firing room momentarily fell silent as Stoney popped the big question: What says the flight crew?

"We're GO," the mission commander would report.

Stoney would relay the message loud and clear: "We're GO for launch!"

The men in the LCC firing room would throw the switches that let the machines take over the countdown, then draw a collective breath as the PAO announced: "This is Apollo/Saturn Launch Control . . . We have just passed the three-minute mark . . . We've had the firing command. That's the signal that the automatic sequence is now in . . . The remainder of the count will be handled by the master computer here in the firing room as various events click off leading up to the ignition of the five engines in the first stage of the Saturn 5 . . . with lift-off coming at the zero mark in the count."

Meanwhile, inside the command module, the astronauts suspended upside down in their padded couches were often surprisingly serene, especially if they had launched into space before. Apollo 11 commander Neil Armstrong's heart rate peaked at 110 beats per minute during the actual lift-off, barely enough to work up a visible sweat. CMP Mike Collins peaked at ninety-nine. LMP Buzz Aldrin, the third Gemini veteran on the crew, was the coolest of all, registering a maximum heart rate of only eighty-eight beats per minute at the climax of the launch.

Even so, every trio of Apollo crewmen realized that the automatic sequence signaled the beginning of the final phase of the countdown, and that they were powerless to do anything except stop the show. They couldn't expedite the process by throwing any more switches. They couldn't ignite the Saturn 5's engines. They couldn't steer the booster safely past the launch tower. All they could do was "punch out" by hitting the abort handle, voluntarily or involuntarily.

Doing the latter—aborting by mistake—would be worse than getting blown to bits by a launchpad explosion, according to the unwritten code of the astronaut corps. As Apollo 11's Mike Collins writes, they had amended the old battle cry "death before dishonor" to "death before embarrassment." No matter how moderate their pulse rates remained, all twenty-four men who launched for the moon were tormented by the same premonition. That fear was summed up in the six words Apollo 16's Ken Mattingly silently repeated to himself throughout his own Launch Day countdown:

"It won't fail because of me . . ."

"It won't fail because of me . . ."

"It won't fail because of me . . ."

Moments later, the PAO at Apollo-Saturn Launch Control would announce that the fuel tanks of the Saturn 5 were being fully pressurized, but most of the VIPs and TV viewers in the audience failed to appreciate the pivotal significance of this news. Only those versed in the mysteries of modern rocketry and jet propulsion realized that the countdown had reached a major turning point, and that the booster's huffing and puffing signaled the true beginning of the final launch sequence.

The PAO would quickly preview the events to come:

"We are now at T minus one minute forty-five seconds, and counting. We'll go on internal power with the launch vehicle at the fifty-second mark. At seventeen seconds in the count, the guidance system goes internal—that's the guidance reference release. We already have the proper flight azimuth in . . . Now ninety seconds and counting . . . The astronauts have turned off their ground communication at this time. However, they are on VHF, and of course, the S Band circuits, as well as the special astronaut communication circuit."

Translated, that meant the three crewmen in the spacecraft could still talk to each other and to Launch Control if they desired. But the words radioed from the spacecraft were usually few and far between. As the mission commander kept reassuring the ground that the crew was still GO for launch, his comrades in the adjacent couches usually said little or nothing out loud, thinking, hoping, praying:

"It won't fail because of me . . ."

"It won't fail because of me . . ."

"It won't fail because of me . . ."

The PAO anxiously watched the ticking of the countdown clock. "We're coming up on the sixty-second mark . . . We're at T minus sixty seconds and counting . . . We are GO for a mission to the moon at this time . . ."

During the final minute of the countdown, time suddenly accelerated. Or so it seemed to the three crewmen who had been waiting and waiting and waiting. Not just since they boarded the spacecraft, but all through flight school, grad school, astronaut training, crew assignments and reassignments, production delays, technical setbacks, and countless make-believe missions in the simulators. Waiting and waiting and now finally getting a chance that might come only once in a lifetime.

"It won't fail because of me . . ."

"It won't fail because of me . . ."

"It won't fail because of me . . ."

At T minus twelve seconds, the swing arms bracing the Saturn 5 to the launchpad scaffolding started to pull away, and the ignition sequence began. The ignition sequence required seventeen major steps starting with the initial spark, but they were all accomplished in three seconds flat.

First, a five-hundred-volt charge was shot through the ground cable on the launchpad, and into the trunk of the Saturn 5, where its spark ignited a mixture of highly flammable turboprop gases. These gases then burned off the igniter links whose meltdown touched off an electrical signal. That signal opened a four-way valve releasing still other gases whose pressure opened the main oxidizer valve. The oxidizer then ignited the hydrogen gases in the combustion chamber.

This high-speed hypergolic chain reaction provided the power to torque up the five F-1 engines in the bottom stage of the booster. The engines' exhaust shot out the rear nozzle into an enormous concave flame deflector composed of volcanic ash and calcium aluminate, which then channeled it down a one-hundred-foot-long, forty-foot-deep concrete trench flooded with 50,000 gallons of cooling water per minute.

At T minus 8.9 seconds, the first fireballs would burst out the end of the trench and billow toward the sky. This fiery spectacle confirmed that the Saturn 5 was indeed lit. But because the booster was still locked tightly to the launchpad by four giant metal hold-down clamps, it did not move upward even a fraction of an inch. Instead, it continued to snort smoke and build thrust.

In essence, the theory of jet propulsion as first divined by Konstantin Tsiolkovsky at the turn of the century was a method for tapping the power of the entire universe. Contrary to popular misconception, the Saturn 5 rocket did not have to "push" against the ground to get itself going. Nor did it have to "push" against the air in order to fly. Such a vehicle would be incapable of propelling itself through the vacuum of space. Rather, the rocket got its power by exchanging the finite momentum generated by its own motors for the infinite momentum generated by the gravitational forces of the solar system.

The Saturn 5 engines started this grand sequence of motion by creating enough heat to convert the fuel into high-velocity gas molecules. As the heat caused the fuel vapors to expand, the pressure built inside the cylindrical fuselage, forcing a powerful stream of exhaust out the nozzle at the bottom of the booster. The combined and concentrated downward momentum of trillions of escaping gas molecules known as thrust was the Newtonian "action" that prompted the "reaction," the equal and opposite upward momentum, required for launch.

One million pounds. Then two million. Then three, four, five, six million pounds of thrust. When the Saturn 5's thrust reached 6.5 million pounds, the equivalent of her total body weight, it would have been theoretically possible, in the absence of the hold-down clamps, to push the booster off the launchpad with no more than a gentle flick of the finger. But in order to build up enough bang to blast well out into space and enough power to blast safely and surely to the edge of the earth's atmosphere, the rocket raised her thrust to a full 7.7 million pounds.

The last five seconds of the countdown always passed in a flash. Or so it seemed to the three astronauts inside the command module. "Five seconds isn't a lot of time anywhere," notes Apollo 14's Stu Roosa, "but that last five seconds before launch after the engines are lit, boy, that's just a blur—it goes by extremely fast."

But all three astronauts heard Stoney's voice droning in their headsets:

"Five . . ."

"Four . . ."

"Three . . ."

"Two . . ."

"Launch Commit!"

Stoney usually skipped right past the one-second mark and blurted a resounding "Zero!"

At that instant, all four hold-down clamps flopped back simultaneously. They were designed to accomplish this task in exactly one-twentieth of a second, and on every Apollo launch they did so. But just in case they did not, a set of explosives was timed to blow them back at exactly one-fifth of a second after the zero mark. Upon seeing the clamps release, Stoney and the PAO would simultaneously cry, "Lift-off! . . . We have a lift-off!"

Oddly enough, to those able to view the launch from a few miles away, the Saturn 5 appeared to be frozen on the pad for a fraction of an instant. This was not an illusion. Although the Saturn 5 was starting to rise, she was still being restrained by a set of eight tapered steel pins wired to the launchpad. The purpose of these pins was to moderate the speed of the booster's initial upward lunge so that it did not shake the launch scaffolding and/or the three crewmen aboard the spacecraft to pieces.

In less than half a second, the Saturn 5 would be about three-quarters of an inch off the ground. The tapered steel restraining pins would slip out, and the service bridges of the launch structure would detach. At one second after lift-off, the booster would begin a preprogrammed yaw maneuver intended to steer it away from a fatal sideswipe of the scaffolding.

The Saturn 5 often appeared to waver in the wind as she climbed past the service bridges. Again, this was not an illusion. The booster was actually designed to do just that to reduce stress on its delicate fuselage. Instead of resisting contrary gusts of wind, it bent with them, and trusted its computer guidance system to calculate, correct, recalculate, and re-correct its vertical yaw trajectory accordingly.

But even from a distance or on TV, the Saturn 5's ascent was awesome to behold. The avalanche of ice crashing and splitting off the sides of its cryogenic tanks and the five-thousand-degree Fahrenheit red orange holocaust pouring out from its exhaust nozzle could be seen clearly in all their frightening glory from the VAB complex, from the VIP grandstands, and from all the spectator-packed thoroughfares and waterways within a five-mile radius. The wail of the booster's F-1 engines and the earth-shaking tremors of her 7.7-million-pound thrust could be clearly heard and viscerally felt for up to ten miles in every direction.

For better or worse, the three crewmen atop the Saturn 5 could not share this apocalyptic vision of the launch, for they had no remote TV monitors on board. Nor could they observe the Saturn 5's climb up the launch tower because all the side windows were covered by the escape booster. Insulated in their "moon cocoons," they couldn't even hear the rocket's full roar. But all three could definitely feel the quake of lift-off and the low-frequency shock waves of the booster's thrust. And as Apollo 16's Ken Mattingly reports, "It feels just like it sounds."

That feeling, Mattingly adds, was a shock in itself. He simply did not expect a booster as big as the Saturn 5 to shimmy and shudder so violently: "Even though I was lying there feeling the booster vibrate back and forth, the very idea was hard for me to accept. Here's this huge thing, and I know intellectually how heavy it is. And yet it's shaking and rattling like a very small, light object on a railroad track."

Fellow Apollo 16 astronaut Charley Duke says he, too, was stunned by the turbulence of the lift-off: "I just wasn't aware that it was going to shake as much as it did. Everybody had told me about the noise and the slow acceleration. But the fact that it was vibrating so violently was a real surprise to me. [Mission commander] John Young was saying, 'We're GO, and Houston's GO, and everything's GO,' and I was just sitting there shaking like crazy. I really wasn't mentally prepared for that. My first reaction was, 'Gosh, this thing can't fly. It's going to shake to pieces.' "

Apollo 17's Jack Schmitt adds that the sound of the launch, almost as much as the physical shaking of the Saturn 5, was what made it feel so violent: "The low-frequency vibration is the unusual thing that you'll never get off the television . . . Some of the films of the crowd that's watching the launch show them breathing very deeply. It's the sound people are reacting to . . . It's that low-frequency sound that's moving your stomach and your innards, your tie and your blouse . . . That's what's different about the launch that just doesn't come out in any audio or video representation . . . It's what makes it an emotional experience even for the viewers."

Schmitt adds that the Saturn 5's shaking also gave him cause for alarm because it blurred his vision of the instrument displays and dials: "It was very hard to monitor the spacecraft systems that you were used to monitoring in simulations. I had a feeling that if we really had a problem, the adrenaline flowing through our veins would have frozen our eyes on the instrument panel very fast. But at the time I remember having the thought, 'With this instrument panel vibrating like this and not being able to read any of the numbers on it, what did we go through all those simulations for? To abort?' "

In addition to blurring the astronauts' vision and rattling their skulls, the Saturn 5's vibrations posed a potentially fatal danger of collision until the booster cleared the top of the launch scaffolding, which took an average of ten full seconds. But even then, the three crewmen faced other equally dangerous obstacles for at least another five heart-throbbing seconds.

"You're most anxious during the first fifteen seconds after lift-off," recalls Apollo 17's Gene Cernan, "because if you lose one of those five engines in the first stage of the Saturn 5 booster, you're going to come right back down on the pad. After that, you've burned off enough fuel to where you're light enough that you can sustain yourself with the thrust of just four engines, and get into Earth orbit. But those first fifteen seconds are very critical. You need all the thrust those five engines can produce for you."

Even those astronauts who claimed to have enjoyed a relatively smooth lift-off say they shared Cernan's fears.

"The launch is the time when you are more conscious than any other time of your dependence on machinery and the frailties of human design and execution," says Apollo 11's Mike Collins. "You worry mightily that the rocket is going to veer off course or stop or explode or do some other terrible thing to you. So you spend your time trying to watch your instruments like a hawk, literally second by second, unable to keep your eyes off them, hoping everything is going to be all right, and trying to remember what you're supposed to do should one of the instruments tell you that things are not all right."

Ten seconds after lift-off, the three crewmen would hear Stoney proclaim, "Tower clear!"

Two seconds later, the Saturn 5 would begin an automatic pitch and roll, spiraling out of her ninety-degree launch trajectory into a graceful upward arch for smooth insertion into Earth orbit.

At first, the three Apollo crewmen barely could feel the rocket's realignment because the sound of Stoney's sign-off was still reverberating in their headsets. In spite of the violent shaking, their blurred vision, and battered nervous systems, almost all the astronauts say they remember the triumph of launch as the most thrilling and uniquely satisfying experience they have ever had.

"There were tears of joy streaming down my face that morning," recalls Apollo 15's Irwin. "There was a moment there of just supreme elation, complete release of tensions, almost the happiest moment of my life . . . To realize that after all these years of training and preparation and education. 'At last it's paying off. At last I'm leaving the earth and I'm destined for the moon!' "

4

THE SPACE RACE

U.S. vs. U.S.S.R.

1957–1967

I looked and looked and looked
but I didn't see God.
SOVIET COSMONAUT YURY GAGARIN,
April 14, 1961

On October 4, 1957, the Russians changed the course of twentieth-century history by putting the first man-made satellite into Earth orbit. Appropriately named Sputnik ("fellow traveler"), it was a 184-pound aluminum ball about twenty-two inches in diameter spiked with four directionally positioned whip antennae. It reached an altitude of 560 miles, and stayed aloft for ninety-two days, orbiting the earth 1,400 times, or roughly once every ninety-five minutes, at an average speed of 18,000 m.p.h. Sputnik did not carry a bomb or a spy camera, but its primitive electronic transmitter emitted a constant stream of staccato "pips" that the Soviets triumphantly rebroadcast all over the globe.

Sputnik alarmed the unsuspecting citizens of the non-Communist world. Noting that the satellite's launch confirmed that the Russians had an ICBM powerful enough to deliver a nuclear warhead, one high-ranking U.S. military officer called it the "Pearl Harbor of the Technology War." Rumors that the Russians intended to launch a military attack from outer space spread throughout the U.S. and her NATO allies with every pip, prompting future British foreign secretary George Brown to ask a visiting U.S. official, "Do you Americans really know what you are doing?"

On November 3, 1957, while the U.S. was still reeling from the shock of Sputnik I, the U.S.S.R. launched Sputnik II, an 1,100-pound satellite carrying a dog named Laika. Though Laika eventually died of suffocation, Sputnik II stayed aloft for over 150 days and 2,370 orbits. Later that same month, the Air Force successfully test-launched the first U.S. intercontinental ballistic missile, the Atlas ICBM, thereby regaining military parity with the Soviet Union. On December 6, 1957, the Navy attempted to put the first U.S. satellite into Earth orbit, but the Vanguard I rocket lurched one inch off the launchpad and exploded. "Washington Humiliated," screamed a front page headline in *The New York Times* the next day. "Congressmen Fear Blow to U.S. Prestige, Especially in View of Advance Publicity."

Fortunately for the U.S., President Dwight D. Eisenhower had authorized Von Braun and the Army to proceed with a substitute satellite project shortly after Sputnik II. On January 31, 1958, the Army launched the 18.13-pound Explorer I. In March of 1958, after two more frustrating launchpad explosions, Explorer III and Vanguard III joined Explorer I in Earth orbit. Eisenhower boasted that the U.S. had now sent more satellites into space than the Russians. He also claimed that the equipment on board, which measured potential obstacles to space travel like cosmic rays, meteorite traffic, and the ionized particles of the Van Allen radiation belt, made the U.S. satellites more scientifically valuable than Sputnik I and Sputnik II.

The U.S.S.R. promptly regained leadership of the space race in May of 1958 by launching Sputnik III, a 2,900-pound vehicle carrying instrumentation equal to that of the next half dozen proposed U.S. satellites combined. Dismissing the much smaller and lighter American space probes as "oranges," Soviet premier Khrushchev alleged that the U.S. was only interested in using satellites for military purposes, and called for the creation of a joint space exploration program aimed at promoting world peace. He also announced that the U.S.S.R. would soon attempt to put a man in space.

Eisenhower simply could not understand why the Soviets were willing to spend billions and billions of rubles to win what he still regarded as basically a propaganda war. Though the president kept insisting that the "space race" was a Communist plot, the Democratic opposition led by Senator Lyndon B. Johnson of Texas and his Defense Preparedness Subcommittee called for an all-out effort to beat the Russians at their own game starting with the formation of a new civilian space agency.

In July of 1958, Eisenhower did an about-face, and signed the bill creating the National Aeronautics and Space Administration. NASA's original mandate was to develop America's aeronautical and space exploration potential "for the benefit of all mankind." The agency charter made no mention of manned space missions or the idea of attempting to land on the moon. But the very fact of NASA's creation signaled that the U.S. was officially entering the space race. In the fall of 1958, NASA announced Project Mercury, a presidentially approved program to launch an American astronaut into orbit.

Although Eisenhower got historical credit as the president who gave birth to the U.S. space program, he continued to doubt the value of the entire enterprise. The man he named as NASA's first administrator was T. Keith Glennan, a veteran of the Atomic Energy Commission and the National Science Board. Glennan considered it his duty to keep a close rein on alleged "space cadets" like Von Braun who dreamed of building giant rockets for manned voyages to the moon and Mars. When Glennan announced the selection of the first NASA astronauts in the spring of 1959, he secretly hoped that the Original Seven would be the last of the breed.

Then, as if on cue, the U.S.S.R. inadvertently gave NASA another shot in the arm. In September of 1959, the unmanned Lunik II probe completed the last leg of a 240,000-mile one-way trip to the moon, and crashed to the lunar surface, a feat comparable to shooting a needle through the eye of a fly six miles away. On October 3, the Lunik III probe successfully achieved lunar orbit, and sent back the first photographs of the far side of the moon. The latest Soviet space triumphs sparked a new round of political hysteria in the U.S. The Democrats declared that the Communist conquest of the moon posed a major threat to national security, and proceeded to make the "space gap" and the "missile gap" major issues in the 1960 presidential campaign.

In the summer of 1960, NASA released a ten-year plan proposing that Project Mercury be followed by Project Apollo, an effort aimed at making a manned circumlunar voyage by 1969. But when Eisenhower's scientific advisory panel submitted a report estimating that a related proposal for a manned lunar landing by 1975 would cost up to $46 billion, the lame-duck president promptly vetoed the idea.

Ironically, President John F. Kennedy did not favor a manned mission to the moon when he took office in January of 1961. At the time, his scientific adviser Dr. Jerome B. Wiesner was urging him to concentrate on cheaper and less risky unmanned space projects. Then a dramatic series of events changed Kennedy's mind. On April 12, 1961, the Soviet cosmonaut Yury Gagarin became the first man in space. After completing one orbit around the globe and parachuting safely back to Earth, Gagarin announced, "I looked and looked and looked, but I didn't see God." Three days later, Cuban premier Fidel Castro decisively repelled the U.S. supported Bay of Bigs invasion.

According to former Kennedy confidant Theodore Sorenson, the combined shocks of the Gagarin flight and the Bay of Pigs debacle made the new president recognize that "prestige was a real, and not simply a public relations, factor in world affairs." Determined to regain the advantage in the so-called cold war, Kennedy ordered Vice-President Johnson to prepare a study answering the following questions: "Do we have a chance of beating the Soviets by putting a laboratory in space, or by a trip around the moon, or by a rocket to land on the moon, or by a rocket to go to the moon and back with a man? Is there any other space program which promises dramatic results in which [sic] we could win?"

Less than two weeks later, Johnson submitted a report urging approval of a project to land a man on the moon. "To reach the moon is a risk," the vice-president noted, "but it is a risk we must take. Failure to go into space is even riskier . . . In the crucial areas of our Cold War, in the eyes of the world, first in space is first, period. Second in space is second in everything."

Kennedy still had serious reservations. Confidential polls showed that a clear majority of the American public opposed the idea of spending billions of dollars to land a man on the moon. Then on May 5, 1961, NASA launched the first American astronaut into space. Alan B. Shepard's fifteen-minute suborbital flight was only a cheap shot compared to Gagarin's orbital mission, but the nation's overwhelmingly enthusiastic reaction convinced Kennedy that it was time to assert the kind of visionary leadership he had promised in his 1960 campaign.

On May 25, 1961, Kennedy issued a historic challenge in a speech before a joint session of Congress: "This nation should commit itself to achieving the goal, before the decade is out, of landing a man on the moon, and returning him safely to Earth. No single space project will be more important to mankind or more important for the long-range exploration of space; and none will be so difficult or expensive to accomplish."

At first, even the NASA "space cadets" were in shock. They realized that this was the chance to make their wildest dreams of space exploration come true. But they also knew how far the U.S. had to go simply to catch up with the Russians in manned Earth-orbiting flights. "When I heard President Kennedy announce that American spacemen would land on the moon, I could hardly believe my ears," Dr. Robert R. Gilruth, the future director of the Manned Spacecraft Center, admitted years later. "I was literally aghast at the size of the project."

Then, as if by magic, the U.S. started closing the gap in the space race. On July 21, 1961, Gus Grissom made the last suborbital Mercury flight in the *Liberty Bell 7.* In late October, just as Arthur C. Clarke was publishing a timely science fiction novel entitled *A Fall of Moondust,* Wernher Von Braun successfully test-fired a Saturn 1 rocket that generated a then unprecedented 1.3 million pounds of thrust. Then on February 20, 1962, John Glenn became the first American to orbit the earth, circling the globe three times. Like his predecessors, Glenn was honored with a ticker tape parade through Manhattan, but his reception was by far the biggest and most enthusiastic to date. In the eyes of most average citizens, Glenn was the first "real" American astronaut, a celebrity status that would later help catapult him to a U.S. Senate seat.

In the wake of Glenn's success, the "space cadets" at NASA began to believe that they might be able to meet Kennedy's challenge after all. But they also realized that it would not be possible to attempt a lunar landing mission until they resolved a fundamental dispute about the best way to send a manned spacecraft from the Earth to the moon. Von Braun advocated a relatively simple but expensive "direct ascent" in which a 600-foot-tall rocket would hurl the astronauts all the way to their touchdown site in one continuous blast. The lesser known but highly respected Dr. John Houbolt proposed a more economical though somewhat riskier "rendezvous" plan in which a two-man landing "bug" would undock in lunar orbit, touch down on the moon, then reunite with the mothership for the return flight to Earth.

In the spring of 1962, NASA announced the creation of Project Gemini, an evolutionary program designed to provide the missing links between Mercury and Apollo. Using two-man spacecraft mated to Agena rockets and other Gemini spacecraft, the astronauts would test rendezvous and docking procedures in Earth orbit. They would also explore an even more important question: Could man endure the weightlessness and cosmic rays of outer space well enough to fly to the moon and back?

Along with new technology, Project Gemini demanded a new complement of spacemen. On September 17, 1962, NASA selected a second class of astronauts destined to become known as "The Nine" to distinguish them from the "Original Seven." Like their predecessors, "The Nine" were all qualified high-performance test pilots under five feet eleven inches tall and less than thirty-five years old; all were married with children. But unlike the Mercury astronauts, the Gemini bunch included the first civilian astronaut, NASA-employed test pilot Neil Armstrong. Six of "The Nine" would one day orbit the moon; three of them—Armstrong, Charles "Pete" Conrad, Jr., and John W. Young—would also walk on the lunar surface.

Not surprisingly, the price tag for landing the first men on the moon turned out to be far higher than the Kennedy administration's original forecast of $20 billion. From fiscal years 1962–1973, official allocations for Project Apollo amounted to roughly $19.4 billion. But that figure did not include the budgets for the Mercury and Gemini programs, without which Apollo could not have gotten off the ground. Nor did it include appropriations for NASA's administrative, overhead, and general operating expenses or the launching of various unmanned flights, most of which directly or indirectly contributed to the success of the three manned programs. The actual cost of "the greatest adventure on which man has ever embarked" could only be measured in terms of the total space agency budget during those years, which was officially estimated to be $39,316,382,000—or nearly $40 billion in round numbers.

However, crafty Vice-President Lyndon B. Johnson made sure that his pet program flourished by spreading NASA contracts from coast to coast. The most visible symbol of LBJ's space patronage was the Manned Spacecraft Center built near Houston (instead of near the launchpads at Cape Canaveral, Florida) on a patch of salt grass prairie donated by Rice University. LBJ's friends at Humble Oil owned most of the surrounding acreage, which they developed into suburbs; Johnson's even closer friends at Brown & Root got the space center construction contract. The most controversial space patronage was the awarding of the Apollo command module contract to North American Aviation, allegedly in connection with a million-dollar service contract the firm gave to Johnson aide Bobby Baker. But Gemini and Apollo plums also went to California-based divisions of Lockheed, Northrup, and General Dynamics; to northeastern-based units of Grumman, Pratt & Whitney, RCA, IBM, Raytheon, and Thiokol; and to midwestern-based firms like Honeywell, Control Data, and Collins Radio.

If the space agency reeked of pork barreling, the Democrats share-the-wealth policy helped make the project to land a man on the moon a kind of middle-class public works program for nearly half a million Americans. No less than 20,000 private contractors and subcontractors participated in the effort. Though a few big corporations grabbed disproportionate chunks, space agency officials boasted that half of all bids were awarded to small independent firms. The Davids and Goliaths of the private sector joined with government in providing over 400,000 jobs. Though shamefully short on minorities, the employee roster included not only computer whizzes and engineers, but also launchpad technicians, secretaries, welders, pipe fitters, plumbers, electricians, carpenters, and old ladies who glued together the linings of the astronauts' pressure suits.

One of NASA's most remarkable accomplishments in the years ahead was managing and coordinating the world's largest organization not dedicated to war. Much of the credit was due the agency's second administrator, James E. Webb. A former executive of Kerr-McGee Oil Company, Webb was a protégé of Senator Robert Kerr of Oklahoma, the man who succeeded LBJ as chairman of the NASA subcommittee. He had also served on the board of Mercury contractor McDonnell Douglas, which also won the Gemini capsule contract. But despite his close ties to industry, Webb institutionalized a "can-do" attitude by insisting that NASA engineers stay at least one step ahead of their private sector counterparts by knowing more about the design and manufacture of spacecraft systems than the contractors. At the same time, he determined to keep NASA rolling full speed ahead without sacrificing quality assurance and safety by establishing the policy of "all-up testing." Instead of separately testing each spacecraft component, NASA would try all the systems at once. The obvious risk was that the failure of one part might destroy the entire vehicle. But barring cataclysmic explosions, the time and money savings would be enormous.

Project Mercury ended with Gordon Cooper's flight on *Faith 7* in May of 1963. That fall, NASA announced the selection of fourteen more astronauts. Celebrated in the media as the first of the Apollo astronauts but known to their peers as the "Third Group," they were like their predecessors in height, weight, age, and most every other way but one. Though each was an experienced aviator, they were not required to be qualified test pilots. Four members of the "Third Group"—Edwin E. "Buzz" Aldrin, Jr., Alan L. Bean, David R. Scott, and Eugene A. Cernan—would one day walk on the moon. Three others—William Anders, Michael Collins, and Richard F. Gordon—would orbit the moon. Three others—Charles A. Bassett II, Clifton C. Williams, Jr., and Roger B. Chaffee—would be killed during training.

In the meantime, NASA became embroiled in a major budget battle in the wake of the Cuban missile crisis. Though JFK won his famous "eyeball to eyeball" confrontation with the Russians, there was bipartisan pressure for more defense spending. At one point, the penny-pinching Democratic senator from Wisconsin, William Proxmire, threatened to chop half a billion dollars from the NASA appropriations bill. Proxmire eventually settled for a $90 million cut, but it was obvious that NASA's honeymoon with Congress was ending. In hopes of mending fences in preparation for the 1964 campaign, Kennedy embarked on a tour of NASA installations in Florida, Alabama, and Texas. On November 22, 1963, he was assassinated in Dallas.

The whole nation mourned, most especially the NASA workers at Cape Canaveral to whom Kennedy had paid an official visit just a few days before his death. Guards had to be posted around the Saturn booster he had inspected to prevent teary-eyed employees from scrawling his name on the fuselage. Kennedy's bold challenge to land a man on the moon before the end of the decade had so transformed and reinvigorated the space agency that many insiders and outsiders wondered if it could go on without him. Newly sworn-in President Johnson tried to allay such fears right away by announcing that Cape Canaveral would be renamed Cape Kennedy.

As it turned out, America's push to reach the moon by the end of the decade continued apace. In February of 1964, the unmanned Ranger IV probe crash-landed on the lunar surface. In late July, Ranger VII sent back the first close-up photographs of the moon—a total of over 4,300 images, some of objects as small as three feet in diameter—before it, too, crashed into the surface. Not to be outdone, the Russians put three cosmonauts in Earth orbit that fall, a feat that seemed to confirm that the Soviets still had their sights set firmly on the moon as well. Then on March 18, 1965, just as the first Gemini crew prepared for launch, cosmonaut Aleksei A. Leonov opened the hatch of the Voskhod II capsule and took man's first walk in space. Furious at being upstaged, LBJ ordered NASA officials to plan a Gemini spacewalk as soon as possible.

In less than two years, the Gemini astronauts completed a total of ten Earth-orbiting missions, an average of one every two months. Gemini 3, the nation's first two-man mission, was launched on March 23, 1965. Though crewman Virgil I. "Gus" Grissom, a Mercury veteran, and John W. Young, a member of the second astronaut class known as "The Nine," did not duplicate the Russian spacewalk, they did something more important, if less spectacular, by becoming the first men to change the trajectory of their spacecraft in mid-orbit. This crucial exercise proved that it was possible to develop the guidance technology needed for orbital rendezvous and docking.

Less than three months later, Gemini 4 astronaut Ed White became the first American to walk in space. In July of 1965, NASA announced the selection of a fourth astronaut class consisting of only five members, all of whom were scientists; the group included Harrison H. "Jack" Schmitt, a Harvard geologist, who was destined to become the twelfth man to walk on the moon. In August of 1965, Gemini 5 crewmen Gordon Cooper and Pete Conrad orbited the earth for over eight days, setting a new space endurance record. In December, Gemini 7 crewmen Frank Borman and Jim Lovell set a fourteen-day endurance record. That same month, Gemini 6-A astronauts Wally Schirra and Tom Stafford guided their spacecraft within a few feet of the Gemini 7 capsule for the first space rendezvous.

On March 16, 1966, Gemini 8 crewmen Neil Armstrong and Dave Scott made the first rendezvous and docking with an Agena rocket in Earth orbit. They also encountered the first emergency in space. Shortly after the two vehicles linked, one of the Gemini capsule's thrusters stuck, causing the spacecraft to spin wildly out of control. With remarkable presence of mind, Armstrong executed an emergency undocking, aborted the remainder of the mission, and managed to pilot the Gemini capsule safely to reentry.

In June of 1966, Cernan made three different rendezvous with an Agena rocket but was unable to dock because a heat shield failed to separate from the target vehicle. Cernan, however, was able to make a spacewalk. In July, Gemini 10 astronauts John Young and Mike Collins made rendezvous with two Agena rockets; Collins also made a successful spacewalk using a hand-held rocket gun for control.

The Gemini program wound up with some Flash Gordon-style stunts. In September of 1966, Gemini 11 astronauts Pete Conrad and Dick Gordon set a new altitude record of 850 miles. Gordon also added a new twist to rendezvous and docking by hooking the capsule to the Agena with a tether so the two craft could separate and spin, creating artificial gravity in outer space. On November 11, 1966, Gemini 12 crewmen Jim Lovell and Buzz Aldrin marched to the launchpad wearing signs that said "THE" and "END" on the backs of their pressure suits. The two astronauts stayed aloft for four days to complete final rendezvous and docking tests; Aldrin also set a new "extravehicular activity" (EVA) record with three space walks totaling over five hours in duration.

Though the Gemini astronauts never left Earth orbit, they proved that a mission to the moon was technically feasible for both men and machines. In the course of their ten orbital missions, they made five spacewalks and, more important, ten rendezvous and nine dockings using seven different approach modes. The long duration flights of Gemini 5 and Gemini 7 indicated that men could survive extended exposure to weightlessness without suffering adverse medical effects, and that the spacecraft could provide sufficient protection from meteorite showers and cosmic radiation.

Now, at last, it was time to turn full attention to Project Apollo, the program that would endeavor to unite science fiction fantasy with scientific fact and land the first man on the moon.

5

SLIPPING THE SURLY BONDS

Into and Out of Earth Orbit

Staging to Trans-Lunar Injection

Oh, I have slipped the surly bonds of earth
And danced the skies on laughter-silvered wings;
Sunward I've climbed, and joined the tumbling mirth
Of sun-split clouds—and done a hundred things
You have not dreamed of . . .
And, while with silent, lifting mind I've trod
The high untrespassed sanctity of space,
Put out my hand, and touched the face of God.

JOHN GILLESPIE MAGEE, JR.
"High Flight"

All twenty-four men who flew to the moon knew the story of John Gillespie Magee, Jr. An American missionary's son born in China, Magee enlisted in the Royal Canadian Air Force at the tender age of nineteen. After putting the finishing touches on "High Flight," he was shot out of the sky during the Battle of Britain. According to Apollo 11's Mike Collins, " 'High Flight' is without a doubt the best-known poem among aviators and can be found, elegantly framed or carelessly tacked, on home and office walls around most military airfields."

Likewise, the three Apollo crewmen atop the Saturn 5 knew that the lift-off, for all its pyrotechnic drama, was merely a prelude to the poetic wonder of their "high flight" into—and then out of—Earth orbit. The three-stage rocket ride from the top of the launch tower to the trans-lunar injection burn was the first stanza, the real "takeoff," of every lunar odyssey. For this was the fiery passage in which the astronauts slipped the "surly bonds" of their mother planet and soared where no man has ever soared before into the "sun-split" blackness of outer space.

"Yahoo!"

"What a ride! What a ride!"

"This really is a rockin', rollin' ride!"

As the mighty booster accelerated through her first-stage pitch and roll, the astronauts on the Apollo 12, 15, and 16 missions let loose cries of boyish delight in testimony to their exhilaration. The astronauts on earlier lunar missions were, understandably, far more restrained. As Apollo 11's Collins recounted in his post-mission autobiography, "All three of us are very quiet—none of us seems to feel any jubilation at having left the earth, only a heightened awareness of what lies ahead. This is true of all phases of space flight: any pilot knows from ready-room fable or bitter experience that the length of the runway behind him is the most useless measure he can take; it's what's up ahead that matters."

The men at Mission Control in Houston had now taken over from the launch controllers at the Cape, and the CAPCOM would let the astronauts know that their guardian angels on the ground were watching out for them.

"Apollo, Houston . . . You're right smack dab on the trajectory."

"Roger, Houston . . . This baby is really going."

"Roger, that."

Of course, there was no guarantee that "this baby" would keep going just because she had scaled the launch tower. Even if the Saturn 5's first-stage engines delivered their full 7.7 million pounds of launch thrust, the booster was still vulnerable to the hazards of wind and weather until she reached the apron of the earth's upper atmosphere. In fact, the crew of man's second lunar landing mission nearly had to abort in midair. Apollo 12 astronauts Pete Conrad, Dick Gordon, and Al Bean launched early on the morning of November 14, 1969, during an electrical storm. Thirty-two seconds after lift-off, their booster was struck by a bolt of lightning that knocked out the main fuel cells and the backup batteries.

"I felt it, I heard it in my earphones, and I saw some light out the front window," recalls mission commander Conrad. "There was a glow inside the cabin. The master caution light was on, and all eleven electrical warning lights were on. Even in the worst simulation, as fiendish as the training people were in trying to slip in multiple failures, they'd never thrown us anything where we'd seen all eleven lights on at one time. My obvious worry was that we were going to get a trip around the world and come on home without getting to go to the moon."

Moments later, the Apollo 12 booster was struck by a second bolt of lightning. The second shock did not trigger multiple alarms, but its concentrated charge did potentially more serious damage by short-circuiting the inertial platform, the spacecraft's primary on-board guidance system. Luckily, lunar module pilot Al Bean had the presence of mind to identify and throw the switch designed to correct such an unprecedented calamity.

"Even with all the training, I wasn't familiar with that switch," Conrad admits. "I made some classic remark like, 'What the hell switch is that?' Al said, 'It's mine,' and let me know that he knew what he was doing. Things settled down pretty good after that."

Apollo 12's brush with disaster turned out to have a silver lining. Bean's quick thinking and grace under pressure impressed both his fellow crewmen and the men back at Mission Control, immeasurably boosting their collective confidence and morale. The special bond created by the bout with the lightning bolts was subtly reinforced throughout the rest of the mission. In a rare departure from traditional protocol, the CAPCOMs and PAOs in Houston stopped referring to the astronauts by formal rank, and started calling them by their first names.

At one minute five seconds after lift-off on all nine Apollo missions, the Saturn 5 would break the sound barrier at a speed equivalent to 738 statute miles per hour. The three crewmen inside the command module couldn't hear a sonic boom, but they all started to feel the rapidly multiplying pressure of a "surly bond" as formidable as gravity—the doubling, tripling, and quadrupling G forces associated with the acceleration of the rocket.

Ironically, the mounting G's were a blessing in disguise, for they overpowered the Saturn 5's postlaunch vibrations and actually helped smooth out the ride. And thanks to the booster's curvilinear pitch and roll, the peak G load was only about half that induced by her more sharply angled ancestors. An astronaut atop one of the Mercury-Redstone rockets was plastered with up to 8 Gs, eight times ground-level gravity. For some two and a half minutes, he could barely lift a finger, much less reach out and touch the control panel. The Apollo astronauts experienced a much safer and far less painful maximum of 4.5 Gs. That was more than enough to keep them securely pinned to their couches, but it did not totally paralyze them even at the peak of the power curve.

"You're psyched up and in good shape," recalls Apollo 14's Stu Roosa, "and you hardly even notice the Gs. Even lifting your arm to move switches is no problem . . ."

Impervious to the astronauts' mounting G load, the Saturn 5 went faster and faster as she climbed higher and higher. At lift-off, the booster had lurched upward at just five feet per second. Upon reaching the top of the launch tower at the ten-second mark, she was traveling at fifty feet per second, or just under thirty-five m.p.h. Less than one minute later, when the Saturn 5 broke the sound barrier at an altitude of one mile, her velocity had multiplied more than twentyfold to over one thousand feet per second. By the two-and-a-half-minute mark, she would be one mile downrange from the Cape and four miles high, clipping along at over 2,100 feet per second, or roughly 1,500 m.p.h.

How was this physically possible? Why was the Saturn 5 able to defy gravity ever more easily even as she ascended? Why did the booster's velocity continue to increase instead of decrease?

The answer never ceased to amaze and delight even the most veteran Apollo astronauts, for it bespoke the technological magic of their flying machine. With each passing foot per second, the Saturn 5 got lighter and lighter. The first stage, officially known as the S-IC, burned propellant at the rate of 3,500 gallons per second. That translated into an ongoing fuel consumption/weight reduction average of roughly two million pounds per minute. But as the booster flew higher, she also got stronger and stronger thanks to the weakening resistance of mother nature. For up where the air was windless and thin, there was less aerodynamic drag on her fuselage, which enabled the five F-1 engines of her first stage to deliver far more punch per pound of propellant.

The Saturn 5's velocity increased exponentially as she lost weight and gained effective boosting power. At lift-off, her sea level thrust of 7.7 million pounds had to push a 6.5-million-pound body. Just sixty seconds later, that same 7.7 million pounds of thrust only had to push 4.5 million pounds. By the two-minute mark, the booster was four miles high and her first-stage thrust was peaking at close to nine million pounds, three and a half times her total body weight, which was now down to 2.5 million pounds. As a result, she could slice through the thinning atmosphere forty times faster than the speed at which she left the launchpad.

The Saturn 5 would have to fly still higher and faster to get into orbit. But because the atmospheric friction—and the undertow of the planet's "surly bonds"—decreased dramatically as the booster got farther away from the gravitational center of the earth, she was able to do the rest of her job with significantly less thrust.

At two minutes fifteen seconds after lift-off, the S-IC's middle engine shut down, and the thrust generated by the four outboard motors plummeted to 7.2 million pounds. Even so, the spacecraft continued to accelerate at an only slightly less dizzying pace.

When the mission clock reached the three-minute mark, the astronauts were hurtling toward the heavens at over six thousand m.p.h., or almost ten thousand feet per second, a fivefold increase in velocity in just forty-five seconds. They were also more than ten times higher and seventy times farther away.

PAO: "We now have an altitude of forty-three miles, downrange distance seventy miles . . ."

The three Apollo crewmen realized that meant the joy ride was about to end. The S-IC, which measured 138 feet from top joint to tail fins, was the biggest and most potent of the three booster stages. During the 180 seconds since lift-off, the S-IC had burned up almost five million pounds of propellant, enough fuel to fly twenty Boeing 707s from New York to Paris and back. But the S-IC had now outlived its usefulness, and the astronauts had braced themselves to absorb yet another violent shock—staging.

At three minutes five seconds after lift-off, a ring of explosive bolts blew off the S-IC. Almost simultaneously the thirty-three-foot-long escape booster was jettisoned. There was a brief, pregnant pause as the five J-2 motors of the eighty-one-foot-long second stage, the S-II, built thrust. According to the astronauts, what happened next was, because of its sheer suddenness, more jarring to mind and body than lift-off.

"The thing rattled and shook," remembers Apollo 16's Charley Duke, "and then the second-stage engines fired—and sort of, 'wong!'—like that—you felt like you were going to get thrown through the front of the spacecraft."

"Staging is an extremely violent maneuver," agrees Apollo 14's Stu Roosa. "One of the reasons why it hits you is because it's so foreign to the way you fly an airplane. In an airplane you can pull Gs, but you've got an onset and a bleed-off. The same is true in any other type of simulation. A centrifuge has got to start slow, build up, and then coast down. But in the spacecraft it's almost four Gs, and then suddenly it's zero. It's like the rocket just hangs there for a second, and then staging hits, and you're slammed forward, and you wonder what the hell is going on."

Staging ended at the three-and-a-half-minute mark. The center motor of the S-II shut down, and the four outboard motors of the second stage began to push the load alone. But instead of weighing 6.5 million pounds as at launch, the spacecraft now weighed around 1.5 million pounds. The J-2 engines of the second stage were only about one-fifth the size of the first-stage engines, and delivered only one-third the thrust, or roughly 2.5

million pounds, at full throttle. But in less than five more minutes, the S-II would double the astronauts' altitude and velocity.

Staging also treated the crewmen to their first outside view since boarding the spacecraft over two hours earlier. The jettisoning of the escape booster unshrouded the side windows, and thanks to the violence of staging, the astronauts were often entertained by a thrilling display of S-IC debris.

"One of the big surprises is all the confetti that's knocked out," remembers Roosa. "All the stuff comes whistling by the window going away from the vehicle . . . I also remember looking back and seeing the earth. You can see the water and the horizon and you know you're getting pretty high. That was a beautiful view . . . But unfortunately, you didn't have much time to sit and watch the view. You had to keep your mind inside in case of an abort."

PAO: "The Apollo-Saturn spacecraft is now downrange 530 miles, altitude 95 miles, velocity 17,358 feet per second—nearly 12,000 statute miles per hour. We are still GO at seven minutes forty-one seconds . . ."

Soon after, a fuel-low sensor in the belly of the S-II started the next staging sequence, igniting the engine of the stage above. At nine minutes eleven seconds, the S-II's outboard motors shut down. A second later, S-II was blown away, destined to plunge into the Atlantic Ocean. Then the Saturn 5's third stage, the S-IVB, took over as the CAPCOM at Mission Control rasped, "Apollo, Houston . . . You are GO for orbit . . . GO for orbit!"

Again, the sudden onset of staging pinned the three crewmen against their couches. But the kick of the S-IVB was not nearly so jarring as the S-II's staging six minutes earlier. Rather, as Apollo 11's Collins puts it, "The third stage is crisp and rattly."

By comparison to the first and second stages, the S-IVB was also a midget. Measuring a little over sixty-two feet in length, it was less than half the size of the S-IC. It was also forty times less powerful, with a single J-2 engine capable of delivering only 200,000 pounds of thrust. But the S-IVB had three things in its favor. One was the velocity and momentum generated by the two previous stages. The second was the fact that the spacecraft was now down to 400,000 pounds, one-sixteenth of her launch weight.

Last but by no means least, the S-IVB had the unique power to shut down and relight its engine, and was the only stage of the booster with two vital roles in the astronauts' "high flight." Its first duty was to get them into a "parking orbit" around the earth. Later, the S-IVB would restart its motor and deliver the all-important trans-lunar injection that blasted them out of Earth orbit on course for the moon.

The first S-IVB engine burn lasted two and a half minutes, and boosted the three crewmen five miles higher and another 250 miles downrange at speeds of up to five miles per second. Then, at nine minutes forty-two seconds into the mission, the S-IVB's engine cut off. Thanks to the instantaneous displays of the on-board computer, the astronauts were apprised of this pivotal event a few moments before their guardian angels on the ground.

"Shutdown!" the mission commander would proclaim.

Houston would officially confirm the event a second or two later via the CAPCOM: "Roger, we have a shutdown."

At that point, the three astronauts suddenly found themselves in the topsy-turviness of Earth orbit. Their altitude was now a hundred miles, they were almost a thousand miles downrange from the launchpad, and their velocity was over 18,000 statute m.p.h. They were also "upside down" with their feet pointing up toward the stars, peering down at their home planet as if through a glass-bottom boat. And as the tension of the launch evaporated, a strange new sensation of wonder would envelop the command module, as the astronauts realized there was no "up" or "down" anymore.

"Things feel sort of funny," recalls Apollo 16's Charley Duke, "and you unbuckle your seat belt and wiggle around in your seat or sort of tap your toes . . . And all of a sudden, your body's just everywhere in the cabin, and you're doing flips and cartwheels and spins . . ."

This "funny" feeling is best known as weightlessness, though the term is really a misnomer. There is no such thing as pure weightlessness, except perhaps in the farthest void of outer space, for the gravitational pull of the planets and major celestial bodies like the sun extends for millions of miles according to their mass and density. The three crewmen were actually in a state of microgravity, not absolute zero G. But even they refer to their condition as weightlessness because that's how it felt. And as the Apollo astronauts unanimously attest, the onset of zero G—or to be strictly precise, the abrupt disappearance of all but a fraction of the G load—always hit with the visceral impact of a punch in the solar plexus.

"Your first feelings of weightlessness as it hits you," reports Apollo 17's Gene Cernan, "is like going down a country road at about sixty miles per hour, and the car goes over a bump and your stomach goes up into your throat, and then you get over the bump and you get a sort of good sensation, and then—'bam!'—you're back down again. Only when you're in space and go into weightlessness, you go over that bump and you never come back down."

Getting over the "bump" and never coming back down could be rough, smooth, or both for different astronauts at the same time or for the same astronauts at different times. According to Apollo 16's Duke, "There are two responses you can have when you first experience weightlessness—'Yahoo! This is great!' or 'Look out, I'm going to get seasick!' My first reaction was, 'Uh-oh, this isn't going to be any fun because I'm going to get sick.' "

Fellow Apollo 16 crewman Ken Mattingly adapted to weightlessness much more quickly and painlessly: "It took me somewhere around ten seconds to get used to zero G. It was as natural as if I had always been there. It made me remember all those years of work in training when I got discouraged and I used to wonder, 'Do I really want to work this hard to do this?' And all of a sudden, in one big moment, the answer was, 'You bet! This is exactly what I was supposed to do!' "

Some of the crewmen on other Apollo missions were not so lucky. Almost all the astronauts admit to feeling at least a touch of "stomach awareness," the accepted euphemism for nausea. Like Apollo 8's Frank Borman, Apollo 17's Ron Evans actually vomited. Even the astronauts who were not overcome with "stomach awareness" or vomiting confess to finding weightlessness more than a little disorienting at first. "It's an unreal world," says Apollo 15's Jim Irwin, "and it's hard to come to grips with . . . There's no sensation of motion. You're floating—and yet you're going eighteen thousand m.p.h."

Be that as it may, the three crewmen now had to attend to the realities of this unreal world. They were in orbit, which meant they were in a geocentric free-fall. At a hundred miles high and 18,000 m.p.h., they had sufficient altitude and forward speed to keep from crashing to the ground. But they had not yet reached the minimum "escape velocity" of 25,000 m.p.h. As a result, the "surly bonds" of the earth's gravitational pull remained just strong enough to bend their trajectory in a gentle orbital circle around the globe.

Theoretically, the astronauts could have been shot straight from the Cape into outer space. But that would have required a much bigger booster than the Saturn 5, perhaps the equivalent of a Saturn 10. Instead, they would employ a much more energy-efficient flight plan using earth orbit as an intermediate staging site from which to blast off for the moon.

The three crewmen also realized that achieving a "parking orbit" before the trans-lunar injection provided an extra margin of safety. They were scheduled to make one and a half revolutions around the globe, a trip that took almost two and a half hours. The astronauts would spend most of that time deactivating the first- and second-stage booster circuitry and checking out the CSM and LEM systems, making sure that propulsion, guidance, environmental control, and communications were all GO before they went for the moon.

"The first few minutes in orbit are busy ones," Mike Collins recalls in his post-mission autobiography, "as a long checklist must be followed to convert the spacecraft from a passive payload to an active orbiter. Between Bermuda and the Canary Islands, I work my way swiftly through a couple of pages of miscellaneous chores, opening and closing circuit breakers, throwing switches, and reading instructions for Neil [Armstrong] and Buzz [Aldrin] to do likewise."

After the three crewmen confirmed that the cabin of the command module was safely pressurized, they proceeded to take off their thick-fingered gloves and unscrew their helmets. This welcome if only temporary release from the total confinement of their "moon cocoons" would allow them more freedom of movement, enabling them to complete their orbital chores on schedule. But because of the unworldly nature of weightlessness, they took care not to go too fast.

"You move very slowly," remembers Apollo 14 lunar module pilot Edgar Mitchell, "and take off your helmet very carefully, and get adapted to their weightless environment with no abrupt movement of any sort. You've still got quite a few tasks to carry out at this point, but you go about them rather methodically in order to make sure, number one, that you don't act like a bull in a china shop, and secondly, in order not to experience vertigo, which would cause nausea."

The next two tasks on the orbital checklist were diametrically opposite procedures for finding out the same essential information, i.e., where the spacecraft was, where it was going, and at what speed. The first procedure, which took the ground-up approach, was a communications drill handled by the commander and the lunar module pilot in cooperation with a worldwide network of no less than seventeen computerized tracking stations.

The three main Apollo tracking stations were strategically located in central Spain (Madrid), western Australia (Canberra), and southern California (Goldstone). In order to maintain round-the-clock electronic surveillance of the astronauts' trans-lunar voyage, each installation was equipped with an eighty-five-foot dish antenna, and each was spaced 120 degrees of global circumference from the other two so that one would always be facing the moon.

In the interim, so as to fine-tune the main tracking antennae and enhance short-range guidance and control, the spacecraft's orbit around the earth was also being followed by twelve smaller ground stations and a navy vessel in the Atlantic. For added safety, six Apollo Range Instrumented Aircraft (ARIA) were standing by in high-altitude holding patterns. In the event of a communications failure, these airborne radio stations would be able to relay messages between the spacecraft and the ground.

As the Apollo spacecraft passed over, each tracking station acquired or reacquired radio contact with the astronauts, and took an independent antenna reading of their present altitude, velocity, and orbital azimuth. The updated telemetry data from each locale was electronically fed into the computers at Mission Control in Houston. The computers then recalculated and refined previous flight path and position estimates, and displayed the results through simulated movements in the orbit of the little green light bug on the main tracking screen up in the control room.

Meanwhile, the command module pilot folded back the center couch inside the cabin of the spacecraft, floated down to the lower equipment bay, and attempted to get an equally accurate position fix from an entirely different angle. Rather than duplicating the efforts of the ground stations, he commenced an exercise in celestial navigation called "platform alignment." This problematic procedure was a vital prerequisite to the trans-lunar injection and to virtually every other major maneuver in the course of the mission.

The inertial platform was the spacecraft's primary on-board guidance system. Located in the lower equipment bay, it had a matrix of three directional axes that measured the acceleration of the spacecraft along each axis with respect to a fourth fixed axis. This "stable member" was, in turn, aligned with a fixed trajectory dubbed REFSMMAT, an acronym standing for "reference to stable member matrix," the term for a kind of imaginary celestial highway. As the spacecraft moved, the platform kept rotating on its gimbals, but it eventually got out of line. The REFSMMAT also changed. At launch, for example, it was a ninety-degree vertical line straight out of the top of the booster. In orbit and during the trans-lunar voyage, the REFSMMAT was a line between two selected celestial navigation stars. Or, to be more precise, a series of lines between several pairs of stars, because, as the mission progressed, a new REFSMMAT had to be selected for each platform alignment.

Just as the tracking stations updated the computers at Mission Control, the CMP's platform alignment updated the electronic brains aboard the spacecraft on the astronauts' position and flight path. There were two computers in the command module, one in the cabin and one in the lower equipment bay, and a third in the lunar module. Made by Raytheon Corporation, each was a rectangular box twenty-four inches long, twelve inches high, and six inches wide weighing a little over seventeen pounds with a sixteen-button keyboard and a twenty-one-digit character display called DSKY (for "display and keyboard").

The astronauts communicated with the DSKY in a 38,916-word vocabulary divided into two principal dialects. Five-digit numbers represented data input and output, such things as altitude, azimuth, and velocity. Two-digit commands preceded by a stroke of the VERB key prepared the computer to execute a program. Two-digit commands preceded by a stroke of the NOUN key told the computer exactly which program and how to execute it.

The accuracy of the platform alignment, however, depended on the human factor, the skill of the CMP, who literally had to eyeball his navigational stars with a one-power telescope and a twenty-eight-power sextant. He generally picked the stars that were easiest to see and yet reasonably far apart, for the wider the angle between them, the more accurate the sextant reading would be. There were over forty stars stored in the computer's memory, but none of the astronauts could remember the pertinent facts about all forty. As a result, each CMP tended to rely on a few favorite pairs over and over again.

"You become pretty good friends with some of these stars," Stu Roosa recalls. "A few of them are very common stars like Sirius, the brightest star in the sky. You also have Polaris, the North Star, which is not necessarily a very bright star, but it is one that's distinctive and easily recognizable."

Once the CMP picked his two stars, he asked the computer to determine their respective locations on the basis of the original platform alignment. Then he fed his current sextant reading into the computer. The sextant data would always differ significantly from the computer's original assessment of the stars' relative positions because of the rotation of the platform and the movement of the spacecraft. But if the CMP did his job properly, his calculations of the fixed angle between the two stars would match the star angle stored in the computer down to a ten-thousandth of a percentage point.

The CMP's goal was to get "five balls," a perfect reading of .00000 on the star angle comparison. But given the manual element involved in sextant sighting, scoring five balls was well-nigh impossible, and the CMP usually counted himself lucky if he got four balls and a low fifth digit. But he also knew that a score of only three balls and change, a variation of only one-thousandth of a percentage point, was an unacceptable indication of human error or a malfunction in the inertial guidance system that might force Houston to abort the mission in mid-orbit.

Assuming the star angle comparisons were acceptable, the rest of the CMP's platform alignment was almost automatic. He simply told the computer that his current sextant readings were accurate within certain parameters between five balls and four balls plus. The computer calculated exactly how much the inertial platform had drifted since lift-off, and realigned it accordingly. Then the CMP floated back up to the cabin to report the results to the ground.

PAO: "This is Apollo Control at fifty-two minutes and the station at Carnarvon, Western Australia, is about to acquire the Apollo spacecraft. We'll stand by for this live pass."

Happily for the three Apollo crewmen, the orbital flight plan included a few officially designated minutes for looking out the window. Most astronauts tried to use every precious second for terrestrial sightseeing from orbit. This was particularly true of the crewmen on the later missions, who got tips on what to look for from the crewmen on earlier flights.

"I kept a shopping list of things to look for that perhaps had nothing to do with the mission," remembers Apollo 16's Ken Mattingly. "One was what people had told me about Africa. There are a lot of nomads out in the desert, and on clear desert nights you see the fires, these little yellow dots that represented fires from all these nomads camping out. And you realized the broad area that you were looking at and the thousands of people that that represented. Each of those little dots represented people, other human beings like you and me, that are out there in an environment which I would consider more strange than they might think about me."

In order to enhance their spotting ability, the Apollo 16 crewmen took along a pair of binoculars. Much like ordinary tourists, they took turns focusing their field glasses on familiar landmarks below. "We sure enough could see Houston when we came over," Mattingly recalls. "We could see boats out in the Gulf or in the ocean. Thunderstorms looked just like thunderstorms do when you fly except that they are miniatures. They're way down there. They're little light bulbs and they flash in the dark like the little winking lights you put on your Christmas tree, like peanuts."

It usually took the Apollo spacecraft about eighty-eight minutes to complete its first Earth orbit. During this first eastbound revolution, the three crewmen witnessed a full cycle of day and night. They also reached a major turning point in their "high flight." If all continued to go well, they would make only one more half circle around their native planet. Then the spacecraft's third-stage engine would ignite for a second and final time, and deliver the trans-lunar injection (TLI).

In the meantime, the astronauts started scribbling down yet another series of numbers relayed by the CAPCOM at Mission Control. These numbers represented navigational coordinates that would supposedly help them find their way home if the TLI burn malfunctioned and they lost radio contact with the ground—at least in theory. As a practical matter the astronauts knew that Mission Control couldn't predict exactly where the command module would go if the engines misfired or failed. In the event of such a mishap, it would be up to the crewmen to pilot themselves safely down to Earth as best they can.

Although the TLI burn was far less violent than lift-off or staging, the checklist also called for the astronauts to put their helmets and gloves back on, and re-strap themselves to their couches. These precautions were intended to protect them if the booster engine exploded and damaged the command module enough to cause a sudden loss of cabin pressurization. But as Apollo 11's Collins observes: "It doesn't make a hell of a lot of sense, really. Should the CM be that badly damaged, certainly our service module engine (which we require to get down out of this orbit) would also be damaged beyond use."

And as Collins notes, the Apollo crewmen probably would be damaged beyond use, as well.

The trans-lunar injection usually began about two hours and thirty minutes after lift-off. More often than not, the three crewmen were making their second pass over the Pacific Ocean and were temporarily out of radio contact with tracking stations in Australia and the U.S. But a faithful squadron of ARIA jets was always hovering below to relay messages between the spacecraft and Houston.

Like staging and every other engine burn during the mission, TLI was always announced by a professionally matter-of-fact exchange between Houston and the Apollo astronauts.

"You are GO for TLI," rasped the CAPCOM.

"Roger," replied the command module pilot. "Thank you."

As Apollo 11's Mike Collins observes, "There ought to be more to it." The three astronauts realized that TLI was not just another engine burn, but the most important burn of their "high flight," the blast that would enable them to slip the "surly bonds" once and for all and set sail for the moon. During their one and a half revolutions around the earth, the spacecraft had established a steady orbital momentum at 25,000 feet per second. The trans-lunar injection would slingshot them out of orbit by increasing their speed by another ten thousand feet per second to the required "escape velocity."

The trans-lunar injection also represented the merging of science and science fiction. In his novel *From the Earth to the Moon,* Jules Verne calculated the "escape velocity" needed to get the three crewmen of the *Columbiad* out of the earth's gravitational grip at twelve thousand yards (36,000 feet) per second. The actual speed of the Apollo spacecraft at the end of the TLI burn would be a little over 35,500 feet per second, roughly 25,022 statute m.p.h., faster than man had ever traveled before.

But most of the astronauts were simply too busy preparing for TLI to ponder the true meaning and significance of this pivotal engine burn until long after it was over. "Before we knew it," recalls Apollo 15's Irwin, "we came around over the Pacific again, and we were given a GO for trans-lunar injection. We fired the third-stage engine, and it was just like we were in a gigantic powerful elevator that was lifted straight up."

The TLI burn was often, in the words of Apollo 11's Buzz Aldrin, "just a tiny bit rattly," but it was nowhere near as bone-jarring as lift-off and staging. Because the trans-lunar injection required much less thrust, the astronauts felt only about one G from the thrust of the S-IVB. And as the burn progressed, they were treated to some breathtaking visuals.

"I just wish I would have had a camera at that point," Irwin exclaims. "I looked out my window, and there, framed beautifully right in the center of the window, was all the Hawaiian islands. It was a clear morning on the Pacific and you could see the volcanoes of Mauna Loa and Mauna Kea. We're being lifted up majestically, accelerating to twenty-five thousand miles per hour, and you see the islands fade, get smaller and smaller."

The TLI burn lasted for five minutes and forty-seven seconds. Then the third-stage engine shut down, the rattling stopped, and the "trans-lunar coast" began.

The three crewmen now found themselves about 180 miles above the earth, only eighty miles higher than their orbital altitude. But instead of circling their home planet, they were now hurling away from the earth and out into space at over four hundred miles per minute. In just three more minutes—less than three hours after lift-off—the astronauts would be over 1,400 miles out. In nine hours, they would be more than 57,000 miles out.

Henceforth, the spacecraft would gradually decelerate until the astronauts left the earth's gravitational "sphere of influence" and entered the moon's "sphere of influence" on day three of their voyage. Then the spacecraft would start accelerating again, and, it is hoped, zoom right into lunar orbit.

In the meantime, the command module pilot had to attend to the first and most important post-TLI maneuver: "transposition and docking." As the name implied, the object here was to separate the command and service modules from the S-IVB, turn around, and dock so that the LM was moved up to the accessway at the nose of the command module. It was, to say the least, a tricky maneuver, but one the astronauts claimed to enjoy.

"Transposition and docking is one of the few times during the mission that anyone ever flies the spacecraft manually," notes Stu Roosa. "That period of time in there is a good one for the command module pilot because you're really out there doing something. Not that you aren't all the time. But during transposition and docking, you've really got your hands on the controls, and you do it."

First, the CMP switched seats with the mission commander, and took over the controls of the command module. Then he threw a series of switches that detached the command and service module (CSM) combination from the rest of the spacecraft. After drifting out about seventy-five feet, the CMP made a 180-degree turn, swinging the nose of the command module around so that it was face-to-face with the top of the lunar module, which still rested within the protective shell of the S-IVB.

Besides being a practical necessity to affect transposition and docking, the CMP's one-eighty provided a unique visual thrill for his fellow crewmen. The command module was now pointed back toward home, and the commander and the lunar module pilot could see the earth not as curving blue horizon but as a majestic blue ball suspended in the blackness of space.

"You could see almost the whole earth by then, in one whole shot, and that is amazing," recalls Apollo 12's Al Bean. "Whenever you see a picture of it, there's always a frame, you know, something that holds the earth there. But in space, there's nothing holding it up. There's no border that a string could come down and hold the earth. There's nothing on either side to hold it magnetically . . . It's not like it's whirling around the sun or anything like that. Relative to you, it's just hanging out there and slowly going away . . . It's one of the few things you see in your life that seems to defy all the normal things you see."

What happened next resembled an in-flight airplane refueling procedure. Slowly, the CMP guided the probe on the nose of the CSM combination into the one-eyed drogue atop the lunar module. Although the CMP could see the LEM through his window, as in most other procedures, he flew by his instruments, aligning the cross hairs on his guidance screen with a three-dimensional cross mounted on the LEM. Then he simply tried to hold that position until the probe of the CSM slipped into the drogue of the lunar module.

When the mothership was securely locked to the LEM, the CMP would float down to the lower equipment bay, and open the hatch to the lunar module. He then conducted a painstaking inspection of the interior to make sure the violence of the launch had not caused any leaks, cracks, or equipment damage.

If the LEM checked out okay, the CMP resealed the hatch, floated back up to the cockpit, and threw a series of switches that jettisoned the S-IVB. He then fired the service module engines for three seconds, further separating the mothership from the booster shell. When the S-IVB got a safe distance away, Mission Control reprogrammed the rocket's remains to veer away from the moon and enter solar orbit, where it would eventually expire.

All three crewmen welcomed the completion of transposition and docking for several very good reasons. The departure of the third stage reduced the original 6.5-million-pound launch vehicle to a lean and agile 96,000-pound spacecraft, and signaled that the command, service, and lunar modules were properly realigned for the undocking in lunar orbit that had to precede their attempt to land on the moon. It also meant that the astronauts could take off their "moon cocoons" and keep them off for the next three days.

Due to the quirks of zero G, trying to undress in the cramped confines of the command module often resembled a slapstick scene in a Three Stooges movie. The three astronauts obligingly tried to help each other unscrew their helmets and unzip their bulky pressure suits, only to find themselves comically inept at performing such a seemingly simple task. Apollo 11's Collins recalls that he and his comrades kept "thrashing around like three great white whales inside a small tank, bashing into the couches and instrument panels . . ."

When the astronauts managed to finish this awkward extraterrestrial striptease, they folded up their pressure suits and stuffed them into storage bags beneath the center couch. Then they changed into their far more comfortable and economically tailored two-piece white nylon jumpsuits, and began their second crucial post-TLI maneuver—the "thermal roll."

Unlike a car or a train, the spacecraft could not simply fly along with forward motion alone. Due to the extreme heat and cold encountered in outer space, it had to keep constantly rotating like a chicken on a barbecue spit so that each half of the fuselage was not overexposed to either sunlight or darkness. If the spacecraft got too hot, the pressure in the service module's propellant tanks could get too high and possibly cause an explosion. If the spacecraft got too cold, its radiators could freeze.

The astronauts always began their thermal roll nice and easy, striving for the optimum rotation of .3 degrees per second, or approximately one full turn every twenty minutes. As Apollo 11's Collins attests: "It requires a very precise sequence of computer-assisted thruster firings to achieve a pure roll motion. If not done properly, pitch and yaw motions will ensue, just like a top wobbling crazily at the end of its spin, and then we must stop and begin all over again."

Once the spacecraft was rolling satisfactorily, the Apollo crewmen finally got their first real chance to take a break. It was now about five hours since lift-off, and about two hours since the trans-lunar injection. The velocity of the spacecraft had plunged from 35,500 feet per second (24,204 m.p.h.) to a little over twelve thousand feet per second or roughly 8,100 m.p.h. But the astronauts were already twenty thousand miles from Earth. And not surprisingly, looking back at Earth was their top recreational priority.

"The first realization is that of leaving Earth," recalls Apollo 17's

Gene Cernan. "You're rendezvousing with someplace, somebody, three days later, but you can't really see it. But the earth is very evident. You could see the earth, plus you know and understand the earth. Although you've never seen it from that spectrum before, that's home you're looking at."

Adds Apollo 12's Al Bean: "Things don't start to sink in until you have time to sit back and look out the window . . . You see that the earth is going away, and you know that you really are on your way to the moon."

What made the astronauts' view truly unique from that of any other mortals was the fact that now, for the first time, they could see the whole earth, not just as an overarching horizon but as it truly was, an island in the sky, a ball surrounded by nothing but blackness.

"When we looked back, the earth was about the size of a basketball," recalls Apollo 15's Irwin. "It was a full earth, fully illuminated. You could see the warm natural colors, the tans of the deserts and the mountains, the blues and greens of the great oceans, and the whites of the few clouds that were on the earth that day . . . Then, as we kept traveling toward the moon, we saw the earth get ever smaller, shrink from the size of a basketball down to the size of a baseball, then a golf ball, then finally the size of a marble . . .

"What a privilege to see the earth in its entirety," Irwin exclaims, "to see it as a very fragile place, to see the earth as God must view the earth when he just wants to see the beauty and not the ugliness that exists. It did remind us of a Christmas tree ornament hanging in the blackness of space. It seemed like there should be a cord attached or something beneath to hold it. But there was nothing. It was just hanging majestically in space. It was so moving to see the earth from that perspective. That view brought me back with a new appreciation for the earth and the things we have on Earth."

Adds Apollo 16's Charley Duke: "It was the most fantastically beautiful thing I had ever seen. I could see Baja, California, the North Pole, Alaska, all of North America, Mexico, Central America, and the northern part of South America . . . The land was an unimpressive brown, but the water was just crystal blue. And all around, there was the blackness of space. The contrast was just so sharp, it was exciting to see . . . The earth had a fragility about it. It looked like a Christmas tree ornament just hanging up there in the blackness, the jewel of Earth hung up in black velvet space."

Seeing the whole earth for the first time inspired an almost opposite reaction in Apollo 17's Jack Schmitt. Unlike Irwin and Duke, who were former fighter pilots turned astronauts and devout Christians, Schmitt was a scientist with a Ph.D. in geology. He therefore viewed his planet with the X-ray eyes of an expert rock cruncher, and saw not fragility or vulnerability but majestic power.

"Within the earth's beauty is a great deal of strength," he observes. "Some aspects of it are fragile, depending on what you try to do. But overall, it's a tremendously resilient and very tough planet. It has a great capacity to absorb punishment that it inflicts on itself, as well as what the cosmos inflicts on it and what man can or has inflicted on it. It heals itself in various ways. So I tended to view the earth not only with that sense of beauty, but as a strong planet that can provide not only sustenance but great pleasure for mankind."

Apollo 11's Mike Collins claims that he was looking forward to being able to see the whole earth for the first time, but confesses that the experience still came as something of a "shock," prompting him to feel a kind of cosmic separation from his home planet.

"Orbiting the earth is an extension of flying an airplane," he notes, "whereas on the trip to the moon where you see the earth start getting smaller and smaller, you really start thinking about space for the first time. You're an extraterrestrial being, for the moment at least, and there is the feeling of remoteness and distance, of having a totally different viewpoint and perspective up there or out there, however you want to describe it, than anyone who is forced to crawl antlike on the surface of that funny little planet over there or down there."

Jim Irwin felt much the same way, but with an added religious dimension. "The fact that you are traveling through space has such an ethereal quality to it," he recalls. "It's different than flying around the earth, what you see, what you feel . . . I guess you feel like an angel."

Apollo 17's Gene Cernan was also moved to contemplate the infinity of space when he saw the whole earth for the first time. But he felt much more at peace with the blackness all around him. "When the sunlight comes through the blackness of space, it's black," Cernan remembers. "I didn't say it's dark, I said black. So black you can't even conceive how black it is in your mind. The sunlight doesn't strike on anything, so all you see is black. You've got this planet called Earth, which is in this blackness but is lit because the sunlight strikes on an object, an object called Earth. And you can feel and see the three dimensions of this Earth within reach, within grasp, not at the end of this blackness, but right out here just a little ways away . . .

"What are you looking at? You can call it 'the universe,' but it's the infinity of space and the infinity of time. I'm looking at something called space that has no end and at time that has no meaning. You can see the blackness and almost focus on the blackness, on that infinity of space, infinity of time. Now that's very difficult to conceive, I know. But that infinity of space and that infinity of time does exist because I've seen it with my own eyes. And it's not a hostile blackness. It's not hostile because of the beauty of the earth, the thing that gives it life."

At the same time, there was also a feeling of solitude and isolation unlike anything the three crewmen had previously experienced on Earth, and understandably so, for the astronauts had truly left their home planet, and were on their own in the infinity of space. If disaster struck, no one could come to their rescue in another spacecraft. They could rely only on the ever more distant computers back at Mission Control and their own limited skills for as long as their fuel supplies and life support systems held out.

"It's a very eerie feeling," admits Apollo 14's Ed Mitchell. "You suddenly start to recognize that, yeah, you're in deep space, that the planets are just that, they're planets, and that you're not really connected to anything anymore, that you are floating through this deep black void . . . The spacecraft really becomes your universe. You're on a little planet. You know that's all there is as far as you're concerned."

But for better or worse, the Apollo crewmen could not dwell on such thoughts or their view through the windows of the spacecraft because they now had to concentrate on the next task essential to survival in zero G—getting themselves something to eat and drink.

6

CHARIOTS OF FIRE

Project Apollo

1967–1969

There appeared a chariot of fire . . . and Elijah went up by a whirlwind into heaven.

Second Book of Kings, II:11.

It was past one o'clock on the afternoon of January 27, 1967, when the Apollo 1 crewmen finally got permission to board their spacecraft. More than a thousand NASA technicians and support personnel were swarming about the giant Saturn I booster at Pad Complex 34 on that cold and cloudless winter day. In keeping with an unwritten custom, virtually everyone stopped to watch in respectful silence as the three astronauts emerged from the elevator at the top of the launch tower and lumbered across the gangplank to the tent-shaped embarkation chamber known as the "white room."

Moments later, after crawling from the "white room" into the cramped interior of the spacecraft, Apollo 1 commander Gus Grissom let launch control know that he was pissed. Problems with the environmental control unit (ECU) had already delayed the countdown more than two hours, but the damn thing still wasn't working right. According to Grissom, the cabin reeked with the smell of "sour milk." To make matters worse, there was so much static on the intercom, he and fellow crewmen Ed White and Roger Chaffee could not tell the boys in the blockhouse just how pissed they really were.

"How the hell can we get to the moon," Grissom roared at the top of his lungs, "if we can't even talk between two buildings?"

Although the Apollo 1 astronauts were not going to the moon on this particular day, they intended to perform a crucial full-scale dress rehearsal for their upcoming Earth orbital mission. This would be the first "closed hatch" countdown demonstration. It was officially designated a "nonhazardous" test because there would be no fuel in the booster, and theoretically, little risk of fire or explosion. But if something went wrong inside the spacecraft, the three crewmen could not simply hit an ejection handle. The outer hatch could be opened only with outside help from the pad crew, a procedure that took at least ninety seconds.

The day before, Apollo 1 backup commander Wally Schirra had warned Grissom of the potential dangers posed by the closed hatch. "If you have a problem, even a communications problem," he cautioned, "get out of the cabin until they've cleared it up." Grissom inexplicably decided not to heed that advice. The countdown demonstration dragged on for five more tedious hours, necessitating a shift change by the overworked pad technicians. But despite the static on the intercom and the stench in the cabin, the Apollo 1 crewmen remained inside the spacecraft with both hatches sealed.

At the time, Project Apollo was between a fiscal rock and a hard place with the Russians in hot pursuit. Back in February of 1966, the Soviets' unmanned Luna 9 probe had made the first "soft" landing on the lunar surface. NASA had duplicated the feat with the Surveyor 1 probe four months later. But the nation's escalating expenditures on the war in Vietnam and LBJ's "Great Society" programs were draining the treasury. NASA officials had insisted on a minimum of $5.8 billion for fiscal year 1967 to have any chance of landing a man on the moon by the end of the decade; instead, they got only $5.02 billion, the first cutback in the annual budget since the space agency's creation.

In the wake of revelations about former Johnson aide Bobby Baker's service contract with North American Aviation, there had also been repeated instances of production slippage by Apollo's prime contractor. Although the first command module had arrived at the Cape on schedule in August of 1966, NASA program managers discovered scores of engineering flaws, including serious problems with the environmental control unit, more than half of which should have been fixed back at the factory. North American had suffered yet another embarrassment in October of 1966 when a service module exploded during a ground test.

NASA's nagging hardware problems had created a host of major headaches for the astronauts. Project Apollo demanded a fundamental change in the preflight training program. Because no one had ever gone to the moon before, there was no way of knowing what it would be like. The Gemini missions hardly offered meaningful comparisons in terms of distance and technical obstacles. The only way the astronauts could hope to prepare themselves for the realities of a lunar voyage was by returning to the realm of make-believe, by flying hundreds of imaginary lunar missions in the flight simulators at Houston and the Cape. But just as the Apollo spacecraft were bigger and more complex than the Gemini models, so were the simulators. What's more, the astronauts had to master the controls of two vehicles—the command module and the lunar module—instead of just one.

Each design and engineering change in the Apollo spacecraft under production required a corresponding change in the simulator. But shortly before the originally scheduled launch date of Apollo 1 in late 1966, director of flight crew operations Donald K. "Deke" Slayton complained that the contractors were so far behind in their modifications that the sims no longer matched the spacecraft. Apollo 1 commander Grissom personally counted 100 significant errors in the command module simulator, and signified his frustration by hanging a lemon over the hatch of the accursed trainer.

Although NASA had eventually decided to postpone the Apollo 1 launch to February of 1967, medical director Dr. Charles Berry was voicing serious concern over potential fire hazards in the command module. Part of his concern centered around the use of 100 percent oxygen in the astronauts' life support systems. But Berry was more worried about the apparent inability to reduce the quantity of flammable materials inside the cabin. He warned that even if the percentage of oxygen was reduced to a bare minimum, a spark from a routine short circuit could ignite the insulation of the electrical wiring and/or the astronauts' pressure suits.

The Apollo 1 crewmen regarded Berry's warning as merely one more last-minute aggravation. They were too gung ho, too macho, too confident in their own abilities to agonize over seemingly unavoidable fire hazards. As Grissom put it, "We're in a risky business, and we hope that if anything happens to us it will not delay the program. The conquest of space is worth the risk of life."

Shortly after 6 P.M. on the evening of January 27, 1967, Deke Slayton entered the blockhouse at the Cape to assist in wrapping up the Apollo 1 countdown demonstration in time for an early dinner. Then at 6:31 P.M. plus 4.7 seconds, he heard a terrifying scream come over the intercom.

"Fire! We've got a fire in the cockpit! . . . We've got a bad fire! . . . Let's get out! We're burning up . . ."

The crew atop the launch tower immediately grabbed the few available fire extinguishers, and charged across the access arm. But by the time they reached the far end of the gangplank, the flames had already spread to the "white room." The pad crew had to put out that second blaze before they could get to the spacecraft itself. At 6:36 P.M., approximately five minutes and six seconds after the first fire report, they finally managed to open the inner and outer hatches, and found that all three Apollo 1 crewmen were dead.

Gus Grissom's wife later sued NASA and North American Aviation (later renamed North American Rockwell and still later, Rockwell International) over the accident, an action that was regarded as tantamount to sacrilege by her late husband's astronaut peers. ("The country hadn't recruited us to take the risk and accept the rewards if successful—and file a lawsuit if it was not," Apollo 7 crewman Walter Cunningham noted afterward in his autobiography *The All American Boys.*) In 1971, North American Rockwell quietly settled Mrs. Grissom's lawsuit out of court for $350,000. Though Martha Chaffee and Pat White never joined in the litigation, the company reportedly paid them similar amounts.

In the meantime, the surviving Apollo astronauts reacted to the tragedy on Pad 34 with a characteristically double-edged mixture of grief and relief. Apollo 7's Cunningham recalls that after the mourning was over, one of his colleagues exclaimed, "Thank God it happened on the pad!"

"That may sound like a cold and strange way to think," Cunningham admits. "But in a time of crisis or tragedy or embarrassment, a convenient process takes hold of men who live by flying. They go on instruments; instincts take over. The crew was dead; nothing could be done for them. The important thing now was to find out how and why—to protect the living. It was that simple. Even as we felt the first dull shock over the deaths of Gus Grissom, Ed White, and Roger Chaffee, we thought, 'What about the program? What about us?' "

Before those questions could be answered, the space race took another life. In April of 1967, the U.S.S.R. announced that Soyuz I cosmonaut Vladimir Komarov had died on reentry when his parachute failed to open. It was later revealed that another Soviet cosmonaut had died some six years earlier in a ground fire similar to the one that had killed Grissom, White, and Chaffee.

Despite all the inquiries, the exact cause of the Apollo 1 fire was never determined. It appeared that the initial spark might have been caused by a short circuit in the electrical system. That theory gained support from the fact that the wiring beneath Grissom's couch was found to be defective. But it was unclear whether poor insulation had caused the short circuit or merely translated it into flames. Even so, investigators were not unclear or uncertain in assigning blame for the Apollo 1 tragedy, criticizing NASA, North American, and even the astronauts for accepting borderline safety margins in their eagerness to launch into space.

Ironically, the Apollo 1 fire did far more good than harm in the long run. "The truth of this is not easily understood," Apollo 7's Cunningham observes. "But the death of Gus Grissom's crew at the Cape made it possible to land a man on the moon on schedule. Indeed, it may have saved America's space program. So we cannot consider their deaths to have been in vain." Cunningham adds that the fire also "reminded the American public that men could and would die in our efforts to explore the heavens."

Project Apollo rose out of the ashes with a new spirit and commitment to reach the moon. The space agency budget was promptly increased by $500 million, roughly eighty percent of which was spent on changes made as a result of the Apollo 1 fire. The most important modifications were in the areas of crew safety. These included the installation of a one-piece hatch that the astronauts could open in ten seconds, and the addition of an escape wire on the launch tower. In order to reduce fire hazards on the launchpad, the astronauts' breathing mixture was switched from 100 percent oxygen to a sixty-forty percent mixture of oxygen and nitrogen. Flammable nylon elements and rubber insulation were replaced with materials like Teflon, fiberglass, and Beta cloth.

Though these changes delayed the next Apollo manned space flight for eighteen months, the astronauts were delighted that the improvements they had urged long before the fire were being made at last. As Cunningham pointed out years later, "From the astronauts' point of view, the changes encouraged the belief that a spacecraft would be built that could, at last, perform its intended mission." Though the Apollo astronauts could not help being overwhelmed by the awesome scale of their quest to land on the moon, their morale improved dramatically as each strove to contribute to the collective effort.

"I clearly could never understand as a crewman how to make it all work," Apollo 16's Ken Mattingly recalled in a recent interview. "I could only learn to operate my share of it. The same held true for other people in other areas. But there seemed to be a common attitude: 'This is such a big thing, but we're all in this together as a team effort, and we're going to make it work. I don't know how to do most of this mission. But I can assure you that my piece of it is going to work, and it won't fail because of me.' "

Although NASA refused to set a firm timetable for the next manned flights, the agency soon formulated the first specific plan for landing a man on the moon. The scenario consisted of seven steps that could be accomplished in an equal, greater, or lesser number of missions. Steps A, B, C, D, and E called for a series of manned and unmanned tests of the newly improved command and lunar modules in Earth orbit. Step F would be a series of lunar orbital missions to evaluate deep space conditions. If all went as scheduled, Step G would feature the first lunar landing attempt.

In early April of 1967, Deke Slayton summoned eighteen astronauts to a historic meeting at the Manned Spacecraft Center in Houston. The roster then read as follows:

Apollo 7
prime crew: Wally Schirra, Donn Eisele, Walt Cunningham
back-up crew: Tom Stafford, John Young, Gene Cernan

Apollo 8
prime crew: Jim Mc Divitt, Dave Scott, Rusty Schweickart
back-up crew: Pete Conrad, Dick Gordon, Al Bean

Apollo 9
prime crew: Frank Borman, Mike Collins, Bill Anders
back-up crew: Neil Armstrong, Jim Lovell, Buzz Aldrin

"The eighteen of you are going to fly the missions that get us to the moon," Slayton announced in a typically matter-of-fact tone, "including the first lunar landing."

At that moment, the Apollo astronauts realized that they were playing the "Big Game" for keeps. Each man present liked to think that objective factors like flying skill and demonstrated competence weighed heavily in his favor. But each man also knew that his chances of making it to the moon would depend on chance itself, unforeseen illnesses, deaths, and technical delays, as well as on byzantine astropolitics and the luck of the draw. As Slayton informed them, "The selection will be made by the normal rotation system."

In theory, the "normal rotation system" automatically determined mission assignments well ahead of schedule by promoting the astronauts from backup crews to prime crews in leapfrog fashion. The operative rule was "back one, skip two, fly one," then start the process all over again. For example, Apollo 7 backup crewman Tom Stafford could expect to sit out the Apollo 8 and Apollo 9 missions, then if all went well, move up to the prime crew for the Apollo 10 flight. In Stafford's case, that was exactly what happened. But as he and his colleagues discovered, there would also be many exceptions to the rule, partly because the rotation system was itself based on an unofficial seniority system known as the "Pecking Order."

The genesis of the "Pecking Order" dated back to the birth of the U.S. manned space program. As Project Mercury evolved into Project Gemini, the Original Seven got preferential status when it came to selecting the mission commanders for future manned space flights. The nine astronauts in the class of 1962 ranked second on the totem pole, followed by the Third Group chosen in 1963, and so on down the line. Thus, it came to pass that Mercury veteran Gus Grissom was picked to command the first Gemini mission and the first Apollo mission, while second- and third-generation astronauts John Young (Gemini 3), Ed White, and Roger Chaffee (Apollo 1) got subordinate crew assignments.

However, as time progressed, the sacred "Pecking Order" started to crumble. Death, retirement, and medical attrition decimated the ranks of the Original Seven. Alan Shepard, the first American in space, would be the only Mercury alumnus to get the chance to walk on the lunar surface. The other twenty-three men who flew to the moon would come from the next four astronaut classes. Thirteen of them, including Apollo 11's Armstrong and Aldrin, were members of the second and third astronaut classes.

It was common knowledge that former Mercury astronaut Shepard, who was grounded by an ear problem until 1971, advised Deke Slayton, who was grounded by a heart murmur until 1975, on crew assignments. But to this day, the other members of the Apollo astronaut corps claim they do not know how or why Slayton and Shepard decided who should be a mission commander, a command module pilot, or a lunar module pilot. "How we were picked," says Apollo 16's Charley Duke, "is a mystery."

"There isn't any big magic selection that goes on for each mission," Slayton would insist. "Everybody wants to be on a crew; that's normal. I can appreciate that more than anyone. I'm not in a job where you worry much about whether people love you or not. You've got a job to do and you hope it's right. This is like handling a squadron of fighter pilots.

You've got a mission to do and you've got so many flights to fly and you assign guys to fly them. Sometimes you ask the commander about the composition of his crew, but there again it goes far back. We know who gets along and who doesn't. If you've got some options, sometimes you give the commander a choice of three or four guys. Sometimes you don't have any choice. We've never pulled a man because of any personality conflict. If that happened, it would mean you had made the wrong selection in the first place."

Although the public tended to perceive the astronauts as interchangeable robots, Slayton knew they were anything but that. The corps included inveterate philanderers and faithful husbands, hot-rodders and born-again Christians. Each astronaut class also had a distinctive character. The first two groups were comprised entirely of test pilots. After the test pilot requirement was dropped, later classes included engineers and scientists. But regardless of class, the men assigned to each respective crew role seemed to share a more than coincidental number of common personality traits.

The commanders, or CDRs, tended to be the strong silent type, stoic to the point of insensitivity, doers not describers. They were usually the best flyers in the group; skilled, confident, experienced pilots, unfazed by crises, accustomed to being in charge, determined to be the hero not the goat. Two of the best exemplars were the laconic and inscrutable Neil Armstrong of Apollo 11, and Apollo 14's Shepard, nicknamed the "Icy Commander" because of his allegedly frigid disposition. And with the exception of Shepard, all came from either the second or third astronaut classes.

The command module pilots, or CMPs, were basically second-string commanders, the men responsible for piloting the mothership while their comrades explored the lunar surface and the only men who would be completely alone in space. In general, they tended to be somewhat more outgoing, observant, and descriptive than the implacable commanders. Though admittedly disappointed about being allowed to come so close without actually touching the moon, they were philosophical about their fates, and often the most lyrical. Appropriately, Mike Collins of Apollo 11 later wrote the best autobiographical account of an Apollo lunar mission, and Ken Mattingly of Apollo 16 provided some of the most quotable post-mission interviews. And like Mattingly, most of the CMPs hailed from the fifth astronaut class.

The lunar module commanders, or LMPs, were literally and figuratively the swing men of each crew, and usually tended to be more sensitive and high-strung than the CDRs or the CMPs. All but one, geologist Jack Schmitt of Apollo 17, came from either the third or fifth astronaut classes. They ranged from extremely observant, visually oriented types like painter Al Bean of Apollo 12 to remote and somewhat esoteric types like the emotionally troubled Buzz Aldrin of Apollo 11 and parapsychologist Ed Mitchell of Apollo 14. Although trained and titled pilots, the LMPs seldom did much flying in the course of a lunar mission. The commanders had the option of taking manual control of the lunar module during the landing approach. The main flying responsibility of the LMP, if the CDR did not co-opt it, was guiding the LEM to rendezvous with the command module for the return trip to Earth.

Meanwhile, all of the Apollo astronauts went through the same basic training. Their regimen combined both fact and fantasy, science and science fiction, including classroom instruction and related field trips, survival courses, practice sessions in the flight simulators, and imaginary lunar expeditions across various terrestrial "moonscapes." But henceforth, their make-believe took on a new reality.

The 240-hour classroom curriculum was designed to bridge the theoretical gap between aeronautics (flying in and around the earth's atmosphere) and astronautics (flying through outer space to the moon). The required courses included astronomy, aerodynamics, rocket propulsion, physics of the upper atmosphere, flight mechanics, guidance and navigation, meteorology, communications, and digital computers, as well as an introduction to rudimentary medicine and first-aid techniques. They were also required to take geology to prepare them for collecting specimens on the lunar surface, though most agreed with Apollo 11's Collins that the rock identification course was "bullshit."

Even more unpleasant from a physical point of view were the astronauts' mandatory survival courses in arid deserts, steamy jungles, and on the high seas. These arduous excursions, which could last several days at a stretch, were not designed to help the Apollo crewmen endure the rigors of outer space, but to survive a post-splashdown crisis on their home planet. Prior to leaving for the moon and upon reentering the earth's atmosphere, the flight paths of the Apollo spacecraft would pass over the equatorial regions of the globe. If the astronauts missed the designated splashdown sites patroled by the recovery vessels, they might find themselves marooned. Among other things, they learned from the air force tropical survival manual that "anything that creeps, crawls, swims, or flies is a potential source of food," including but not limited to "grasshoppers, hairless caterpillars, wood-boring beetle larvae and pupae, ant eggs, and termites . . . hedgehogs, porcupines, pangolins, mice, wild pigs . . . bats, squirrels, rats, and monkeys."

Oddly enough, the astronauts had no formal physical training program. There was a gym with squash and handball courts at the Manned Spacecraft Center in Houston, but no required exercise regimen. Even so, they found that it didn't hurt to stay in shape if only for the sake of enduring the rigors of the make-believe moon missions that consumed more and more of their training schedules as time progressed.

In order to prepare for their extravehicular activities on the lunar surface, the astronauts had to make rigorous treks across various artificial moonscapes not so affectionately known as the "rock piles." Apollo 11's Neil Armstrong and Buzz Aldrin, who would attempt man's first lunar landing in the Sea of Tranquility, practiced on the relatively flat rock piles in Houston and at the Cape, which were modeled after the boulder fields of the lowland maria. The crewmen on subsequent missions to the lunar highlands rehearsed their EVAs at much more rugged locales like Craters of the Moon National Monument in Idaho and the volcanoes of Hawaii. All of these terrestrial training sites were subject to the full gravitational burden of Earth. Walking even a few feet in a 185-lb. pressure suit was like climbing a mountain in a coat of medieval armor.

The astronauts found that the most enjoyable way to experience the one-sixth gravity of the moon without leaving the earth was by hitching themselves to the "Peter Pan rig," a pole-tied harness equipped with pulleys that offset five-sixths of the passenger's weight. Although the Peter Pan rig did not enable them to fly all around the globe like the fairy tale hero for which it was named, the contraption provided a very realistic simulation of lunar buoyancy within the reach of the tether cable. "You had the feeling of being able to jump very high—a very light feeling," Armstrong recalls. "You also had the feeling that things were happening slowly, which indeed they were. It was sort of a floating sensation."

The gymnastic routines performed on the "slippery table" and in the underwater submersion tank were far more arduous and exhausting. The slippery table, also known as the boxing ring, was a steel-walled enclosure with super-slick surfaces that made it impossible to stand or walk in normal fashion; the creators of this diabolical little torture chamber claimed that it simulated the frictionless environment of outer space. The submersion tank was a fancy swimming pool which supposedly provided yet another opportunity to experience the one-sixth gravity of the moon. But according to the astronauts, the only lasting effects of their antitraction and aquatic drills were bruised bodies and wounded egos.

Perhaps the most disorienting training exercise was going for a ride on the "zero G plane," a KC-135 with a custom-padded cargo bay. By flying up and down in steeply angled parabolic loops, the zero G plane could induce twenty-second intervals of weightlessness. During the dive toward Earth, the astronauts and everyone else on board not strapped in a seat belt would bounce off the cabin walls like lighter-than-air balloons. When the plane reversed direction and nosed upward, they would be smacked against the floor by a force equivalent to twice the pull of gravity.

"There is something extremely unsettling, even for the experienced aviator, about repeating forty or fifty parabolas," Apollo 11's Collins would report. "A typical planeload included cameramen, engineers, and various support personnel—most of whom were not accustomed to this sort of thing and who were almost sure bets to throw up during the first half dozen parabolas . . . Some astronauts threw up. I never did, but I was close enough at times to be utterly miserable."

The antithesis of the zero-G plane was the centrifuge, also known as "the wheel," which simulated the added gravitational force the astronauts would experience during both launch and reentry. The centrifuge was basically a one-man merry-go-round. It consisted of a single-seat cockpit stuck on the end of a fifty-foot-long metal arm; when the arm began to rotate at ever-increasing speeds, the gravitational force increased proportionately. The veterans of Project Gemini had experienced up to four Gs, or four times the force of Earth gravity, on their missions. The Apollo astronauts expected to encounter about seven Gs. But the centrifuge could increase the load to as much as fifteen Gs, a sensation that made the rider feel as if he were being smashed flat as a pancake.

By far the most important part of the Apollo astronauts' training was the time they spent in flight simulators, the delicate and often moody computerized trainers designed to replicate the cockpit and instrument performance of an actual spacecraft. The three main flight simulators were in Houston, the Cape, and St. Louis. Each consisted of two or three seats (depending on whether it was a command module or a lunar module sim) surrounded by steel panels crowded with a mind-boggling assortment of handles, levers, buttons, switches, and lights. Each of the astronauts spent literally hundreds of hours inside the flight simulators practicing imaginary launches, engine burns, orbital flight patterns, dockings, undockings, lunar landings, lunar lift-offs, and Earth reentries.

Flying a simulator was decidely more difficult than flying an actual spacecraft, and intentionally so. The simulators provided the only means for preparing the astronauts to handle unexpected malfunctions and in-flight crises. The sims were the places to make mistakes—and to learn how to avoid making mistakes—before it was too late. As Apollo 11's Collins would observe, "Running on the beach is grand; learning geology is commendable; jungle living is amusing; the centrifuge hurts; the spacecraft ground tests are useful; but one is not ready to fly until the simulator has told him he is ready."

If, as Apollo 16's Ken Mattingly maintains, the astronauts had "no idea how to make the whole thing work," each became an expert in an assigned area of lunar mission technology suited to his individual background and interests. Apollo 11's Buzz Aldrin, known as "Dr. Rendezvous" by virtue of his MIT doctoral thesis, got mission planning. In addition to helping decide how many flights to send to the moon and in what order, he helped devise the delicate maneuvers needed to get the lunar module back to the mothership once it lifted off from the moon. Apollo 12's Al Bean, a former carrier pilot, was assigned responsibility for the recovery systems required to pick up the spacecraft after splashdown in the ocean. Other astronauts concentrated on areas ranging from guidance and navigation to pressure suit assembly.

The Apollo astronauts later praised NASA for having the wisdom to include them in the design phases of the program, but the attendant responsibilities did have their downsides. Because JFK and LBJ had spread the space program's political patronage from coast to coast, the astronauts had to fly from coast to coast to visit NASA installations and contractors. Although most of their training equipment was in Houston or at the Cape, they had to go to the Grumman plant in Bethpage, Long Island, to inspect the lunar modules. The command modules were built in the North American Aviation (née Rockwell) plant in Downey, California. The three main sections of the Saturn 5 boosters were built by three different contractors in three different locales. The first stage, known as the S-IC, was assembled at the Boeing plant near New Orleans. The second stage came from the North American facility at Seal Beach, California. The third stage, aka the S-IVB, was test fired by McDonnell Douglas in Sacramento. The guidance system for the Apollo spacecraft was developed by engineers at the Massachusetts Institute of Technology in Cambridge.

Getting back and forth between these widely dispersed sites could be delightfully quick but exceedingly dangerous. The astronauts were provided personal transportation in the form of Northrup T-38s, the most advanced jet trainers. A T-38 could fly faster than the speed of sound, but it was extremely rough at low velocities, especially before landings, and its high-performance engines were easily damaged by ice. Before the first Apollo mission launched for the moon, four astronauts—Ted Freeman, Elliot See, Charlie Bassett, and C.C. Williams—were killed in T-38 crashes.

Despite the hazards of the T-38s, the astronauts appreciated the importance of visiting NASA contractors and subcontractors far more than spending time in geology class. It was reassuring for them to know that the people who were putting the spacecraft together knew that the astronaut who would be on board was a real live human being just like them. More important, most of the private sector executives and engineers seemed to be seriously interested in what the astronauts had to say about particular problem areas, and made numerous modifications as a result. Best of all, as Collins points out, "No one ever told us we were running the price up too high."

The astronauts, being astronauts, naturally felt the need to let off steam. But here again, thanks partly to scandalmongers in the national media and partly to the astronauts' lust to display what author Tom Wolfe termed "the right stuff," fact and fiction became indistinguishable.

Besides drinking, the favorite extracurricular activities involved fast planes, fast cars, and fast women, not necessarily in that order. A classic example was the day Apollo 7's Wally Schirra and Walt Cunningham challenged Apollo 9 astronauts Jim McDivitt and Rusty Schweikert to a race from El Paso to Los Angeles. After spotting the competition a head start, the Apollo 7 crewmen pushed their T-38 up to the breakneck speed of Mach 1.2, only to lose by a good five minutes. Other astronauts, most notably Apollo 12's Pete Conrad, liked to drag race their customized Corvettes on the runways at Ellington and on the heavily congested Gulf Freeway linking the Manned Spacecraft Center with Houston and Galveston. As Cunningham confesses, "We got high on such encounters."

Sex, by far the most common vice, claimed more casualties than speed. Astronaut groupies euphemistically referred to as "specialists" patrolled the hotels, motels, bars, and beer joints at the Cape, and infiltrated the shopping centers, subdivisions, and apartment complexes around the Manned Spacecraft Center in Houston. Unfounded rumors of the astronauts' sexual escapades abounded in the scandal sheets. Apollo 11's Neil Armstrong was falsely reported to be having an affair with singer Connie Stevens. Two different astronauts, Apollo 7's Walt Cunningham and Apollo 12's Pete Conrad, were simultaneously linked to singer Keely Smith, ex-wife of Louis Prima; both Cunningham and Conrad denied having any such relationship. Apollo 7's Wally Schirra learned that he had a "second wife" he had never even seen before who checked into a Cocoa Beach, Florida, motel part-owned by the Original Seven, and announced that she was the real Mrs. Schirra. NASA security eventually exposed the woman as a fraud, and she was never heard from again.

Of course, NASA officials knew that not all of the rumors were unfounded, a fact that made them extremely nervous. While chief astronaut Deke Slayton did not expect his men to behave like a flock of angels, he determined to protect the space agency's squeaky clean public image at all costs. "Most of you are bright enough to know that if you are going to be screwing around you'd better be damn discreet about it," he told one group of new recruits on their first visit to the Cape, adding, "Anytime your outside activities are more important to you than this job, just let me know and we'll arrange for a change."

The indiscreet suffered swift retribution. Dr. Duane Graveline, one of the five scientists in the class of 1965, was the first to get the ax. Shortly after he joined the astronaut corps, Graveline's marital problems reportedly became "public knowledge." When his wife filed for divorce, NASA officials asked for and got his resignation before Graveline ever got the chance to fly in space.

Apollo 7's Donn Eisele, who commenced a rather open extramarital affair well before his mission left the launchpad, was the first astronaut to get a divorce after returning from outer space, but he was by no means the last. Fellow Apollo astronauts Young, Aldrin, Mitchell, Worden, Cernan, and Bean were also destined to part with their wives in the years to come.

By any objective standard, the astronauts' domestic difficulties were par for the course. Of the forty married men who flew in the Mercury, Gemini, and Apollo programs, a total of eight, or roughly twenty percent, got divorces. That was actually less than half the divorce rate for the general population. But as the astronauts knew, NASA and the public held them to a higher standard. Following the breakups of their marriages, none of the divorcés ever flew in space again except Young, who later qualified to fly on the space shuttle.

Even those marriages that survived Project Apollo felt the strain of the astronaut life. Most of the wives had small children to raise, and were understandably lonely, exhausted, and nervous when their husbands were gone all week to visit contractors or practice simulations at the Cape. One evening when a group of astronaut wives were bemoaning their hardships at a party for the Apollo 7 crew, Apollo 13's Jack Swigert, one of a handful of bachelors in the astronaut corps, told them off.

"You wives are foolish as hell if you take this attitude every time your husband comes home," Swigert declared. "Yeah, we've been off all week in the land of milk and honey and beautiful secretaries. But we're also working our tails off. Your husbands look forward to coming home for a little loving relaxation, not a lot of nagging. You should make them feel great when they get back to town, not like they're taking a dose of medicine. If you don't, some weekend when he's faced with a choice of staying where he is or flying home, he isn't going to make the right decision."

Though Jack Swigert earned a well-deserved reputation as a tireless skirt chaser, not all the other bachelors affected the stereotypical astronaut swagger. Apollo 17's Jack Schmitt was one who recoiled from his new-found celebrity status, and reportedly told more than one aggressive female seductress, "I don't just tumble on the first date. It means more to me than that." Likewise, the other Apollo astronauts knew they were much more—and much less—than their official and unofficial public images suggested.

"I can see where we developed a kind of split personality," Apollo 7's Cunningham wrote later. "There was the astronaut image; the pursuit of a life-style that in other places or other times would never have been accessible to us. Then there was the side that shrank from the image and the fanfare, when we reminded ourselves that at the bottom we could be dull men, sent to do a technician's job."

The sheer frustrations of the astronauts' daily routines battered their self-images and self-esteem. Buzz Aldrin later recounted his sudden outburst one day while he and Gene Cernan were in the submersion tank. As the two men practiced tedious exercises of screwing, unscrewing, clipping, unclipping, connecting, and unconnecting various nuts, bolts, wires, and hoses, Aldrin realized, "The monkey I had bought for Joan [his wife] could've performed these jobs, probably not underwater, but at least out in the garage." When Cernan interrupted his partner's musings to ask about the next item on the underwater checklist, Aldrin suddenly retorted with a high-pitched shriek: "Shut up, and pass me a banana!"

The astronauts also had problems on the financial front. When the Apollo 11 crewmen flew to the moon, they had already been wined and dined by presidents, kings, and multimillionaires, but their base pay was only $17,000 a year. And contrary to popular misconception, they were not getting rich from endorsements. The Original Seven had signed a highly publicized—and highly controversial—$500,000 personal story contract with *Life* magazine which paid each of them approximately $71,000 over the duration of the Mercury project. In 1963, the astronauts signed a second personal story contract for $1,040,000 with *Life* and the Field newspaper syndicate, but this time the money had to be divided among twenty-nine instead of just seven astronauts, so each man wound up netting about $16,250 over four years. Field dropped out of the deal in 1967, and *Life* eventually renegotiated a contract which paid each astronaut about $3,000 a year through 1970. The following year, the magazine folded.

Although the astronauts were offered an endless and invariably tempting array of private deals, they were constantly reminded by NASA officials and each other to be wary of exploitation. And yet almost each and every one of them wound up getting bamboozled in some way. A classic astronaut rip-off involved an electronics company that invited Apollo 7's Walt Cunningham to join the board of directors. When news of Cunningham's deal got out, several of his fellow astronauts including Neil Armstrong, Mike Collins, Al Bean, and Bill Anders invested in the company's stock, a move that only made Cunningham more anxious. In the end, the company folded. Cunningham later reported that he wound up with "ten thousand dollars invested in worthless stock, plus a bucketful accepted in lieu of salary, plus the pinched conscience of knowing that friends and relatives had blown some dough as well."

By the fall of 1968, the overworked, underpaid, henpecked, hamstrung, and generally unappreciated Apollo astronauts also realized that the rules of the "Big Game"—and everything else—had changed dramatically. The American body politic was in an uproar. Race riots had burned the ghettos of Detroit the previous summer. In January, LBJ had all but admitted that his Vietnam policy was a disaster; he later announced that he would not run for reelection. In April, a sniper had assassinated civil rights leader Martin Luther King. In June, an Arab fanatic had assassinated Democratic presidential candidate Robert F. Kennedy.

Meanwhile, as Richard M. Nixon and Hubert H. Humphrey fought for control of the White House, the nation's college campuses were erupting in bloody antiwar demonstrations. Even the children of the proudly patriotic "silent majority" were among the draft dodgers and bra burners who marched in the streets or simply tuned in, turned on, and dropped out to rural communes devoted to peace, free love, drugs, and/or esoteric Eastern religions.

Oddly enough, the astronauts—not the hippies or the yippies—were the true idealists of the era. They did not share the disillusionment, cynicism, and profound alienation that afflicted the younger generation. Nor did they sympathize with increasingly shrill demands to terminate Project Apollo so the budget could be spent on domestic social programs. The astronauts still believed that what they were doing was noble, just, and justifiable, that the space program was not just a multi-billion-dollar boondoggle for the infamous military-industrial complex, that going to the moon really would reap enormous benefits for all mankind—provided that the U.S. got there first.

At the time, NASA officials believed that the U.S.S.R. still had a very good chance of landing the first man on the moon. Although the Soyuz 1 catastrophe had caused costly revisions and delays similar to those resulting from the Apollo 1 fire, the Soviets had reportedly developed a rocket booster even more powerful than the Saturn V. In early September of 1968, an unmanned Soviet probe orbited the moon, and for the first time, returned to Earth intact.

The first series of unmanned Apollo missions, three far less ambitious Earth orbital flights, were officially hailed as successful, but critics claimed that they were nearly catastrophic embarrassments. Apollo 4 suffered a fuel spill and a major computer malfunction. Apollo 5, the inaugural test flight of the lunar module, experienced two equally serious problems. Prior to lift-off, the windows of the LEM shattered to pieces for no apparent reason. When the hastily refurbished spacecraft finally got off the ground, the LEM's engine, which was supposed to fire for thirty-eight seconds at full power, only managed to burn for four seconds at ten percent thrust. Apollo 6 failed even more miserably. Due to a sequence of booster engine malfunctions, the spacecraft was catapulted into the wrong orbit before it got the chance to show its stuff.

Then the seemingly unspectacular Apollo 7 mission got the U.S. space program back on track. Launched from the Cape on October 11, 1968, it was the nation's first manned space shot since the Apollo 1 tragedy. It was also the maiden voyage of the new and improved command and service module (CSM) combination, and one of the most ambitious test flights ever attempted. Apollo 7 astronauts Wally Schirra, Donn Eisele, and Walt Cunningham orbited the earth for eleven days, coming within seventy-two hours of the American endurance record set by the

Gemini 7 astronauts. Though all three crewmen came down with head colds in space, they completed their operational and experimental assignments on schedule, a feat that provided a direly needed boost of confidence for their embattled colleagues at the Manned Spacecraft Center in Houston.

But unbeknownst to the general public, the Apollo 7 astronauts were flirting with disaster from the moment they left the launchpad. While the mass media celebrated the fact that America had at last gotten back in the space race, NASA quietly compiled a list of no less than fifty malfunctions that had occurred during the mission. The most ominous included repeated errors by the spacecraft's guidance and control systems, inexplicable surges in orbital velocity, a nine-minute communications blackout, and the loss of three days worth of biomedical monitoring data. Luckily, these and other deliberately unpublicized problems had not derailed the Apollo 7 flight, but if left uncorrected, any one of them could create a life-threatening crisis for the first astronauts who dared to venture out of Earth orbit.

Nevertheless, the widely heralded successes of Apollo 7 and the perceived threat of Soviet one-upmanship inspired space agency officials to make a bold revision in the flight plan for Apollo 8. According to the original scenario, the next alphabetized item on the agenda was step D, which called for the first combined test of the command, service, and lunar modules in Earth orbit. But because of unforeseen production delays, the Apollo 8 lunar module was not ready for launch. Rather than postponing the mission (and thereby incurring cost overruns of up to $30 million per month) or merely retracing the Apollo 7 flight plan (which might bore American taxpayers and incite political opponents), NASA announced that Apollo 8 crewmen Frank Borman, Bill Anders, and Jim Lovell would attempt something man had never done before—a round trip voyage from the earth to the moon.

Although the other Apollo astronauts remained dutifully tight-lipped in public, most of them privately agreed that the revised flight plan for Apollo 8 was, as Walt Cunningham put it, "spectacular but superfluous." Lacking a lunar module, Borman, Anders, and Lovell could not land on the moon. Nor could they test the crucial procedures required for a rendezvous in lunar orbit. And given the fact that they would be orbiting over sixty miles above the lunar surface, they could not improve on the previous photo reconnaissance of lower-flying unmanned probes. In essence, the mission was a grandiose public relations venture that promised no tangible scientific or technological rewards—except for proving that man could fly around the moon and return safely to Earth.

Apollo 8 turned out to be an extraterrestial spectacle that was well worth watching. The astronauts launched from the Cape on December 21, 1968. Three days later, on Christmas Eve, they became the first men to circle the moon. During the prelaunch conferences on eliminating unnecessary weight, Borman had opposed the idea of carrying a television camera aboard the spacecraft. Fortunately for him, his fellow crewmen, and the citizens of Earth, his objections were overruled. According to a survey by *TV Guide,* nearly one billion people in sixty-four countries watched the first live TV broadcasts from lunar orbit.

As the Apollo 8 spacecraft rounded the moon, Borman offered a brief prayer. Then he and the other astronauts took turns reading the biblical version of creation in the first ten verses of the Book of Genesis. On Christmas Day, after completing ten lunar orbits, the astronauts fired the engines of their spacecraft and headed back to Earth. "Please be informed," Lovell radioed with glee, "there is a Santa Claus!"

The Apollo 8 astronauts' Bible reading did not receive universal applause. Outraged atheist Madalyn Murray O'Hair filed suit against NASA and the three crewmen for allegedly violating constitutional prohibitions against the establishment of religion. The mission also raised some ominous questions about man's ability to adapt to weightlessness. During the first day of the translunar voyage, Borman suffered severe nausea and began to vomit. Though he bravely concealed his illness from Mission Control until he had almost fully recovered, NASA doctors naturally worried that air sickness might also afflict the crewmen on future lunar missions. But as far as NASA officials were concerned, Apollo 8 was a smashing triumph from a propaganda point of view. "I think the Russians would have gone all out to orbit the moon if we hadn't done it when we did," Manned Spacecraft Center chief Robert Gilruth later told an interviewer. "We preempted that. If they had orbited the moon, our press would have said they had won."

On January 6, 1969, Deke Slayton summoned the Apollo 11 crew, and informed astronauts Neil Armstrong, Buzz Aldrin, and Mike Collins, "You're it." Three days later, NASA announced the news to the public. Assuming that the Apollo 9 and 10 missions succeeded as planned, Apollo 11 would attempt mankind's first lunar landing in July of 1969.

At first, the Apollo 11 astronauts reacted to their extraordinary good fortune with surprisingly mixed emotions. All three crewmen publicly professed to be elated at winning the plum flying assignment of all time. But in private, they harbored some quite serious doubts and reservations.

"The LM hadn't even flown men yet," Collins noted in his post-mission autobiography, "and if it acted up on Apollo 9, the landing would probably slip to Apollo 12. Apollo 10 was to be a dress rehearsal for 11, and again it could develop problems which would require delaying the landing . . . If I had been trying to establish odds at the time, I think I would have given Apollo 10 a ten percent chance, 11 a fifty percent, and 12 or subsequent a forty percent chance to attempt the landing."

Much to the relief of the Apollo 11 astronauts, the next two missions proved to be unqualified successes. Apollo 9, which launched on March 3, 1969, was the first real-life test of the lunar module/command module combination. Although astronauts Jim McDivitt, Dave Scott, and Rusty Schweikart never left Earth orbit, their ten-day space voyage was of vastly underrated significance. The Gemini flights had pioneered space rendezvous and docking. Apollo 9 demonstrated that similar maneuvers could be executed with the lunar module and the command module—at least in Earth orbit. The only disconcerting aspect of the mission was the fact that Schweickart suffered nauseating air sickness for four days. Though he nevertheless managed to perform his scheduled space walk, Schweickart's illness, which proved even more debilitating than Borman's, again raised the spectre that future astronauts might not be able to cope with long-term exposure to weightlessness.

Apollo 10 had the bittersweet task of playing bridesmaid for Apollo 11. Like the Apollo 8 astronauts, the crew of Apollo 10 went all the way to the moon, but they were not allowed to touch down on the lunar surface. Their assignment was to test rendezvous and docking procedures in lunar orbit, an exercise crucial to the success of man's first lunar landing mission. On May 22, 1969, astronauts Tom Stafford and Gene Cernan flew the Apollo 10 lunar module Snoopy to within 50,000 feet of the moon. Their flight path took them right over the approach Apollo 11 planned to use to get to the Sea of Tranquility, a route they dubbed "U.S. 1." Then Stafford and Cernan pulled up, reunited with crew mate John Young in the command module Charlie Brown, and returned to Earth.

A few days later, the Apollo 11 astronauts got formal approval to proceed with man's first lunar landing mission. NASA officials announced that the launch was scheduled for July 16, 1969. That date happened to be the twenty-fourth anniversary of the first nuclear explosion in Alamogordo, New Mexico. It was also five and a half months before the deadline set by the late President John F. Kennedy.

But despite all the advance publicity, the world public never appreciated the magnitude of risks the Apollo 11 crewmen were about to face. That was partly because the astronauts themselves invariably minimized the dangers of their mission at every prelaunch press conference. Neil Armstrong even declared with characteristic confidence that the Apollo 11 mission was "not truly a flight into the unknown," adding that he and his crew were "prepared to go to the moon."

Armstrong had a point. By the summer of 1969, man had compiled almost as much basic astronomical and geophysical data about the moon, one of the largest planetary satellites in the solar system, as he had about the Earth. Centuries of telescopic observations had enabled him to estimate with remarkable accuracy the diameter of the moon (2,155 miles), its circumference (6,790 miles), its mass (1.23% of the Earth's mass), and its weight (81 billion billion tons). Having charted the moon's elliptical orbit, he had determined that her relative position was constantly changing, so that her distance from the earth varied from 221,643 miles to 252,710 miles. Likewise, the irregular speed of the moon's monthly revolutions could only be calculated in comparison to the positions of other celestial bodies. Her sidereal month, which was measured with respect to a relatively fixed background of stars, lasted 27.32 days. Her synodic month, which was gauged by her location relative to the sun, lasted 29.53 days.

Man had also discovered the existence of two other more subtle but equally influential forms of lunar motion. Contrary to popular belief, the moon was not frozen in space. Like the earth, the moon rotated on its own axis, but it did so in monthly cycles, steadily turning the far side away as it proceeded through her orbit. That was why she always kept one side hidden from view. Yet, the moon still revealed up to fifty-nine percent of her total surface area when full due to a peculiar rocking motion known as "libration." These lunar vibrations alternately exposed the edges of her "limb," the twilight zones on her eastern and western horizons.

Thanks to recent unmanned probes, the Apollo 11 crewmen were supplied with similarly detailed data about conditions on the lunar surface. That data included tens of thousands of photographs and innumerable meteorological and topographical readings transmitted by the Ranger and Surveyor spacecraft. Among other things, the astronauts knew that surface temperatures could range from a low of minus 250 degrees Fahrenheit to a scorching high of plus 250 degrees Fahrenheit. They also knew that the topography ranged from 30,000-foot peaks higher than the Himalayas to flat desertlike lowland maria similar to their intended landing site in the Sea of Tranquility and cavernous rilles deeper than the Grand Canyon.

In order to reach their destination and return to Earth, however, the Apollo 11 astronauts would also have to confront assorted hazards of unknown origin and nature. Not the least of these was the great unknown of infinity itself. The astronauts did not expect to find life on the moon or do battle with armies of alien beings on their voyage through space. But one did not have to be a lunatic to believe with reasonable certainty that just about anything could happen in an infinite universe—modern science said so by definition.

One of the unknowns the Apollo 11 astronauts had to face was the question of whether it was even possible for a manned spacecraft to touch down on the moon. Although a dozen American and Soviet probes had already crashed or soft-landed on the lunar surface, the heaviest unmanned probe was less than half the weight of the 33,205-pound Apollo 11 lunar module. The available remote instrument data indicated that the moon consisted of a layer of granular topsoil buttressed by a firm foundation of semi-porous bedrock. But no one knew for sure if the lunar surface would support a manned spacecraft or dash it to pieces or smother it in a bottomless pit of ashes as foretold in Arthur C. Clarke's 1961 science fiction novel *A Fall of Moondust.*

The geological composition of the lunar surface also posed serious guidance and navigation problems of an unknown nature. Previous unmanned probes had revealed that the moon was dotted with invisible magnetic potholes called "mascons." Deeply buried and widely dispersed, these treacherous mass concentrations of iron ore exerted a much greater gravitational pull than surrounding rock and crater formations. The mascons' unusually powerful G force had already played havoc with the guidance and navigation systems of the Ranger and Surveyor probes. The Apollo 11 astronauts feared that the mascons might cause their lunar module to overshoot or undershoot their landing site, forcing them to abort their mission before they reached the lunar surface.

Even if Armstrong and Aldrin managed to touch down safely, they would have to face the unknown dangers of cosmic rays and micrometeorites on the lunar surface. Cosmic rays were believed to be ultrahigh velocity bolts of energy capable of penetrating hard metal and human tissue with or without harmful effect. Micrometeorites, on the other hand, were potentially destructive by definition. They consisted of microscopic particle fragments that rained down from outer space at blinding speed and with more than enough force to tear holes in spacecraft, space suits, and spacemen.

Assuming that Armstrong and Aldrin survived their excursion on the lunar surface, they were by no means home free. In order to rendezvous with the mothership and return to Earth, the astronauts would have to lift off at precisely the right time and in precisely the right trajectory. And unlike the command module, the lunar module did not have a backup motor. There was only one ascent engine. It had to work right the first time, for there would be no second chance. If the ascent engine did not fire, the astronauts would be doomed to die on the moon as soon as their oxygen supplies were depleted.

And even if the ascent engine worked, all three Apollo 11 astronauts would have to make the crucial trans-Earth engine burn without the help of their on-board computer guidance system. This unenviable predicament resulted from the controversial prelaunch decision to delete the "Return to Earth" program in an effort to streamline their computer's already overloaded memory bank. Before the spacecraft began each lunar orbit and disappeared out of radio contact behind the moon, the astronauts would have to hand-copy key navigational data relayed from the computers back at Mission Control so they could execute their own manually punched-in return program.

Meanwhile, as the launch date grew nearer, the Apollo 11 astronauts began to have grave doubts about whether they would get a chance to practice their lunar landing maneuvers, much less get final approval to attempt the real thing. Along with ominous slippage in the production schedule of the Apollo 11 lunar module, there were annoying delays in completing corresponding changes in the lunar module simulator. "The LM sim did not become operational for our mission until late March," Aldrin complained later, "leaving us little time for the myriad experiments and training tests we needed to conduct."

There were also continuing problems with supposedly improved models of the LLRV, the "Flying Bedstead" that Armstrong had crashed at Ellington AFB in May of 1968. Seven months after that mishap, Joseph S. Agranti, head of the Aircraft Operations Office in Houston, had been forced to bail out of a similar Lunar Landing Training Vehicle (LLTV) at Ellington, which also crashed and burned. It was not until March of 1969 that Gilruth approved the resumption of LLTV flights, and then only by nonastronaut test pilots.

Armstrong didn't get another crack at practicing free flight lunar landings in the trainer until June, barely a month before Apollo 11's lift-off. Even then, the LLTV was considered so dangerous that Gilruth refused to let anyone but Armstrong go up in it. Though Buzz Aldrin had the title of lunar module pilot, the mission commander would do the piloting during the actual lunar landing attempt. Gilruth saw no reason to risk two lives at such a late date.

Gilruth and Slayton were also becoming concerned that the hectic pace of the astronauts' preflight training schedule was taking its toll. The three crewmen were constantly on the go and under stress. They were each spending ten to fourteen hours per day in the command module and lunar module simulators at the Cape five days a week. On Fridays, they flew home to Houston in their T-38s, tried to spend time with their wives and children like normal family men, then hopped back in their planes early Monday morning and returned to their literal and figurative rock piles.

Jan Armstrong, whose husband had the added duty of practicing in the LLTV, grew increasingly anxious about the mental and physical condition of her spouse and his fellow crew mates. "Neil used to come home with his face drawn white," she recalled afterward. "I was worried about him. I was worried about all of them. The worst period was in early June. Their morale was down. They were worried about whether there was enough time for them to learn the things they had to learn, to do the things they had to do, if this mission was to work."

Fearful that the Apollo 11 crewmen might wear themselves out before the mission even got off the ground, Slayton asked the three astronauts if they wanted to postpone the launch for a month. All three voted no.

Then, almost miraculously, Apollo 11's preflight depression bottomed out, and problems started to resolve themselves on all fronts. On June 14, Armstrong made his first successful practice landing in a rocket-powered free-flight training vehicle at Ellington Air Force Base in Houston. Two days later, he repeated the landing drill eight times in a row.

By the time the official countdown began in mid-July, there were even a few moments of unintentional humor amid the Apollo 11 prelaunch frenzy. Careful to a fault, NASA medical director Dr. Charles Berry insisted that the three astronauts be virtually quarantined for the week preceding their flight, and required them to wear surgical masks when they appeared in public for their final prelaunch press conferences.

Berry also insisted on canceling the astronauts' scheduled dinner with President Nixon the night before launch, a precaution all three crewmen regarded as well beyond the boundary between the sublime and the ridiculous.

Following Apollo 11's triumphant touchdown in the Sea of Tranquility on July 20, 1969, the world public got the impression that the next six lunar landing missions would be a technological fait accompli with little or no danger to the astronauts. As the near fatal explosion on Apollo 13 demonstrated, nothing could have been further from the truth. Although the Apollo 11 crewmen faced the great unknown at its greatest and most unknown, the astronauts who flew on subsequent lunar voyages faced at least equal if not greater risks.

The success of man's first lunar landing by no means guaranteed the success of the second, third, fourth, fifth, or sixth. As NASA safety director Jerry Lederer had pointed out in a speech prior to the Apollo 8 launch, the spacecraft and booster consisted of over 5.6 million parts and 1.5 million subsystems. Even if all these components functioned with 99.9 percent reliability, the spacecraft could be expected to suffer a statistical minimum of 5,600 malfunctions, almost any of which, if left uncorrected, could lead to disaster.

NASA administrator Dr. Thomas O. Paine later reported that he and Gilruth feared that the likelihood of tragedy increased exponentially with each mission to the moon. Those fears eventually convinced them that they should terminate the program after the Apollo 18 mission instead of after the scheduled Apollo 20 mission. As it turned out, federal budget cuts forced them to end the program with Apollo 17. But as Paine later admitted, "Every one of [the nine lunar missions] was a hazard to the three astronauts involved . . . eventually we would have lost a mission."

Even so, the Apollo astronauts all seemed to share the sentiments Neil Armstrong expressed shortly before lifting off on man's first lunar landing mission. Asked to explain why he and his crew mates were embarking on such an audacious undertaking, he replied, "I think we are going to the moon because it's in the nature of the human being to face challenges. It's by the nature of his deep inner soul. Yes, we're required to do these things just as salmon swim upstream."

7

LIFE IN ZERO G

Cislunar Space

Day One–Day Three

When it is dark enough, men see the stars.
RALPH WALDO EMERSON (1840)

Day One

All twenty-four men who flew to the moon ate lunch on Day One and every other in-flight meal out of red, white, or blue plastic bags. Each bag was color coded according to crew member and labeled with the intended date and time of consumption. But almost all the solid food and flavored drinks came in dehydrated form. And except for slight variations in color and consistency, each packet of pre-frozen granules looked the same as every other packet of pre-frozen granules. According to Apollo 16's Charley Duke, "You really couldn't tell what you were eating unless you read the menu or read the instructions on the side of the bag."

Although the Apollo astronauts were usually big eaters back on earth, most did not enjoy their first meal aboard the spacecraft. That was primarily because many of them were still reeling from the tumultuous events of the launch, staging, and TLI, and were not yet fully adapted to weightlessness. Even those who did not actually feel nauseated complained that their newly induced "stomach awareness" suppressed their appetites.

But almost all the astronauts also complained about the unappetizing content of their in-flight menu. Apollo 11's Buzz Aldrin later gave one of the most balanced critiques in his post-mission autobiography *Return to Earth:* "It wasn't long before I developed my first affinity for food in space: the shrimp cocktails. The shrimp were chosen one by one to make sure they would be tiny enough to squeeze out of the food packet, and they were delicious. There was some fairly good soup and some cheese and meat spreads for crackers, but most of the rest of the food was bland."

Aldrin also debunked a popular commercial myth about the Apollo 11 crew's preferred in-flight drink: "As anyone who watches television knows, the people who manufacture Tang make a big thing out of the fact that astronauts drink it for breakfast while they're traveling in space. I can't speak for other flights, but before ours, the three of us sampled the orange drink, supposedly Tang, and instead chose a grapefruit-orange mixture as our citrus drink. If Tang was on our flight, I wasn't aware of it."

Before digging in, the three Apollo crewmen had to mix the contents of their plastic food bags with hot or cold water as specified by the accompanying instructions. The instructions on a packet of pea soup, for example, might read, "Add five ounces of hot water and wait ten minutes." A packet of solid food typically required less water and a certain amount of kneading in order to become edible.

"To get the water in the bag," explains Apollo 16's Charley Duke, "we had a water pistol which we loaded from a spigot in the command module. Every time you squeezed the trigger, it squirted out one ounce of water. You stuck the barrel of the gun into the top of the bag through a little valve, and squeezed the trigger four times or whatever [the recipe] called for. Then you shook up the bag, kneaded it, and just sorta let it float while you went on to [prepare] the rest of the meal."

The three crewmen usually discovered that they could get the food out of the bags in much them same way they might do on Earth. First, they snipped off the bottoms of the bags with a pair of surgical scissors. Then they took spoons and scooped out the contents one mouthful at a time. The mundane simplicity of this process was amazing given the fact that the food, like the astronauts, was weightless. "I don't know how it works," says Duke, "but you can reach in and pull out one spoonful, and the rest of it stays in the bag, I guess because of surface tensions and things like that."

Getting the food from spoon to mouth could be even more amazing thanks to the benign influences of zero G, according to Duke. "You've got a spoonful of pea soup that's the shape of the spoon, and you can take your spoon and smooth it around in any set direction and the soup stays on the spoon. But if you take your spoon and pop it, the soup separates. When the soup comes off, instead of [remaining in] the shape of the spoon, at a snap of a finger the molecular force is equalized and pulls it into the shape of a perfect sphere, a little green marble of pea soup floating around in space. And when you reach up and touch it, it just migrates down the spoon and takes the shape of the spoon again. You can eat it that way or you can just float over next to this little marble and—swoooop!—suck in a mouthful of pea soup."

Cleaning up after a meal was as time-consuming as preparing one. All the food bags had to be collected, spiked with pills that dissolved the residue, then rolled up as tightly as possible and deposited in a plastic garbage bag. Periodically, the astronauts would vent their food wastes into outer space, but they did not have time to perform this disposal procedure after every breakfast, lunch, or dinner. In the interim, they stashed their garbage in cabin storage compartments that filled to capacity as Day One wore on. "The trash you accumulate is just unbelievable," recalls Apollo 14's Stu Roosa.

Disposing of human wastes posed even greater difficulties than eating and drinking. In order to urinate, the crewmen had to float over to a tiny urinal compartment in the lower equipment bay, and try to brace themselves in a steady position until they were done. In order to defecate, they had to completely undress and suffer through a procedure many astronauts remember as the most uncomfortable aspect of life in zero G.

"We had a really crude system for [disposing of] your feces," attests Apollo 16's Duke. "It was a bag. The only thing is that nothing goes to the bottom of the bag in zero gravity. Everything floats. In some cases with a loose [bowel] movement, it was really unmanageable. It just didn't work."

Conducting the otherwise ordinary routines of personal hygiene was equally problematic, especially when it came to shaving. Most of the astronauts found that their standard issue electric razors mysteriously began to malfunction after a day or two in space. According to Duke, "It felt like it was pulling out your beard instead of cutting it."

The crew of Apollo 11 steadfastly endured the aggravation of shaving because they were constantly reminded that the whole world was watching man's first lunar landing mission. A few of the astronauts on later lunar missions brought along conventional safety razors, and wound up making an even worse mess of their faces and the cabin environment than the crewmen who stuck with the stubble-collecting electric models. But most eventually quit shaving altogether.

Although many astronauts were glad to forego the inconveniences of shaving and let their beards grow for a week or so, none of them were happy about the fact that the command module did not have room for a shower. Their only means of washing their bodies was by taking a sponge bath. As Duke describes the ordeal, "The other two guys would float up to the other side of the spacecraft, and you'd get naked to take your sponge bath. We had washrags and towels, and you could get hot water out of the water gun. You'd put your washrag right up against the water pistol, and hit it a couple of times just to get the towel damp, and then start rubbing down."

Unfortunately for all three crewmen, the sponge baths did not do the job. No matter how often the astronauts bathed, no matter how methodically they sponged, they could not really get themselves clean. As the mission progressed, their bodies got dirtier, and the human aromas inside the spacecraft grew increasingly pungent.

"The odor was pretty bad," Duke confirms. "But at the time you didn't really notice it because you're in there smelling right along with it . . . I remember when we got back to the aircraft carrier after splashdown, and got out, we were out for a few hours, had a physical and washed up, then went back inside the spacecraft to retrieve our little personal articles, the odor almost knocked you down. I thought, 'God, how could I stand this?' "

Apollo 11's Buzz Aldrin reports that the cabin odor was further exacerbated by the unavoidable onset of flatulence after meals: "It wasn't long before we discovered that the little device designed at the last minute to ventilate hydrogen from the water, as it passed from the gun to the food bag, was not always the success its designers had hoped. Frequently, the hydrogen tended to do what it had done on previous flights: It stayed in the water and was swallowed by us. The result was stomach gas. At one point on the trip back to Earth it got so bad it was suggested we shut down our altitude-control thrusters and do the job ourselves."

Meanwhile, the peculiarities of the spacecraft's weightless environment created a general sense of disorientation far beyond what the astronauts had previously experienced in their brief simulations in the zero G airplane or even in previous Earth orbital space flights in the Gemini program. In zero G, anything that was not tied down would float. What's more, everything and everyone was subject to unceasing demonstrations of Newton's law that every action produces an equal and opposite reaction.

"I'd turn a valve, and the valve would turn back," recalls Apollo 17's Gene Cernan. "I'd get the momentum of my one hundred-eighty-pound body moving and the only way I could stop it was with the strength of my arms. Then I'd have to start over and wrap my legs around something so I could secure a stance from which to work."

Cernan says he suffered nagging physical pains when he tried to move about the cabin's weightless environment: "I didn't have any problem on my Gemini flight, but once we got into an Apollo spacecraft with a larger volume to move around in, then the disorientation becomes greater. I found that the symptoms I felt were essentially those that I would work up after forty-five minutes or so of heavy aerobatics in a T-38, slight headache and a little bit of stomach awareness. As soon as I started to feel those, I would just quit moving around, and get back in my seat and look out the window for a while, and then in a few minutes, it would go away."

Almost all the other astronauts admit that they, too, were afflicted with some recurrent physical discomfort on Day One no matter how quickly they adapted to zero G. These symptoms could be mild or temporarily immobilizing, similar in form or totally different. Cernan claims he felt a certain "light-headedness" during his initial adjustment period. Other astronauts say they were beset with a diametrically opposite feeling.

"I felt a fullness in my head," says Apollo 14's Roosa. "It was like hanging upside down on a gymnastic bar. Your heart is used to pumping that strong blood to your brain going against gravity. So if you suddenly take the gravity away, the heart hasn't realized it yet, and it's pumping too much blood to your head."

The physical disorientation of zero G was exacerbated by the spacecraft's thermal roll. Although the astronauts had no sense of forward motion except through second-hand instrument readouts, they could see—and therefore claim to feel—their continuous turning motion by looking through the windows of the command module. The ever-changing view dramatically reminded the astronauts that they were rotating like chickens on a barbecue spit, increasing their sense of vertigo. The only consolation was the magnificence of the visual spectacle that paraded past their portals during every roll, what Aldrin calls "an incredible panorama every two minutes as the sun, moon, and Earth appeared in our windows one at a time."

The three crewmen's initial disorientation and discomfort often continued right through their first "night" aboard the spacecraft. Their Day One rest period was usually scheduled to start about fourteen hours into the mission. Depending on the time of the launch, it might have been morning, noon, or night back in Houston. But the astronauts could not tell what time it was by looking out the windows because there was no sunset or sunrise to mark the passage of the day. According to Gene Cernan, "We thought in terms of daytime and nightime in Houston, but we really lived by our own clock. Our sleep cycles and work cycles were all based on total ground elapsed time (GET) from the commencement of the mission."

And yet, as Apollo 11's Mike Collins observed, time itself no longer had any meaning except as a measure of space: "What do we call this strange region between Earth and moon? Cislunar space is the most common term, but it doesn't say much. Is it day or night? Since we humans generally define night as that time when our planet is between our eyes and the sun, I suppose this must be considered daytime, but it sure looks like night out of several of my windows . . . My wristwatch knows best, or at least better than my eyes do."

The astronauts had to depend on their natural circadian rhythms to make them drowsy at the appropriate hour. But if they had to re-synchronize their biological alarm clocks to conform with the countdown schedule, nature didn't always cooperate. Apollo 12 crewmen Pete Conrad, Alan Bean, and Dick Gordon, who launched well before dawn, had to stay awake for the first twenty-one hours after lift-off in order to get their sleep cycles in sync for the rest of their lunar voyage. As Bean recalls, "By that time, we were all ready to go to bed."

After tidying up the cabin, the three crewmen fastened covers on the windows to keep out the sunlight as the spacecraft continued its thermal roll. Then the commander and the lunar module pilot floated down to separate sleeping compartments behind the left and right couches. Each compartment was actually a rectangular shelf about six feet long and one foot wide, with just enough overhead clearance to accommodate a body-size mesh hammock. But many astronauts complained that getting to sleep in their hammocks was not as easy as getting into them.

"That first night was miserable for me," Charley Duke reports, "because I felt like I was going to get seasick. You close your eyes and your head doesn't want to go anywhere, it just sorta stays there. I kept wondering, 'Where's my pillow? Where's my head?' So to feel some pressure, I wedged my head up under the couch and the strut, but then I floated loose."

While his crewmates tossed and turned, the command module pilot remained topside and strapped himself into the left couch in front of the control panel. Many astronauts claim that was by far the most comfortable berth, but the couch also had its inconveniences, for the CMP had to keep his headset on in case of an overnight emergency. Such occurrences proved to be thankfully few and far between during the nine Apollo lunar missions. But even if Houston kept the chatter to a minimum, there was always a dull roar of radio static to disturb the soundest sleeper.

At the end of Day One, the CMP and the two insomniacs in the hammocks below were also more than slightly overwhelmed by how far they had gone and by how far they still had to go. Since the wake-up call back at the Cape, they had soared from their home on Earth into the blackness of cislunar space, but they were only a quarter of the way to the moon. It would take three more days to get close enough to enter lunar orbit, and another half day after that to prepare for their lunar landing attempt, which meant there was still much to do and plenty of time for all sorts of things to go wrong.

PAO: "This is Apollo Control at fourteen hours six minutes into the mission. The spacecraft is now 66,554 nautical miles out. Speed is 7,095 feet per second, about 4,800 statute miles per hour. All three crewmen are sleeping . . ."

Day Two

The morning reveille was always a rude awakening, similar in form and content to the one that roused Apollo 11 command module pilot Mike Collins and the crew of man's first lunar landing mission.

"Apollo 11, Apollo 11, this is Houston. Over."

There was a brief pause as Collins blinked his bleary eyes and cleared his throat.

"Good morning, Houston, Apollo 11."

"Roger, Apollo 11, good morning."

That was the extent of the pleasantries. Without further ado, the CAPCOM at Mission Control got right down to business.

"When you're ready to copy, 11, I've got a couple of small flight plan updates and your consumables update, and the morning news, I guess."

The three crewmen often felt a mite hung over, as if they had spent the previous night in the bars of Cocoa Beach rather than in the void of cislunar space. Their first in-flight rest period was officially scheduled to last for six to seven uninterrupted hours, but most of the astronauts got only four to five hours of fitful sleep. Several also developed mysterious lower back pains in the middle of the night.

Houston's morning news broadcast was always a carefully censored upbeat report intended to boost the astronauts' morale as they tried to regain their bearings for the second day of their trans-lunar voyage. The briefing typically consisted of several major headlines, and selected stories about the mission from the international media, which consciously and unconsciously reflected the interests and ideology of the crew and the tenor of the times.

The Apollo 11 astronauts, for example, were updated on the status of the Soviet Union's unmanned Luna 15 probe that launched for the moon several days before their lift-off from the Cape on July 16, 1969. Initial concerns that the Luna 15 might somehow interfere with the Apollo 11 flight path proved to be unfounded, but the Soviets' customary veil of secrecy left U.S. officials in an agonizing state of doubt. As a result, the CAPCOM at Mission Control was forced to rely on second-hand foreign press reports.

"Okay, from Jodrell Bank, England, via A.P. Britain's big Jodrell Bank radio telescope stopped receiving signals from the Soviet Union's unmanned moon shot at 5:49 EDT today. A spokesman said that it appeared the Luna 15 spaceship 'has gone beyond the moon.' Another quote: 'We don't think it has landed.'

"The House of Lords was assured Wednesday that an American submarine would not 'damage or assault' the Loch Ness monster. Lord Kilmany was told it was impossible to say if the 1876 Cruelty to Animals Act would be violated unless and until the monster was found . . . Mexican immigration officials say they are refusing entry to American hippies unless they bathe and cut their hair . . . Vice-President Spiro T. Agnew says the U.S. should attempt to land a man on Mars before the end of the century . . ."

Subsequent wire service bulletins carried the official Soviet update on the Luna 15 probe and a special announcement from the White House concerning Apollo 11's upcoming lunar landing attempt.

"Latest on Luna 15. Tass reported this morning that the spacecraft was placed in orbit close to the lunar surface, and everything seems to be functioning normally on the vehicle. Sir Bernard Lovell [at the British radio telescope station] said the craft appears to be in an orbit of about sixty-two nautical miles . . .

"And also, President Nixon has reported—has declared a day of participation on Monday for all federal employees to enable everybody to follow your activities on the surface [which were scheduled for the preceding Sunday] . . . It looks like you're going to have a pretty big audience for this EVA . . ."

The Apollo 11 astronauts, however, were not informed of a major scandal that broke while they were en route to the moon: Edward M. Kennedy's auto accident at Chappaquiddick, Massachusetts. That act of censorship may have been motivated by some judicious political restraint by space agency officials, for Kennedy had recently called for a NASA budget cut so more funds could be allocated for federal antipoverty programs.

The Apollo 11 crew was also spared the latest news on what many Americans believed was the most important story of the time: the war in Vietnam. Instead, Armstrong, Aldrin, and Collins were treated to an uplifting editorial comment on what they believed to be the most important story of all time from the Paris daily *Le Figaro,* which declared, "The greatest adventure in the history of humanity has started."

Many of the Apollo astronauts treated the morning news briefings as something of a joke, especially because they had firsthand experience acting as the voice of Mission Control before flying on their own lunar missions. Apollo 16's Charley Duke, who served as CAPCOM for the Apollo 11 lunar landing, kept asking for stock market quotes on "Consolidated Jackpine." His requests had the intended effect of sending the press corps into a frenzy in a vain attempt to dig up information on a company that existed only in Duke's imagination.

Such pranks illustrated the new mood that would envelop the Apollo spacecraft in the course of Day Two. The work load was relatively light, and the three crewmen rapidly got over the initial discomforts of weightlessness and started to appreciate its special joys. Even Apollo 16's Duke, who was troubled by severe "stomach awareness" the first day and slept fitfully the first night, found himself becoming ever more enthralled with life in zero G. As he puts it, "Zero G is a lot of fun once you get past the nausea."

Apollo 17's Jack Schmitt had an even easier time adapting to zero G: "It took me about twenty-four hours to feel completely comfortable, but I never felt uncomfortable. It was surprising to the medical profession that human beings would adapt so completely to a new environment."

Adds fellow Apollo 17 crewman Gene Cernan: "God created a wonderful thing when he created the human being because we do adjust physiologically and psychologically to new environments very quickly. The whole system powers down when you get up in zero G. There's no big barriers or space euphorias or physical handicaps you have to overcome. It takes less effort to move in zero gravity than it does to walk on Earth. It takes less effort to pump blood around the human body because you're not pumping uphill to the head. So the heart gets smaller. The mechanism of the human body says, 'I don't have to work as hard, therefore I don't have to be as big. Therefore I can shrink in size and I can slow my heart rate down and keep this human being operating very capably.' "

Most other astronauts also wax poetic about the newfound joys of weightlessness. "The biggest one is the freedom of movement," claims Stu Roosa. "If you were suddenly put in the command module and launched there, you wouldn't appreciate it as near as much as you do after all of that scrambling you've done over hundreds of hours in the simulator. Now you can just get around so much easier."

Adds fellow Apollo 14 astronaut Ed Mitchell: "You feel light, you feel no pressures on the body. You can just move around into any position you want to move into . . . float around in space, hover above the couch, or go down wherever you want . . . We've all had dreams of flying just like a bird, hopping off and kind of sailing through the air like Superman . . . It's a little like those dreams. I guess some people are terrified of it, but to me it was a very euphoric feeling, very freeing."

Thanks to their newfound freedom of movement, the crewmen would suddenly discover that the cabin of the command module seemed to have expanded as if by magic. Ceilings, tunnels, nooks, crannies, and corners that were once physically inaccessible because they were overhead or out of reach were now just an easy float away. The crewmen could roam throughout the entire cabin at will, using virtually every cubic inch of the interior.

"It's much roomier, and this is something that you really appreciated," Roosa remembers. "After a thousand hours [of ground training] in the simulator, you've hit your arms and you've hit your shins I don't know how many hundreds of times. In the weightless state, you could just float down and float up, and you didn't run into things like you did in the simulator. So this was a real plus, and you really enjoyed it."

Day Two was all the more enjoyable for the astronauts because they didn't have to worry about "flying" the spacecraft. Or at least not much. As Apollo 11's Mike Collins wryly observes, their pilot was Sir Isaac Newton, the "fourth" member of the crew. In accordance with Newton's law, the astronauts were actually sailing through space on the wind of their initial momentum. The centrifugal force created by the trans-lunar injection combined with the gravitational forces of the sun, the moon, and the earth to navigate, control, guide, and propel them via a cosmic sequence of actions, reactions, and invisible magnetic attractions.

During the night, the three crewmen had allowed the Newtonian engines of the universe to pilot them another 30,000 miles through the cosmos. At the twenty-four-hour mark, just one day after leaving the Cape, they were over 100,000 miles away from Earth and almost halfway to the moon. Their speed had gradually decreased 5,400 feet per second (3,700 m.p.h.), one-seventh of the post-TLI peak and 1,000 m.p.h. slower than they were going the night before. But just as Sir Isaac could have predicted, nature's guidance system had also permitted a certain amount of directional drift that could only be offset by the human beings aboard the spacecraft.

As a result, the biggest item on the Day Two checklist was a mid-course correction at about twenty-six hours three minutes into the mission. This deceptively difficult maneuver required the official skipper of the spacecraft, the command module pilot, to make a three-second engine burn. That wasn't much "flying" for a veteran fighter jock, but the CMP's brief turn at the controls was as crucial as every other manual and automatic thruster firing from launch to lunar orbit, for it had to be precisely timed and executed in order to keep the spacecraft on course for the moon.

Houston always helped the CMP and his fellow crewmen prepare for the mid-course correction by beginning the countdown well in advance. But like staging and TLI, the burn often came as an unexpectedly violent shock, especially to those astronauts who were already enthralled with the wonders of weightlessness.

Apollo 14 crewmen Alan Shepard, Edgar Mitchell, and Stuart Roosa got so comfortable with life in zero G that they forgot to secure the loose objects floating about the cabin before ignition of their mid-course correction burn. According to Roosa, "We just weren't prepared. Things went flying back by our heads that we just hadn't fastened down. A Snoopy cap flew off a couch . . . a checklist went that way . . ."

Once the fury and confusion of the mid-course correction burn abated, the three crewmen would attend to the next biggest item on the Day Two checklist: a series of televised performances for the tax-paying public back on Earth. Most consisted of brief tours of the command module cabin, demonstrations of how to float around, eat, sleep, and shave in zero G, or narrated views of the solar system as seen through the windows of the spacecraft. As Apollo 11's Aldrin later revealed in his post-mission autobiography: "We went to great lengths to make the telecasts seem improvised, and we apparently were successful; we received compliments on them when the flight ended. The fact is they were carefully planned in advance and for me the exact words were written down on little cards stuck onto the panel in front of us. Spontaneous I was not."

Nevertheless, the telecasts from cislunar space were always spectacular by virtue of their unique origin and nature. Apollo 11's first in-flight performance was no exception. While command module pilot Mike Collins steadied the TV camera on Earth, commander Neil Armstrong began describing the ice cap around the North Pole, which appeared at the top of the frame, and the cloud cover around the equator. But in the midst of Armstrong's monologue, Collins suddenly blurted: "Okay, world, hang on to your hat . . . I'm going to turn you upside down."

"Roger," replied the deadpan voice of Mission Control.

With a flick of his wrist, Collins flipped his camera 180 degrees, and made the earth stand on its head.

"You don't get to do that every day," Collins chortled with glee.

Apart from that pre-rehearsed TV prank, the Apollo 11 astronauts tried to remain professionally calm and businesslike throughout most of their trans-lunar coast, as if man's first moon landing mission was just a routine dress rehearsal. Now that the impossible dream was coming true, the only way to cope with the incomprehensible reality was by retreating to make-believe.

"We must pretend that the flight has not begun yet," Collins would write, "that it will not begin until we get to the moon and begin our preparations for landing . . . It really helps that all three of us have been in space before. I sense a quiet awareness inside the command module, rather than the jubilation and apprehension which I am sure would have infected three rookies."

However, the wonders of weightlessness inspired the astronauts on later Apollo flights to indulge in all sorts of spontaneous playfulness. This was especially true of the crewmen on the last two lunar landing missions whose confidence was so fortified by previous successes that handling the technical chores of a voyage to the moon really did seem routine, and at times, almost an afterthought to the extraordinary recreational activities afforded by life in zero G. "We had a hard time keeping our minds on our jobs," confesses Apollo 16's Ken Mattingly.

One of the favorite pastimes of the Apollo 16 and Apollo 17 crews was turning inanimate tools into zero G toys. They delighted in making tops, tape recorders, clipboards, and pencils sail back and forth across the cabin in crazy cartwheels, parabolic loops, centripetal spins, and slow motion swan dives. They also loved to show off their newly acquired acrobatic prowess by performing midair somersaults and "upside down" juggling routines.

Apollo 17 commander Gene Cernan claims that at least some of this playfulness was deliberate and purposeful: "One of the things we made sure to do was have fun. We were a very light-hearted, jovial, and yet very professional crew, and we had fun doing what we did. I think having fun allowed us to do well."

On most Apollo flights, the three crewmen could not see the moon after the transposition and docking maneuver on Day One. That was partly because the LM attached to the nose cone further restricted the forward view through the five windows of the command module. But the real culprit was the trajectory of the spacecraft, which effectively aligned the moon so close to the sun that it was lost in the blinding light of the solar glare.

The astronauts also had a hard time seeing the stars even with the help of a special "monocular" (half a binocular) used to supplement the scanning telescope and the sextant. Due to the absence of an atmosphere to refract and filter light, the stars do not twinkle in cislunar space. Rather, as Stu Roosa puts it, "The stars look like little points of light or fuzzy little dots."

The three crewmen, however, were always in sight of land. That's partly because the "albedo" or reflected light of the earth was seven times brighter than that of the moon, which made their home planet clearly visible despite the solar glare. Most of the astronauts say they took advantage of every opportunity to look homeward.

"Looking back at Earth was a pastime I never got tired of," Ken Mattingly recalls. "One thing every spacecraft ought to have is a big bay window to sit in and be able to appreciate the view. Already I was getting the impression, 'This is such an amazing thing that I'm gonna forget most of it. I wish I had time to sit back and write the thoughts that come to me, but I don't.' Each image was there for a flash to be appreciated and savored for its content, and then reluctantly let go, because you know it's going to be replaced with other images in just a few minutes."

"The view of Earth is without a doubt the strongest magnet of the whole mission," says Roosa. "You get to thinking of the earth as an entity. You don't think of it as Texas or the United States. You really think of it as Earth."

The Apollo 11 astronauts also saw something decidedly unearthly on their trans-lunar voyage—an unidentified flying object (UFO)—that they did not report to Mission Control in Houston or anyone else until long afterward. In his post-mission autobiography *Return to Earth,* Aldrin described the sighting as follows:

"In the middle of one evening, Houston time, I found myself idly staring out the window of the *Columbia* and saw something that looked a bit unusual. It appeared brighter than any star and not quite the pinpoints of light that stars are. I pointed this out to Mike and Neil, and the three of us were beset with curiosity. With the help of the monocular, we guessed that whatever it was, it was only a hundred miles or so away. Looking at it through our sextant, we found that it occasionally formed a cylinder, but when the sextant's focus was adjusted, it had a sort of illuminated 'L' look to it. It had a shape of some sort—we all agreed on that—but exactly what it was, we couldn't pin down. We asked Houston some casual questions: "How far away is the Saturn third stage?" The response was in the vicinity of six thousand miles. That wasn't it.

"It could possibly have been one of the panels of the Saturn third stage which fly off to expose the LM and cannot be traced from Earth. We could see it for about forty-five seconds at a time as the ship rotated, and we watched it on and off for about an hour. We debated whether or not to tell the ground we had spotted something, and decided against it. Our reason was simple: the UFO people would descend on the message in hordes, setting off another rash of UFO sightings back on Earth. We concluded it was most likely one of the panels. Its course in no way appeared to conflict with ours, and it presented no danger. We dropped the matter there."

Sometime in the late afternoon or early evening of Day Two, the crewmen on each lunar mission would pass the halfway point in their voyage. The exact moment varied according to the particular time of the lunar month, for the moon's elliptical orbit around the earth could be as close as 220,000 miles or as far as 244,000 miles. But as the second day of the trans-lunar voyage drew to a close, the astronauts always realized that they were more than twice as far from home as they were the day before.

PAO: "This is Apollo Control at forty hours fifty-eight minutes into the mission. The Apollo spacecraft is now 146,300 nautical miles from Earth, velocity is 3,917 feet per second, approximately 2,700 statute miles per hour. The astronauts are four hours into their scheduled ten-hour rest period . . ."

Day Three

The wake-up call on Day Three was both a pleasant surprise and a sleepy shock. Having tossed and turned throughout their first in-flight rest period, most of the astronauts enjoyed eight to nine hours of uninterrupted slumber during the second night of their trans-lunar voyage. Those quickest to adapt to weightlessness the day before claim that they found themselves able to sleep even more soundly in zero G than on Earth.

"It was fun getting used to," says Charley Duke. "I could go to sleep with my chin on my hand, and just sleep and sleep. You don't get cricks in your neck and things like that. You wake up refreshed."

The astronauts were also pleasantly surprised to find that the lower back pains they suffered following their first night in space had disappeared. The crewmen of Apollo 14, man's fourth lunar landing mission, were the first to discover the cause and the cure of this mysterious sacroiliac ailment. "In the zero G environment, you can never be perfectly stable," says lunar module pilot Ed Mitchell. "There's always movement in some direction. After experimenting a little bit, we realized that our lower back pains resulted from trying to clench our toes while we slept in weightlessness, which is an interesting anthropological artifact—trying to stabilize yourself by clenching your toes. We found that if we unclenched our toes and just relaxed, our back pains would go away."

As Houston read the morning news, however, the astronauts also realized they were about to reach a major turning point in their trans-lunar coast. The wake-up call usually found them about 160,000 miles from Earth. By late afternoon or early evening, they would be over 186,000 miles out, and the spacecraft would have slowed to just three thousand feet per second (approximately two thousand m.p.h.), less then one-tenth the velocity of the trans-lunar injection. That night, the astronauts would cross the invisible boundary between the earth's sphere of influence and the moon's sphere of influence. As the pull of lunar gravity gradually overpowered the reverse tug of terrestrial gravity, the spacecraft would start to accelerate once again. From that point forward, the astronauts' position would no longer be calculated relative to their distance from the earth but in terms of the total miles remaining between them and the moon.

In the meantime, they often began to feel, see, smell, and hear the long-term effects of life in zero G. One of the most noticeable was the change in the faces of their comrades. "You look at your colleagues, and they don't quite look the same," recalls Apollo 15's Irwin.

Apollo 11's Collins explains why: "It's their eyes. With no gravity pulling down the loose fatty tissue beneath their eyes, they look squinty and decidedly Oriental."

As Day Three wore on, the astronauts would also become anxious about a much more serious physical effect of weightlessness. The "shrinking" of their hearts and the "powering down" of their metabolic systems was sapping their overall strength and stamina. And because of their relative inactivity, they were also gaining weight, in some cases as much as a pound per day.

The problem was not whether to exercise but how? Besides lacking the room to do extensive calisthenics, the astronauts lacked the means to wash off a heavy sweat except by taking a laborious sponge bath. They also had to guard against overexertion, for they couldn't request an extra day of R&R before attempting to land on the moon. As a result, the most favored conditioning exercise was running in place in the lower equipment bay, a regimen pioneered by Mike Collins on man's first lunar landing mission.

The stress of living in the close confines of the command module was even more bothersome. "Housekeeping takes on monumental importance because of the smallness of the spacecraft," notes Stu Roosa. "There are only three of you, but it seems more like eight people in a rowboat . . . It doesn't take long before everything is in shambles."

It was up to command module pilots like Roosa to serve as chief cook and bottle washer for their fellow crewmen. In addition to preparing the food packets and cleaning up after meals, the CMP's duties included monitoring the current supplies of consummables like fuel and oxygen, changing the lithium dioxide canisters that purified the cabin air, charging batteries, dumping waste water, dumping urine, disposing of solid wastes, and purging fuel cells of overheated chemical residues.

In between housekeeping chores, the command module pilot also had to keep realigning the inertial platform at regular intervals throughout the day. In order to enhance his dominant eye's view through the sextant, the CMP usually covered his outside eye with a black patch that made him look like a pirate. But his comical appearance was the least of his worries, for he quickly became preoccupied with the difficulties of working in zero G.

"In the course of a day, you're literally on the sextant, it seems like, all the time," Roosa recalls. "I never had any problem doing platform alignments in the simulator on Earth. But in zero G, you're floating, and it seems like you're always being pushed up while you're trying to hold yourself down."

As the work load piled up, the astronauts' lengthening time in space became divided into shorter and shorter spaces in time. According to Roosa: "You're always budgeting time because you're on a time line and that clock never stops. You're always thinking ahead because if you ever get behind that time line, you don't get a second chance. When you're doing one task, you're thinking, 'What am I going to be doing when I come out of this?' because in many cases you didn't have any time between tasks. And once you start shuffling your time line, it's like pushing dominoes: Everything has to move and give and you're canceling things. It takes a lot of time just to read up all the changes in your flight plan."

The relentless ticking of the clock put a premium on cooperation and compatability. Given the lack of breathing room between particular tasks and time lines, the crewmen had to be ever ready to help each other complete assignments. And given the sheer length of elapsed time they spent together in the spacecraft, they also had to be able to understand, appreciate, and tolerate each other regardless of their respective idiosyncrasies.

Inevitably, some crews got along much better than others. Apollo 12 astronauts Pete Conrad, Al Bean, and Dick Gordon had a particularly close and warm relationship even before their Apollo mission. They had been classmates at the navy's advanced aviation school, and Conrad and Gordon had flown together on Gemini 11.

"We were all good friends . . . We ate, slept, and worked together for so long during training that we knew each other's thoughts and reactions in every situation . . . It was a comfortable feeling."

The Apollo 12 astronauts celebrated this "comfortable feeling" by tormenting each other with devilish practical jokes. "We played games on each other all the time," Gordon recalls. "For example, when you defecate, you have to take care of it: put it in a bag, put in the antibacterial pills, wrap it up, and tape it. But we didn't have anyplace to dump the feces bag, so we just put it in the food storage pantry and hoped that somebody else would find it when they got up for a midnight snack."

Apollo 14 astronauts Alan Shepard, Edgar Mitchell, and Stuart Roosa went through a similarly long and arduous training period before lift-off, but they did not develop the same personal bond. As Mitchell admits, "We were not close friends in any sense."

The same was true of Apollo 11 crewmen Neil Armstrong, Buzz Aldrin, and Mike Collins, who hailed from widely divergent backgrounds and represented sharply contrasting personality types. According to some reports, Armstrong and Aldrin were barely on speaking terms, allegedly because of lingering ill-will over the decision about who would be first to step out onto the lunar surface.

However, every trio of Apollo astronauts managed to submerge their individual differences in the interests of their mutual survival. Apollo 16's Charley Duke admits that he and his otherwise amicable crewmates John Young and Ken Mattingly "had a little spat or two" on the way to the moon, but insists that most of their arguments concerned technical points and operational alternatives that were quickly resolved. "We knew that if you let your personalities go, you could be in trouble. We had had enough of these problems in simulation to know that we had to watch ourselves. There wasn't enough time to behave like we were on a fishing trip together."

The bond between Apollo 17 crewmen Gene Cernan, Jack Schmitt, and Ron Evans was strengthened by the otherwise unfortunate intestinal virus that afflicted Evans on the first two days of their trans-lunar voyage. As Cernan attests, "That's when you really become buddies. When you're sick, you're sick. You've got to rectify it. You've got to help the guy because that's your home, your environment. And if you're not careful, you'll literally have shit all over the spacecraft."

Fortunately, the Apollo astronauts could also dissipate the tensions of the trans-lunar voyage by listening to recorded music. Each crewman was allowed to take along three cassette tapes of his own choosing. The astronauts' tastes ranged from classical music to pop and country and western, and included such whimsical selections as the theme from *2001: A Space Odyssey,* the Berlioz "Symphony Fantastique," "The Impossible Dream," Frank Sinatra's version of "Everyone's Gone to the Moon," a folk song entitled "Riding Old Paint," and the gospel favorite "How Great Thou Art."

A few of the astronauts' cassettes were specially recorded by well-known artists, and featured original compositions and personalized messages. Stu Roosa listened to a particularly appropriate custom-made cut by country and western star Buck Owens that began, "Now when you get back from the moon, we know they're probably gonna put you in the movies, so here's a little something to help you get ready . . ." Owens and his group then played a rousing country and western rendition of "Act Naturally," crooning, "They're gonna put me in the movies . . . They're gonna make a big star out of me . . . Yes, they're gonna put me in the movies . . . And all I have to do is act naturally . . ."

"The personal touch really helped," says Roosa. "You can see the earth getting smaller and smaller, and the tapes really gave you a touch of home . . . But I suspect that no one ever listened to all the music they took with them. Nobody ever had enough spare time . . . I know I never did."

Oddly enough, most of the Apollo astronauts say they did not suffer pangs of homesickness during their trans-lunar voyage. They were able to communicate with their loved ones back on Earth via the CAPCOM at Mission Control, but they seldom did so, simply because they were so busy. "Once you get into the mission," notes Roosa, "it's just all-engrossing. You know there are people back there on Earth leading lives, but it doesn't concern you. What concerns you is that clock going around and the flight plan that you are matching to that clock, and that's it."

Apollo 15's Jim Irwin explains the underlying reasons for this seemingly callous attitude: "I don't think that I missed my family as much as if I had been traveling somewhere on Earth because when you are traveling on the earth, if you see an interesting place, an enjoyable place, you sometimes wish your family could be with you. On a trip to the moon we of course realized that it was quite impossible for them to join us, and so that thought was quite remote."

But the three crewmen always maintained an extremely close and warm working relationship with Houston, for they knew they were even more dependent on the men at Mission Control than they were on each other. "We were very conscious that we were not about to get in any flap with the ground," says Apollo 12's Dick Gordon. "Hell, you can't get along without them. They're the bosses, they're the leaders. You're doing essentially what they want you to do. It's kind of a business arrangement. You're talking to the bosses on the ground, and you're passing vital information to them, and they're passing vital information to you."

Adds Apollo 16's Charley Duke: "There's a sense of officialness about it, but there's a lot of friendship, too. You knew the guys on the consoles. You knew the team members, you knew the positions, you knew their first names. Through training and simulations and the fact that you worked there before [during previous Apollo flights], you knew their problems and you knew what they were facing. You had a lot of confidence in those guys. You knew what they said went, and that it was right. So while it sounded official, it was really more than that."

The men at Mission Control dramatically demonstrated their competence and grace under pressure following the near fatal oxygen tank explosion during Apollo 13's trans-lunar voyage. As the astronauts' families, friends, and a worldwide TV audience hoped, prayed, and agonized, Houston coolly advised crewmen Jim Lovell, Jack Swigert, and Fred

Haise how to get home safely by using their lunar module as a "lifeboat." (See Chapter 11, "Unlucky Thirteen.") But Houston also played a heroic role of major or minor proportions at some point on virtually every moon shot, including the historic (and nearly aborted) Apollo 11 lunar landing.

Apollo 15's Jim Irwin recounts a few of the untold crises that transpired during his trans-lunar voyage: "There were times where we had doubts as to whether we would be able to complete the mission. A circuit breaker popped that affected the firing of the service module engine that would put us into a lunar orbit and bring us back to Earth. We had breakage of glass on a key instrument in the lunar module. We also had some water leaks in the command module that we didn't know whether we would be able to contain. It would be horrible if you had to be drowned in space. But with the help of the people on the ground, we were able to stop the leak and fix the other problems so we could continue."

Even so, the inherent danger of each and every manned lunar mission was fully appreciated by all the astronauts both before and after Apollo 13's calamity shattered the complacency of the general public on Earth. As Apollo 12's Al Bean attests, "You see the risk you're taking. When you're out there in this little command module, you realize that if the glass breaks or the computers quit working or the electrical system quits working, you're not gonna get back. And you have time to contemplate this, you have time to think about it and run it through your mind a lot of different times, play it over and over again."

However, as Apollo 15's Irwin hastens to note: "Our emphasis was always on mission success. We knew where we wanted to go. We knew what we wanted to do when we got there. We knew clearly the purpose of our exploration . . . And we always had the feeling that somehow we would come back to Earth."

Adds Apollo 12's Dick Gordon: "If you thought you were never going to get back, you'd never go. Believe me, there's not anybody that would do that, regardless of what they might say. If they knew they weren't coming back, they wouldn't be going, I don't care what bullshit anybody puts on you. You're gonna come back. You know you're gonna come back."

The astronauts were further reminded of the dangers ahead when they tackled the key item on the Day Three checklist—preparing the lunar module for their upcoming landing attempt.

The command module pilot would begin this solemn ritual by removing the probe and drogue apparati blocking the accessway between the mothership and the attached landing craft. Then the lunar module pilot opened the hatch of the LEM, and conducted a meticulous inspection of the interior, checking for launch damage, leakage, and loose wiring, verifying switch positions, testing manual levers, etc. The mission commander, who would actually fly the LEM, usually got in on the act by offering to assist in the LMP's inspection and/or by filming the procedure for the TV audience back on Earth.

Last but not least, the commander and the LMP would try out their assigned "seats." Unlike the command module, the lunar module did not have room for conventional couches. The LEM's twin flight stations were really small metal platforms equipped with shoulder harnesses that would keep the astronauts standing upright like trolley car conductors during the landing attempt. "The cabin of the lunar module is about the size of two phone booths standing side by side," notes Charley Duke, "and it starts to feel pretty cramped when the time comes to get inside, close the hatch, and undock."

But that time was still two full days away, and for all three crewmen, there was still a bewildering sense of unreality about the trans-lunar voyage even after they finished prepping the LEM for lunar landing. Their disorientation was compounded by the lack of familiar milestones to mark their progress through the uncharted frontier of cislunar space. As Apollo 12's Al Bean observes, "One of the things different about a lunar trip from a trip that you take in other vehicles is that you don't really pass anyplace on the way. It looks like these guys are rotating around you and that you are stationary, and that has a pleasant effect as far as feeling that you are not going anywhere. If you're going from New York to Los Angeles, for example, you maybe pass Chicago, Denver, and whatever else. Going to the moon, you leave the launchpad, then you leave Earth orbit, then about a couple of days later, after passing nothing and just essentially floating along and watching the earth get smaller, all of a sudden you're at the moon. That lack of way points made it seem a little magical or mystical getting there. All of a sudden, you just showed up."

"I know it sounds kind of screwy," admits Stu Roosa, "but what really made it dawn on me that we were getting a long way from home, even more than seeing the earth getting smaller, was when you start to pick up the delay in radio transmissions from Mission Control. All the time since we left Earth orbit, we had that delay building, but it just wasn't as obvious. There is an immediate reply to any early transmission. But as you get closer to the moon, you're looking at a five- to six-second delay. Six seconds is a long time when you're sitting there waiting for an answer."

At approximately sixty-three hours into the mission, the astronauts entered the lunar sphere of influence. They were now over 200,000 miles from home, and less than 40,000 miles from their intended landing site. The moon's newly dominant gravitational pull caused the spacecraft to speed back up to a forward velocity of about 3,800 feet per second or roughly 2,600 m.p.h., but the acceleration was so smooth that it was virtually imperceptible. And by this time, the three crewman were usually sound asleep.

8

HELLO, MOON

Lunar Orbit

Day Four

The small moon lightens more;
And as I turn me home,
My shadow walks before.

ROBERT BRIDGES
The Clouds Have Left the Sky (1876)

All twenty-four men who flew to the moon arrived in lunar orbit on the same day. The wake-up call usually came at about seventy hours into the mission, which was technically the tail end of Day Three. But by the time the astronauts finished breakfast, the morning news briefing, and the housekeeping chores, Day Four had officially begun.

"Day Four has a decidedly different feel," recalls Apollo 11's Mike Collins. "Instead of nine hours' sleep, I get seven—and fitful ones at that. Despite our concentrated attempt to conserve our energy on the way to the moon, I feel the pressure is overtaking us (or me at least) and I feel that all of us are aware that the honeymoon is over and we are about to lay our little pink bodies on the line."

Day Four was the day the three Apollo crewmen would attempt to orbit the moon. As far as they were concerned, the launch, Earth orbit, the trans-lunar injection, and the trans-lunar coast were all prelude. This was the true beginning of their mission.

"You don't have anything until you insert into lunar orbit," notes Apollo 14's Roosa. "All the other things except reentry to Earth you could shuffle around. But if you want to have a mission, you've got to make your lunar orbit insertion burn. You have one shot. You don't get a second chance."

And that one shot demanded split-second accuracy. During the night, the three crewmen had traveled more than 20,000 miles, and their velocity had climbed to 4,047 feet per second or roughly 2,700 m.p.h. The moon was still some thirteen thousand miles away, and as the astronauts got closer and closer, their velocity would nearly double. But in order to insert into lunar orbit, they had to slow down enough to allow the spacecraft to be captured by the moon's gravity.

The first "lunar orbit insertion" burn (LOI #1) was actually a braking maneuver, the reverse of the trans-lunar injection that blasted the spacecraft out of Earth orbit. During TLI, the spacecraft had been pointed "forward," with the exhaust nozzle at the rear, so firing the S-IVB's engine made the ship accelerate to the required "escape velocity." But the astronauts approached the moon "backward," with the exhaust nozzle of the spacecraft pointing forward. Firing the service module engine during the LOI #1 maneuver, therefore, produced a reverse thrust, which caused the spacecraft to slow down to the "capture velocity" required to enter lunar orbit.

LOI #1 was much trickier and much riskier than TLI, however, because it always took place amid a temporary loss of signal (LOS) period as the astronauts rounded the back side of the moon. Each Apollo spacecraft was equipped with a Very High Frequency (VHF) radio for close-range transmissions and an S-band radio that could broadcast hundreds of thousands of miles through space. But both the VHF and the S-band required "line of sight" reception.

As a result, the three crewmen had to insert themselves into lunar orbit without the benign guidance of their guardian angels back on the ground. Houston would have no way of spotting any problems that might occur in mid-burn, no means of telling the astronauts what to do if something went wrong. The men in the trench at Mission Control could only sit and wait with bated breath to learn the outcome of LOI #1 until the astronauts were already safely inserted in lunar orbit and reemerged on the front side of the moon—if they ever did.

"The moments leading up to the LOI burn were pretty nervous," remembers Apollo 14's Ed Mitchell. "For the first time in our entire flight we would be entirely on our own. We had to execute the burn correctly, and come out on the other side of the moon—yea or nay."

Instead of continuing to float around the zero G environment of the command module in their long johns and jumpsuits, the astronauts helped each other get back into their "moon cocoons." Then they returned to their couches, and strapped on their shoulder harnesses to prepare for the impact of the burn and the centripetal tug of the moon's one-sixth gravity.

There was no way to prepare for what happened next. Just as the three crewmen settled into their couches, the spacecraft abruptly stopped her thermal roll, and they got their first good look at the moon since the trans-lunar voyage began. No longer blinded by the glare of the sun, the astronauts could see the lunar surface illuminated by the glow of reflected light from their home planet appropriately known as "earthshine."

The crew of Apollo 11 was treated to a particularly dazzling view, for the moon appeared to be back-lit by the corona of the sun as if in a solar eclipse. Buzz Aldrin tried to describe the scene for the men back at Mission Control:

"It's quite an eerie sight. There is a very marked three-dimensional aspect of the corona coming from behind the moon glares . . . And it looks as though—I guess what gives it that three-dimensional effect is the earthshine. I can see Tycho [a large crater] fairly clearly—at least if I'm right-side up, I believe it's Tycho in moonshine, I mean in earthshine. And of course I can see the sky is lit all the way around the moon . . ."

Apollo 11 commander Neil Armstrong, by far the most laconic member of the crew, was also moved to comment: "Houston, it's been a real change for us. Now we are able to see stars again and recognize constellations for the first time on the trip. The sky is full of stars, just like the nights out on Earth."

While Apollo 11 command module pilot Mike Collins was delighted to be able to see stars again, he candidly admits that the "eerie sight" of the moon illuminated in earthshine startled him. "It was a totally different moon than I had ever seen before," he recalls. "The moon that I knew from old was a flat yellow disc, and this was a huge three-dimensional sphere, almost a ghostly view, tinged sort of pale white. It was very, very large, and very stationary in our window, utterly silent, of course, and it gave one a feeling of foreboding. It didn't seem like a very friendly or welcoming place. It made one wonder whether we should be invading its domain or not."

The astronauts on subsequent Apollo flights also saw "a totally different moon." But unlike Collins, they did not feel the same sense of foreboding.

"I didn't find the moon hostile," says Gene Cernan, who commanded man's last lunar landing mission. "I found it bland in color but majestically beautiful. The moon's been waiting for us for thousands of years, millions of years, maybe, unless someone else has already been there before us sometime. That's possible, although we didn't see any evidence of that. I felt very welcome there."

But despite their diametrically opposite emotional reactions, both Collins and Cernan agreed with Apollo 11 commander Armstrong, who exclaimed to Mission Control, "It's a view worth the price of the trip!"

PAO: "This is Apollo Control at seventy-five hours twenty-six minutes into the mission . . . The astronauts are 966 nautical miles from the moon, velocity 6,511 feet per second, more than 4,400 statute miles per hour . . . We're fifteen minutes away from loss of signal, twenty-three minutes away from the lunar orbit insertion burn . . ."

The PAO did not mention the far more sobering fact that the astronauts were also more than three full days and over 200,000 miles away from home. Nor did he remind the worldwide TV audience that if the spacecraft did not slow down enough to be gravitationally curved into the proper lunar orbit, they would hurtle right on past the moon toward the outer limits of the solar system with no hope of turning around and flying back to Earth before their fuel and oxygen supplies ran out.

Mission Control always bid the trio one last formal farewell before the communications blackout. The dialogue during Apollo 11's approach to lunar orbit went as follows:

CAPCOM: "Two minutes to LOS . . . Apollo 11, this is Houston. All your systems are looking good going around the corner and we'll see you on the other side. Over."

Armstrong: "Roger. Everything looks okay up here."

CAPCOM: "Roger, out."

Invariably, there was almost total silence inside the command module as the three crewmen barreled toward the moon. At the moment of losing radio contact, they were supposed to be a good three hundred nautical miles "above" and to the "left" of the hemisphere visible from Earth. But due to the exponentially increasing pull of lunar gravity, their forward velocity had climbed to over 5,200 statute m.p.h., and due to the closely angled trajectory of the spacecraft, they appeared to be plunging in a suicide dive toward the lunar surface.

"You're coming in awful fast," recalls Apollo 12 command module pilot Dick Gordon, "and the moon is growing—it's visibly getting larger by the minute—and it looks like you're going to run into the middle of the damn thing. You know intellectually that you're not gonna hit it, but what your eyes see and what your intellect tells you are two different things."

"It was the closest thing on the entire flight to what you would classify as a '2001' type of scene," says Apollo 14's Roosa. "You're drifting in, rocking on ready, and that old moon is growing magnificently fast, just filling up the window—pale brown, pale gray, turning black . . . Then you drift into the shadow of the moon and turn around the back side, and pretty soon you're drifting in total darkness . . ."

Roosa says he celebrated this dramatic event by listening to a specially recorded tape of "How Great Thou Art" by Sonny James. "I thought 'How Great Thou Art' was a very appropriate song to play when you are about to drift in behind the moon."

"When we went out of radio contact behind the moon, I had a sense of, 'Okay, we're really here,' " remembers Apollo 14 lunar module pilot Ed Mitchell, "You are in darkness, so you can't see the surface of the moon at all, but you know it's there—at least your instruments tell you it's there . . . It was a poignant moment, one of the most dramatic moments of the entire flight. You're coming up on the lunar orbit insertion burn, and you know the whole success of the mission depends on that burn coming off . . ."

LOI #1 was usually scheduled to commence about eight minutes after loss of radio contact with Earth. Because the astronauts could no longer hear the familiar voice of Mission Control, the commander and the lunar module pilot monitored the spacecraft systems and called out the final countdown. But the CDR and the LMP were merely informative passengers faithfully preparing to go along for the ride. The command module pilot was in the driver's seat all by himself. Only the CMP knew exactly how to throw the switches that would ignite the service module engine and insert the spacecraft into lunar orbit.

Tapes of the dialogue among the Apollo 11 crewmen re-create the subtle drama of the countdown of lunar orbit insertion.

Aldrin: "We're coming up on T minus thirty seconds to LOI . . ."

Armstrong: "All systems are GO for LOI."

Collins: "Roger, we're GO for LOI."

There was a brief shower of radio static.

Aldrin: "T minus ten seconds . . . eight seconds . . . six seconds . . . five . . ."

Collins reached for the control panel in front of him.

Aldrin: "Three seconds . . ."

Armstrong watched wordlessly as Collins fingered the ignition switches.

Aldrin: "Two . . . one . . ."

With a well-timed flick of his wrist, Collins lit the single engine of the service module, and all three crewmen were instantaneously plastered against their couches. The recoil from the LOI burn measured only half a G, but after three full days of life in zero G, that relatively mild kick could feel as bone jarring as the 4.5 Gs the astronauts experienced during lift-off and staging. And yet, according to Collins, it also felt "good."

Apollo 11's first lunar orbit insertion burn lasted five minutes and fifty-nine seconds. Then the service module engine shut down, and the astronauts anxiously checked the telemetry data displayed on the DSKY screen. According to their on-board computer, the velocity errors amounted to no more than one-tenth of a foot per second in three directions.

Armstrong: "That was a beautiful burn."

Collins: "Goddamn, I guess."

Aldrin, who had worried that the spacecraft might crash into the moon before, during, or after the burn, sighed with relief: "At least we haven't hit that mother."

Collins, who couldn't help but have pride in the accuracy of the burn, ignored Aldrin's remark, and shouted triumphantly:

"Hello, moon! How's the old back side?"

The answer was utterly dark, utterly silent, and as far as the astronauts could make out, rougher than a target range for B-52 bombers. Because of the moon's tilted axis, almost sixty percent of the lunar surface was visible from Earth, with the hemisphere bordered by western and eastern shadow lines known as "terminators."

On their first pass around the back side of the moon, the Apollo astronauts were usually far too preoccupied with the outcome of the LOI #1 burn to do much sight-seeing or systematic ocular reconnaissance. And even if the crewmen wanted to check out the features of the back side, their visibility was restricted to the twilight zones around the terminator lines.

The initial flight into and through the darkness was made even more eerie by the almost total absence of sound. The astronauts had prepared themselves well in advance to lose radio contact with the familiar voice of Mission Control. But they were seldom prepared for the absolute quietude on the back side of the moon, especially after the service module engine shut down following LOI #1.

"After we burned into lunar orbit, I had the feeling, 'I wonder why we keep orbiting, I don't hear the sound of any motor,' " recalls Apollo 12's Al Bean. "When the motor quits running in an airplane or a car, you have the feeling that your plane's gonna land in a minute or your car's gonna coast to a stop. I frequently had the feeling that I was gonna land on the other side of the moon because nothing was gonna keep me moving."

Apollo 12 command module pilot Dick Gordon claims he nevertheless enjoyed each and every pass around the back side of the moon precisely because he was out of radio contact with Houston. "It's relaxing to get out of their control and go on the back side of the moon. You don't have to answer any questions. You don't have to talk to anybody. It's the old fighter pilot's syndrome: It's that solitude they enjoy. Finally, I'm alone in my airplane. I can look out the window, star gaze, or do whatever the hell I want to do."

The astronauts' first loop around the back side of the moon usually took about thirty-four minutes. Just as they approached the eastern terminator, the LOS period ended with acquisition of signal (AOS) from Mission Control. And even Apollo 12's Gordon admits, "It was nice to hear talking again."

Houston always felt the same way. While the three crewmen were on the far side of the moon, the men in the trench had been figuratively in the dark about the progress of the mission, holding their collective breath and hoping for the best. They were even more relieved to hear the astronauts talking than vice versa, and understandably anxious to learn the outcome of the lunar orbit insertion burn. The Apollo 11 mission commentary tapes provide the typical flavor of the AOS event.

PAO: "This is Mission Control, Houston . . . It's very quiet here in the Control Room. Most of the controllers are seated at their consoles, a few are standing up, but very quiet . . . We are four minutes away [from acquisition of signal] now . . . There are a few conversations taking place here in the Control Room, but not very many. Most of the people are waiting quietly, watching and listening, not talking . . ."

A sudden humming, whirring, and clicking interrupted the narrative in the Control Room.

PAO: "That noise is just bringing up the [ground tracking] system. We have not acquired a signal [from the spacecraft]. We're a minute and a half from acquisition time . . . Thirty seconds . . ."

Half a minute later, the giant dish antenna at the Apollo tracking station in Madrid reacquired radio contact with the spacecraft.

PAO: "Madrid AOS, Madrid AOS . . . Telemetry indicates the crew is working on the antenna angles to bring the high-gain antenna to bear . . ."

Then Houston heard a barely audible squeak broadcast from Apollo 11.

CAPCOM: "Apollo 11, this is Houston. Are you in the process of acquiring high-gain antenna? Over."

There was no answer. Or at least not one that could be heard right away. Apollo 11 was over 230,000 miles away, and the lag in radio signal transmission was a full six seconds even from the most advantageous line of sight position in lunar orbit.

CAPCOM: "Apollo 11, Apollo 11, this is Houston. How do you read me?"

Six more seconds passed. Then Apollo 11 answered.

COLLINS: "Read you loud and clear, Houston."

The entire Control Room heaved a collective sigh of relief.

CAPCOM: "Roger. Reading you the same now. Could you repeat your burn status report? We copied the residuals' burn time [via the Madrid station] and that was about it. Send the whole thing again, please."

ARMSTRONG: "It was like—like perfect. DELTA T-O, burn time 557, ten values on the angles, BGX minus .1, BGY minus .1, BGZ plus .1, no trim, minus 6.8 DELTA V-C, fuel was 38.8, OX 39.0, plus 50 on balance, we ran an increase on the PUGS, NOUN 44, show us in a 60.9 by 169.9 . . ."

CAPCOM: "Roger, we copy your burn status report, and the spacecraft is looking good to us on telemetry."

Translated from the NASA-ese, Armstrong's report confirmed that Apollo 11's lunar orbit insertion burn had lasted exactly five minutes and fifty-seven seconds. Although propellant consumption ("fuel" and "Ox") was slightly greater than anticipated, CMP Mike Collins's velocity errors on the burn amounted to only one-tenth of a foot per second in three directions ("BGX minus .1, BGY minus .1, BGZ plus .1"). Best of all, the spacecraft had safely maneuvered into a 60.9-mile-by-169.9-mile elliptical orbit around the moon.

Upon completion of LOI #1, the Apollo crewmen know they "have a mission"—or at least a circumlunar mission. But before they could think about descending all the way down to the surface, the astronauts immediately prepared for the "trans-Earth injection" (TEI) that would blast them out of lunar orbit and, it is hoped, all the way back to their home planet, should they have to abort the mission before their landing attempt.

Unfortunately, the Apollo spacecraft's on-board computer was not prepared for such a perilous eventuality because of the fateful decision to purge the "Return to Earth" program from its already overloaded memory bank. The astronauts therefore had to pull out their notepads and manually record a new set of TEI guidance, navigation, and control data radioed up from the computers at Mission Control each time they came back around the front side of the moon.

"There were fifteen to twenty such (TEI) possibilities," recalls Buzz Aldrin," and one by one as they came up from Houston, we copied them down and checked them, and the data was then ready for insertion into our computer in case of any emergency. This system replaced the Return to Earth computer program we had decided to discard a year earlier."

No one understood the vital importance of copying the TEI data better than the three crewmen whose lives were at stake, yet all three always found it almost impossible to keep their minds on their stenographic jobs. For as the astronauts reacquired radio contact with Houston, they commenced their first extraordinary pass through the lunar daylight zone at altitudes as low as sixty miles above the surface. As Apollo 12's Dick Gordon recalls, "We were all off on one side of the spacecraft goggling at the moon."

"You make the LOI burn in the darkness, and then without any warning, you pop out into the light, and you look out the window, and you get your first good close-up view of the moon," remembers Apollo 14's Roosa. "At that point, we were still about a hundred twenty miles above the surface, but it looked so close you thought you could just reach out and touch it."

The first human beings treated to this "first good close-up view of the moon" were Apollo 8 astronauts Borman, Lovell, and Anders, who inserted into lunar orbit on Christmas Eve, 1968. Although all three rejoiced over the successes of man's first lunar mission, they did not have many nice (or reassuring) things to say about the moon from sixty miles away. CMP Jim Lovell commenced man's fist eyewitness description of the lunar surface shortly after copying the TEI data.

Lovell: "Okay, Houston, the moon is essentially gray, no color—looks like plaster of paris or a sort of grayish deep sand. We can see quite a bit of detail . . . The Sea of Fertility doesn't stand out as well here as it does back on Earth. There's not much contrast between that and the surrounding craters. . . . The craters are all rounded off. There's quite a few of them; some of them are newer. Many of them—especially the round ones—look like hits by meteorites or projectiles of some sort."

The other two Apollo 8 crewmen had harsher comments about the moon. Frank Borman called the lunar surface "a great expanse of nothingness that looks rather like clouds and clouds of pumice stone." Bill Anders described it as "a dirty beach." According to Anders's post-mission account: "It looked much like a battlefield, hole upon hole, crater upon crater. There was a total lack of sharp definition. It was completely bashed."

The astronauts on Apollo 10, man's second voyage to the moon, and most of the crewmen on the next seven lunar landing missions, Apollo 11–Apollo 17, would provide much more detailed complimentary descriptions as they got closer and closer to the moon. And for all but the unlucky threesome of Apollo 13 who had to return to Earth prior to lunar orbit insertion, getting closer to the still hazy lunar surface was the next major item on the checklist.

As the astronauts completed their first pass around the front side of the moon, their velocity had already decreased to about 4,200 statute m.p.h., roughly 1,000 m.p.h. slower than before LOI #1. But their altitude could vary by as much as a hundred miles between the apogee and perigee of the orbital ellipse. In order to safely execute the undocking maneuver prerequisite to a landing attempt, they had to "circularize" their orbit with a second manually controlled lunar orbit insertion burn (LOI #2).

LOI #2 was always scheduled to begin at the beginning of the astronauts' third orbit around the moon. And unlike LOI #1, it always occurred when they were on the front side, not the back side, in clear radio contact with Houston. But even though the burn itself lasted only seventeen seconds, LOI #2 was no less crucial—and in many ways, far trickier and more hazardous—than the first lunar orbit insertion.

"The second burn to place us in closer circular orbit of the moon, the orbit from which Neil [Armstrong, the Apollo 11 commander] and I would separate from the [mothership] *Columbia* and continue on to the moon [in the lunar module *Eagle*], was critically important," recalls Apollo 11's Buzz Aldrin. "It had to be made in exactly the right place and for exactly the right length of time. If we overburned for as little as two seconds we'd be on impact course with the other side of the moon."

Fortunately, the Apollo 11 trio did not overburn, underburn, or crash into the moon. Neither did the astronauts on Apollo 10, the "dress rehearsal" for man's first lunar landing attempt, or those on Apollo 12, 14, 15, 16, and 17, who made man's last five lunar landings. On each and every mission, the second lunar orbit insertion burn always duplicated the success of the first.

By the end of LOI #2, the astronauts were flying "lower" and "slower" than at any time since lift-off three and a half days earlier. They were orbiting the moon in a nearly perfect sixty-mile-by-sixty-mile circle, approximately half the altitude at which they had orbited the earth before the trans-lunar injection. Their forward speed had slipped to 5,300 feet per second, which was roughly 3,600 statute m.p.h. or barely one-fifth of their velocity before the trans-lunar injection.

Even so, the astronauts suddenly began to appreciate the true relativity of time and space. It took less than ninety minutes to make one complete revolution around the earth. It took at least two full hours to circle the moon. But because the moon was only one-sixth the size of their home planet, it seemed as if they were actually speeding up instead of slowing down.

"If you looked out the window," says Apollo 12's Al Bean, "you felt like you were moving very fast. When you're in Earth orbit, you can see the curve of the horizon, but the ball of the earth is so big you can't visualize it. Around the moon, since it's one-sixth the diameter of the earth, you're mind's eye can almost extrapolate the ball. It's a little bit like those animations you see before you go where they show you a moon that's about four inches around. You feel like if you looked down just a little bit further, you could see the other side of the ball."

But before the astronauts could see "the other side of the ball," they had to attend to the next major item on the Day Four checklist. If all went well, the CDR and the LMP would attempt to undock from the mothership and begin their descent toward the surface of the moon in less than twenty-four hours during the spacecraft's thirteenth lunar orbit. Although the LMP and the CDR had checked out and tidied up the LEM on Day Three, the landing craft had remained a dormant beast ever since leaving Earth. Now they had to bring the LEM to life by powering up the electrical systems and pressurizing the cabin.

Despite the countless hours of ground simulations, the astronauts never ceased to be amazed by the smallness and peculiar shape of the spacecraft that would supposedly land them on the moon and then return them to the mothership. Not counting the spindly landing legs, the two-stage LEM measured less than twenty-three feet from top to bottom. Because of the absence of atmospheric friction in the lunar environment, the LEM did not have to be streamlined, a luxury that enabled its designers to produce a vehicle worthy of the nickname "bug."

The lunar module's cylindrical "descent" stage measured 10.5 feet high and fourteen feet in diameter, and consisted almost totally of fuel and engine. The upper stage, which included the cabin and the single ascent engine, was also fourteen feet in diameter and about twelve feet high. But rather than preserving the cylindrical form of the lower half, the "ascent" stage bulged out at the top like a bruised and blackened pomegranate stuck with all sorts of prongs, probes, and protruding radio antennae.

Charley Duke's description of the LEM's interior as "about the size of two phone booths" was generous. The volume of the command module measured about 2,200 cubic feet. The interior of the LEM abided only 160 cubic feet, less than one-tenth the space available in the mothership, and was cramped with all sorts of levers, handles, dials, switches, warning lights, and circuit breakers. Because the LEM was designed to be flown "standing up," there was no place for the two passengers to sit down. There were only armrests, restraints, and Velcro footpads.

The process of powering up the LEM marked another important turning point in the mission, for the Apollo spacecraft was now about to become two vehicles instead of one.

PAO: "This is Apollo Control at eighty-three hours and two minutes into the flight, apolune 65.3 nautical miles, perilune 53.9 . . . Transfer to lunar module power due momentarily . . ."

Then the Control Room would fall silent as all awaited the first words from the LEM in lunar orbit. On the Apollo 11 mission, the mothership was named *Columbia* after the fictional spacecraft *Columbiad* in Jules Verne's *From the Earth to the Moon.* The lunar module bore the proudly patriotic handle *Eagle.* The self-effacing crew of Apollo 10 named their sibling craft after the cartoon characters *Snoopy* (LM) and *Charlie Brown* (CM). The all-navy crew of Apollo 12 christened theirs *Intrepid* (LM) and *Yankee Clipper* (CM). Apollo 14 became *Antares* (LM) and *Kitty Hawk* (CM). The all-air force crew of Apollo 15, the first moon rover mission, christened their lunar module *Falcon* after the official air force mascot and called their mothership *Endeavor.* Apollo 16 was *Orion* (LM) and *Casper* (CM). The crew of Apollo 17, man's last lunar landing mission, revived the patriotic tradition of Apollo 11 by calling their mothership *America* and their lunar module *Challenger,* a name that would later become synonymous with the worst tragedy in the history of the manned space program.

After powering up the LEM, the astronauts once again lost radio contact with Houston as they slipped around the back side of the moon for the fourth time. The first thing they saw when they emerged from the darkness over half an hour later was a phenomenon unique to lunar orbit, a silently stunning spectacle they were too busy to notice on previous transitions from LOS to AOS—earthrise. Apollo 11's Mike Collins describes the sight:

"[The earth] pokes its blue bonnet over the craggy rim [of the moon's eastern terminator] . . . and surges up over the horizon with an unexpected rush of color and motion. It is a welcome sight for several reasons: It is intrinsically beautiful; it contrasts sharply with the smallpox below; and it is home and voice for us. This is not at all like sunrise on Earth, whose brilliance commands one's attention; it is easily missed and therefore all the more precious."

As the astronauts began their fourth pass around the front side of the moon, however, their attention quickly shifted from their home planet to their intended destination as the lunar daylight and their lowered orbit allowed them to examine the surface more closely than ever before. Unlike their predecessors on Apollo 8, who had frowned upon the moon, Apollo 10 astronauts Tom Stafford, John Young, and Gene Cernan called the lunar surface "incredible," "fantastic," and "unbelievable."

Apollo 11's Neil Armstrong happily reported: "The pictures and maps brought back by Apollo 8 and Apollo 10 give us a very good preview of what to look at here. It looks very much like the pictures but like the difference between watching a real football game and watching it on TV—no substitute for actually being here."

At first sight, all the Apollo astronauts also found the lunar landscape forbidding, even those who arrived two and three years after Armstrong. Apollo 15's Jim Irwin, who inserted into lunar orbit on July 30, 1971, explains: "I think it's primarily the color. When you go into orbit around the moon, you are looking at a place where life has never existed. It's a barren desert. There is a great contrast between the moon and the earth, where life does exist, where you have greens and blues and depths of color. If some alien or extraterrestrial life were to move into the vicinity of our solar system, I think that they would immediately recognize there is a great possibility of life on the earth because water implies life. But you see no water and none of those colors on the moon."

Although none of the Apollo astronauts saw the "yellow" or "silvery" moon visible from Earth, they disagree on what colors they did see on the surface. According to the crew of Apollo 8, the lunar landscape appeared to be a colorless wasteland of whites, grays, and blacks. The Apollo 10 crewmen claimed to have seen numerous intermediate shades of browns and tans. Apollo 11's Mike Collins later tried to resolve the controversy by taking both sides:

"It seems to depend on the sun's angle. At dawn and dusk [when the spacecraft is flying over the eastern and western terminators], we have to vote with the Apollo 8 crew. It is dark gray and some white, but no other colors—a darkened monochromatic plaster of paris. On the other hand, near noon [when the spacecraft is flying over the middle of the front side] the surface assumes a cheery rose color, darkening toward brown on its way to black night. We vote with the Apollo 10 crew in the late morning and early afternoon."

Whatever colors the astronauts saw always paled by comparison to the truly amazing and multivariate topography of the moon. Some of the most prominent features were those that formed the face of the "Man in the Moon": the Sea of Serenity *(Mare Serenitatis)* and the Sea of Rains *(Mare Imbrium),* his right and left eyes, located above the lunar equator in the middle of the front side; the Sea of Cold *(Mare Frigoris),* his furrowed brow near the North Pole; and the Sea of Clouds *(Nubium)* south of the lunar equator, which formed his mouth.

These seas or maria did not actually contain water, just miles and miles of desertlike flatlands. But in between the Seas of Serenity and Rains, the "Man in the Moon" had a giant "nose" formed of the Apenninus, Causcasus, and Alps mountain ranges, which towered up to six miles above the surface. There were also all manner of cliffs, faults, plateaus, escarpments, boulder fields, highlands, lowlands, canyons, valleys, cracks, and crevices.

And then, of course, there were craters. Craters in the marias. Craters in the highlands. Craters in the sides of mountains. Craters in the valleys. Craters in rows like strings of pearls. Craters in circular, crescent, and scattershot patterns like the pinholes in a giant dart board. Craters inside of craters inside of still other craters. Craters in every imaginable size, some smaller than a man's fist; others bigger and broader and deeper and higher than the Grand Canyon, like the giant Newton Crater, eighty-five miles wide and thirty thousand feet deep with a rim thirteen thousand feet high.

And in between, around, over, and through the craters, mountains, and maria were millions of razor-thin "ray systems." Lacking in depth and varying in width from less than a millimeter to several meters, these mysterious lines of lighter or darker soil crisscrossed the entire lunar surface like the web of a giant spider. Some measured only a few inches in length. Others ran for thousands of miles. But exactly what they were and what caused them, no one knew for sure.

The same held for all the features on the lunar surface. The "vulcanists" insisted that the moon was once an extremely active "live" planet whose topography was sculptured by massive volcanic eruptions and lava flows. The "impact" theorists argued that the surface was formed mainly by meteorite bombardment. But neither the vulcanists nor the impact men could fully account for such features as the maria, which were often far removed from volcanic regions and far too large and flat to be caused by meteorite bombardment.

The origins of the moon were also subject to debate. Some scientists believed that the moon might have been ripped from the womb of the earth at the dawn of the solar system. Others claimed the moon was born independently in the far reaches of the solar system and inadvertently "captured" by Earth gravity. But neither could completely explain the differences between the front side of the moon, which boasted an impressive and varied landscape, and the back side, which, as best could be discerned through the darkness, was uniformly battered into no particular major regions or distinct features.

Soviet unmanned probes had sent back the first photographs of the back side of the moon nearly a decade before the first American manned lunar mission. But Apollo 8 astronauts Borman, Lovell, and Anders were the first three human beings to see the back side firsthand. As Anders reported, "It certainly looks like we picked the most interesting places on the moon to land in. The back side looks like a sand pile my kids have been playing in for a long time. It's all beat up, no definition. Just a lot of bumps and holes."

Passing from the light of the moon's front side into the darkness of her back side only highlighted the contrast between the two halves. "It was eerie," recalls Apollo 14's Roosa. "You could see the craters and everything in the Earth light, and as you approached the western terminator, the area behind the moon where you don't have any Earth light, you see these big long shadows bringing out the relief of the ridges . . . And then total darkness—everything gone . . . It was kind of like a giant inkblot test that the shrinks give you."

Each time they came back around the front side, the astronauts were naturally most interested in spotting their assigned landing site. The Apollo 11 crew, who were targeted for the presumably flat eastern edge of the Sea of Tranquility near the lunar equator, was shocked by how rugged the place looked from sixty miles up. Mike Collins, who would remain in *Columbia* while comrades Armstrong and Aldrin tried to land the *Eagle,* recalls: "The sun was hitting the moon at a very, very shallow angle, and the craters were throwing very long shadows, and the area looked so chopped up and so sharp, I thought, 'My God, there's no way they're going to find a place smooth enough to put the spacecraft down in that kind of terrain. There's just no way.' Then I said to myself, 'Aw, come on, you've seen the photographs from some of the unmanned flights up close, and you know there really are smooth spots down there.' "

Apollo 11 commander Neil Armstrong, who would have the responsibility for finding that smooth spot, if one existed, seemed undaunted. As the spacecraft passed over an exceptionally large crater near the Sea of Tranquility, he coolly exclaimed into his radio mouthpiece: "What a spectacular view!"

The astronauts on the five subsequent Apollo landings enjoyed some equally spectacular but dramatically different views. Apollo 12's Pete Conrad and Al Bean hoped to make man's second lunar landing in a similarly flat region known as the Ocean of Storms nearly one thousand miles west of the Sea of Tranquility. Apollo 14's Al Shepard and Ed Mitchell were targeted for the rugged highlands of the Fra Mauro region just south and west of Tranquility, and would have to guide their LEM through a corridor of sharp peaks guarding the approach path.

Apollo 15's Dave Scott and Jim Irwin planned to make the first landing in the lunar mountains amid the jagged Apennine range. Apollo 16's John Young and Charley Duke intended to explore the crater-pocked Descartes region south of the lunar equator. Apollo 17's Gene Cernan and Jack Schmitt were bound for the Taurus-Littrow region in the northeast quadrant of the front side, and would have to pilot their LEM through a narrow valley to reach the intended site of man's sixth and final landing on the moon.

The astronauts' visual inspection of the lunar surface was not simply leisurely sight-seeing. On every pass at least one of the three was busily running Program 22, the auto optics photo-mapping program that would help chart the topography of the lunar landscape for their successors. They were also trying to get a precise fix on the spacecraft's position relative to the lunar surface, which fluctuates in small but significant degrees with each orbit due to the treacherous magnetism of invisible mass concentrations of iron ore known as "mascons."

Yet, most of the astronauts were surprised to find themselves adapting much more quickly to lunar orbit than they did to Earth orbit or to life in zero G. "Amazing how quickly you adapt," Apollo 11's Collins informed Mission Control. "Why, it doesn't seem weird at all to me to look out there and see the moon going by, you know . . ."

But by the end of Day Four, as the three Apollo crewmen embarked on their seventh orbit around the moon, the anxiety level inside the spacecraft began climbing in quantum leaps. The undocking and separation maneuver would occur on the thirteenth orbit, which was now less than half a day away. Touchdown (or at least the officially scheduled time for touchdown) was only fifteen hours away.

The astronauts realized that Day Five would be the biggest day of their entire lives. But in the meantime, the only thing they could do to enhance their chances of success was perhaps the hardest thing of all—try to get some sleep. Not surprisingly, the crewmen on the last five successful lunar landing missions had a much easier time than the pioneers of Apollo 11. According to Apollo 15 lunar module pilot Jim Irwin, who would become the eighth man to walk on the moon in the summer of 1971, he and commander Dave Scott approached the final rest period before their landing attempt with the same businesslike attitude that had carried them through every preceding item on the checklist. As Irwin recalls: "I determined to get a good night's rest because I knew the next day would be very busy."

Apollo 12's Pete Conrad and Al Bean, who flew on man's second lunar landing mission in November of 1969, were a great deal more keyed up, not so much from unspoken fears but from nervous anticipation. But they were also confident enough to settle down. As Conrad recalls: "You settle down for the night bright-eyed and bushy-tailed because you know, 'Tomorrow, I'm going to land!' "

The Apollo 11 astronauts retired to their hammocks with much greater trepidation. Buzz Aldrin, who would be the second man to set foot on the moon, later described his last night in lunar orbit as follows: "Although it was not in the flight plan, before covering the windows and dousing the lights, Neil and I carefully prepared all the equipment and clothing we would need in the morning, and mentally ran through the many procedures we would follow. I was so anxiously anticipating the next few days' events, I completely forgot to look for the odd flashes I had seen the two previous nights en route to the moon. I slept fitfully."

Neil Armstrong, who would be the first man to make the "giant leap" onto the surface of the moon, was the first member of the Apollo 11 crew to fall asleep. Unfailingly cool and taciturn, he would also be the last one up the next morning. Command module pilot Mike Collins, the only member of the crew who would not get the chance to land on the moon, was the last one to fall asleep, for he had to keep the night watch. But Collins was the only one who fully articulated the thoughts running through his mind and the minds of his soon-to-be immortal comrades. As he would write in his post-mission autobiography *Carrying the Fire:*

"Everything we have done so far amounts to zilch if we make a serious mistake, and we all know it, although Neil and Buzz seem less prone to admit our vulnerability than I do . . . My wristwatch tells me it's already a few minutes past midnight in Houston. That makes it July 20, lunar landing day. If we were bullfighters, we would call it the moment of truth, but all I want is a moment of no surprises."

STAGE 2:

MEN WALK ON MOON

Houston, Tranquility Base, The Eagle has landed.
APOLLO 11 COMMANDER NEIL ARMSTRONG
July 20, 1969

9

A GIANT LEAP

Man's First Moon Landing

Apollo 11: July 20–21, 1969

That's one small step for man, one giant leap for mankind.
APOLLO 11 COMMANDER NEIL A. ARMSTRONG

"Apollo 11, Apollo 11 . . . Good morning from the Black Team."

The wake-up call came at 6 A.M. Houston time on July 20, 1969, a typically hot and humid Sunday morning that also happened to be the date when men would attempt to walk on the moon.

"Good morning, Houston," mumbled command module pilot Mike Collins after a pregnant pause of close to ten seconds.

"Good morning." repeated the voice of the Black Team.

The voice belonged to astronaut Ron Evans, the night shift CAPCOM for Apollo 11 who would become command module pilot on the last lunar landing mission nearly three and a half years later.

"Got about two minutes to LOS here, Mike," Evans reported.

During their rest period, the Apollo 11 crewmen had completed their seventh and eighth lunar orbits. They were now approaching another loss of signal (LOS) period when they would be out of radio contact with Houston for the duration of their ninth pass around the back side of the moon.

Collins, who had spent the night in the left couch in front of the control panel, reported getting a full six hours' sleep. But even though he seemed to have gotten his circadian rhythms in sync with the flight plan, he still felt drowsy and more than a little fatigued when Houston sounded reveille.

"Oh, my," he groaned, "you guys wake up early."

"Yes," confirmed CAPCOM Evans, "you're about two minutes early on the wake-up. Looks like you were really sawing them away."

"You're right," Collins admitted. Then he dutifully inquired about the status of the spacecraft systems.

"Looks like the command module's in good shape," Evans assured. "The Black Team has been watching it real closely for you."

"We appreciate that," Collins replied, "because I sure haven't."

Half a minute later, the spacecraft disappeared behind the moon's western horizon, and Collins realized that what he called "the moment of truth" was fast drawing nigh.

The Apollo 11 mission clock showed a ground elapsed time (GET) of ninety-three hours and twenty-nine minutes or nearly four full days since lift-off from the Cape. In about six more hours, during orbit number twelve, commander Neil Armstrong and lunar module pilot Buzz Aldrin were scheduled to undock from the mothership. About one hour after that, while Collins piloted the command module *Columbia* through her thirteenth lunar orbit, his comrades in the *Eagle* would begin their descent toward the lunar surface.

Aroused by the crackle of the radio dialogue, Armstrong and Aldrin tumbled out of their hammocks beneath the left and right couches to help Collins fix breakfast. Both men were starting to feel the pressure of their historic mission. Armstrong had gotten less than six hours of sleep (5.5 to be exact); Aldrin had managed to get only five. Despite the fact that they were beginning Day Five in space and were also veterans of Gemini space flights, they fumbled with their color-coded food packets like nervous rookies. As Aldrin recalls:

"By now we had worked out various routines for living together, but in our excitement this particular morning the system came unglued. For eating we had a system where one guy pulled the food out, another snipped the packages open while the third liquefied his food and passed the water gun to the man waiting. The rhythm got slightly out of whack this last morning, and once we finally got it going properly, we all three bemoaned the fact that the simple act of eating was something there was no training for. It was one of those things we had to learn by doing."

The same held for the "giant leap" the Apollo 11 astronauts would try to make later that day. There was no way to train for man's first foray onto another planet—except through computerized simulations and other elaborate forms of make-believe. Landing on the moon was something they would have to learn by doing. But then, all three crewmen had spent their entire adult lives doing exactly that.

Apollo 11 commander Neil Alden Armstrong, the sole civilian member of the trio, was the best pilot in the predominantly military astronaut corps and by far the most inscrutable. Blond and blue-eyed, five feet eleven inches tall, and a surprisingly flaccid 165 pounds, he was pleasant but aloof, boyishly enthusiastic and yet stoic as the statuette of Buddha on display in his living room. Though he was destined to attempt one of the most extraordinary feats in human history, he came across as being ordinary as every man, and at the same time profoundly unlike other men.

Born on a farm near Wapakoneta, Ohio, on August 5, 1930, Armstrong was the son of a state auditor whose job kept him moving from one town to another as he reviewed each county's tax records. Young Neil was an itinerant bookworm who learned to play the piano and the bass horn, but his lifelong passion was flying. In addition to building all sorts of model airplanes, he worked part time in local groceries, pharmacies, hardware stores, and bakeries at forty cents an hour to earn money for nine-dollar-an-hour flying lessons. He got his pilot's license on his sixteenth birthday before he was even licensed to drive a car.

After flying seventy-eight combat missions in the Korean War, Armstrong resigned from the navy and returned to Purdue University to get his engineering degree. Shortly after graduation, he was hired by NASA as a civilian experimental test pilot at Edwards Air Force Base in California, where he flew every type of high-performance aircraft from the X-15 on down. Along the way, Armstrong married college sweetheart Jan Shearson, who gave birth to two boys and a girl while living in a mountainside cabin with no hot water. The couple later suffered the tragic loss of daughter Karen, who died of a brain tumor before her third birthday.

On September 17, 1962, Armstrong was accepted into NASA's second astronaut class, known as "The Nine" so as to be distinguished from the "Original Seven." Unlike most of his peers, Armstrong was neither a heavy drinker nor a physical fitness fanatic, but it was he who rose to the top by demonstrating remarkable grace under pressure. In 1966, he had confronted a most precarious crisis aboard Gemini 8 when the spacecraft suddenly began gyrating out of control during docking with an Agena rocket. According to manned spacecraft program chief Robert Gilruth, Armstrong's success in safely steering the command module back to Earth "left a very great impression" on the NASA officials who selected the crews for the first moon landing missions.

Armstrong nearly lost his life before he ever got the chance to make his "giant leap" onto the moon. On May 6, 1968, he was at Ellington Air Force Base in Houston practicing in a free-flight trainer known as the Lunar Landing Research Vehicle (LLRV). Irreverently nicknamed the "Flying Bedstead," the LLRV was a wingless jet-and-rocket-powered craft with a single seat-box-shaped cockpit, a girdered tail section, and four spindly landing legs. Because of the vehicle's ungainly body and the inability to simulate one-sixth gravity at Ellington, handling the LLRV was far more difficult than maneuvering a real LEM, and in many ways, more dangerous.

"I reserve my God-given right to be wishy-washy," Armstrong had informed fellow Apollo 11 astronaut Buzz Aldrin a few weeks before his LLRV exercise.

When Armstrong was in the terminal phase of his descent less than 200 feet above the landing site, the LLRV suddenly began bucking and backfiring, spewing a thick cloud of engine exhaust. Then, as if struck by some invisible arrow, the vehicle pitched nose up, rolled over to the right, and tumbled toward the earth. After exercising his "God-given right to be wishy-washy" for several breathless seconds, Armstrong finally hit the ejection handle. The LLRV crashed to the pavement and exploded in flames. Instead of perishing in the fiery wreckage, Armstrong managed to parachute onto a patch of grass between the runways. Then he walked directly to the hangar, where he apologized for losing his ship and blamed the accident not the LLRV or the ground crew but on "pilot error"—i.e., himself. "The only damage to me," he reported afterward, "was that I bit my tongue."

However, Armstrong seemed to take his historic Apollo 11 assignment with remarkably unflappable professional modesty bordering on nonchalance. During the crew's final prelaunch press conference, he gave advance credit for the mission to the thousands of support personnel on the ground, insisting that "It's their success more than ours." Asked if he had ever dreamed about going to the moon, he claimed he had not, but admitted, "I guess after twenty-four hours in a simulator, I guess I have dreams of computers." Asked if he would have done anything different in preparing for his lunar mission, he replied, "If I had a choice, I guess I'd take more fuel." Although Armstrong had no way of knowing it at the time, all of his ostensibly innocent comments would echo with prophetic resonance on the fateful afternoon of July 20, 1969.

So would an anecdote Armstrong shared with the Apollo 11 crew's official biographers shortly before lift-off. The story involved his recurrent boyhood dream of being able to float in mid-air. "I could, by holding my breath, hover over the ground," he recalled. "Nothing much happened; I neither flew nor fell in those dreams. I just hovered. They can't have been bad dreams. But the indecisiveness was a little bit frustrating. There was never any end to the dream."

Edwin Eugene "Buzz" Aldrin, Jr., Apollo 11's balding but broad-shouldered thirty-nine year-old lunar module pilot, was, in sharp contrast to the Zen-like Armstrong, a man in intense inner conflict with himself. Though he, too, claimed in public that he had not dreamed of going to the moon during his prelaunch training, he was not being completely candid. As he would write in his 1974 post-mission autobiography *Return to Earth,* "I remembered one [science fiction] story about a voyage to the moon during which a great deal of trouble was encountered and, once the moon had been reached, the space travelers departed for Earth, returning home insane. It had given me nightmares as a youngster, and had secured an odd corner in my psychic life."

In that same autobiography, Aldrin revealed that for many months before lifting off for the moon, he had secretly been repressing the realization that he was a borderline alcoholic and manic depressive, and attributed his psychic turmoil to a classic Oedipal conflict. Born in Montclair, New Jersey, on January 30, 1930, he was the son of a hard-driving former air force colonel turned Standard Oil executive, and a mother whose maiden name was Marion Gladys Moon. In addition to getting a Ph.D. at MIT, Edwin E. Aldrin, Sr., had studied under rocketry pioneer Robert Goddard, and prided himself on his close friendships with military heroes like General Billy Mitchell and General Jimmy Doolittle. And according to Aldrin, Jr., his father, who had "skirted the fringes of fame most of his adult life," proceeded to plant "his own goals and aspirations in me."

After graduating third in his class at West Point, where he was also a star pole vaulter, Aldrin was twice rejected for a Rhodes scholarship. Undaunted, he went on to fly sixty-six combat missions for the air force in the Korean War, then enrolled in the Ph.D. program at MIT where he wrote his 1962 doctoral thesis on what would prove to be one of the key maneuvers needed to get a man on and off the moon—"Line of Sight Guidance Techniques for Manned Orbital Rendezvous." On October 17, 1963, after being rejected for the second class of NASA astronauts to which Armstrong had been admitted, he was accepted in the class of fourteen, known as "The Third Group."

Nicknamed "Dr. Rendezvous" after his thesis topic, Aldrin first flew in space in November 1966 aboard Gemini 12, the last of the two-man flights, during which he set a new American spacewalking record of five hours and twenty-six minutes. But the mission took its toll on him and his wife and three children. During the Christmas holidays of that year, he suddenly felt "an overwhelming sense of fatigue mixed with a vague sadness," which prompted him to stay in bed for five days. Only after he returned from his trip to the moon three years later did he realize that his nervous system had been "sending out a distant early warning . . ."

Unbeknownst to his astronaut peers, Aldrin was so bitterly ambivalent about his role on the Apollo 11 mission—which he believed would not allow him sufficient time on the lunar surface to use his scientific skills—that he and his wife privately discussed the idea of declining the assignment in hopes of being shifted to one of the later longer-duration missions. "We kept this secret," Aldrin would claim in his post-mission memoir, "because in my business such an idea is tantamount to sacrilege. No one had ever refused a flight. If I, as one individual, refused, both Mike [Collins] and Neil[Armstrong] would likely be taken off the flight. And if I did such a thing, I would so impair my position that I'd probably never be assigned to a subsequent flight."

Aldrin's ambivalence was compounded by the fact that he would not be permitted to take man's first step onto the moon. He later maintained that he had simply assumed the lunar module pilot would exit first since the commander had customarily remained on board the spacecraft during the EVAs on Gemini missions. But shortly before the Apollo 11 lift-off, he heard through the NASA grapevine that Armstrong, by virtue of being a civilian astronaut as well as the mission commander, would go first. Aldrin decided to confront Armstrong face-to-face.

"Neil equivocated for a moment or so," Aldrin recalled afterward, "then with a coolness I had not known he possessed, he said that the decision was quite historical and he didn't want to rule out the option of going first."

NASA officials would also equivocate when explaining how Armstrong got the option to make his "giant leap." Some claimed he was the logical choice all along because his seat in the lunar module was closest to the exit hatch. Others reported that Armstrong simply exercised the "commander's prerogative." In any event, tensions were further strained when Armstrong "crashed" a LEM simulator shortly before the crew was scheduled to launch from the Cape. He and Aldrin reportedly stayed up into the wee hours of the morning arguing about the incident until they finally declared a truce.

Though still rankling about having to be "only" the second man to walk on the moon, Aldrin had all but rationalized his fate by the time the Apollo crew arose on the morning of July 20, 1969. As he saw it, the landing phase was the trickiest and most hazardous aspect of the mission;

the EVA portion promised to be a relative snap. And though Armstrong might be the first man to plant his boots in the lunar soil, he and Aldrin would actually touch down on the moon at exactly the same moment. What's more, Aldrin had a few surprises of his own in store, one of which involved a holy secret cargo stashed away in his personal preference kit.

Michael Collins, Apollo 11's amiable and athletic thirty-nine-year-old command module pilot, was the only member of the crew who would not get the chance to land on the moon. Most of the general public pitied him as the odd man out, but Collins didn't feel sorry for himself at all. As he declared at the astronauts' final prelaunch press conference, "I don't feel in the slightest bit frustrated. I'm going 99.9 percent of the way there, and that suits me just fine . . . I couldn't be happier right where I am."

Collins, who had nearly lost the chance to fly to the moon due to a prelaunch back problem, was not just telling a white lie. Born in Rome, he was the dutiful and patriotic son of a well-connected army brigadier general, a former polo player and horse cavalryman, who moved his wife and children from Italy to Oklahoma, New York, Puerto Rico, and finally, back to Virginia. After prepping at prestigious St. Albans School in Washington, D.C., where he captained the wrestling team, Collins attended West Point, then joined the air force to win his wings as a fighter pilot. In the summer of 1953, he was accepted for advanced day fighter training at Nellis Air Force Base near Las Vegas. "In the eleven weeks I was there, twenty-two people were killed," Collins reported afterward. "It was a hectic time, and I'm surprised to have survived. I have never felt quite so threatened since."

Collins emerged from the Nellis experience sobered but unscarred, with a wife and three children and the still somewhat heretical ambition of becoming a space pilot. He was denied admission to the second NASA astronaut class, Armstrong's group, but was later accepted along with Aldrin in the "Third Group" selected in the fall of 1963. Three years later as a crewmen aboard Gemini 10, he became the third American to walk in space. Wide-eyed, optimistic, and dutifully trim at five feet eleven inches and 163 pounds, he also became the best handball player in the astronaut corps.

Then, in the summer of 1968, Collins was dismayed to find that he had a bone spur in his spine that required immediate surgery. Though the operation forced him to be bumped from his assignment as command module pilot for Apollo 10, the "dress rehearsal" for the first lunar landing, it proved to be a stroke of priceless good fortune. Collins managed to make a miraculously quick recovery from his back operation, and was promptly reassigned to Apollo 11 at Armstrong's request.

Despite being denied the privilege of walking on the moon, Collins deeply appreciated the indispensable nature—and potential anguish—of his role as command module pilot. It was up to him to keep the mothership in proper orbit while his comrades explored the lunar surface, a task of solo space pilotry that was in many ways more difficult and demanding than landing on the moon. It would also be up to him to effect the rendezvous with the lunar module when Armstrong and Aldrin were ready to return to Earth. And if, for some unforeseen reason, Armstrong and Aldrin were unable to get off the moon and rendezvous with the command module, it would be Collins's unenviable duty to fly home alone.

In the meantime, Collins could look forward to being only the second man after Apollo 10's John Young to be totally cut off from Earth in outer space. After his comrades undocked in the lunar module, he would be completely alone in an LOS period each time his spacecraft circled the back side of the moon, a potentially fearful experience he welcomed as a unique opportunity for solitary observation and reflection. As he later demonstrated in his post-mission autobiography *Carrying the Fire,* Collins was the most articulate and poetic of the twenty-four men who flew to the moon. But he was also one of the most wary. And on the morning of July 20, 1969, the words he had uttered at the crew's final prelaunch press conference were still racing through his mind: "The most dangerous items are the ones we've overlooked."

When the spacecraft emerged from behind the moon fifty-two minutes later, flight director Gene Kranz and his White Team were filing into the Control Room down in Houston to relieve the Black Team. CAPCOM Ron Evans relayed one more update on remaining supplies of oxygen, fuel, and other consummables. Then, before passing the mike to the White Team, Evans startled the still bleary-eyed astronauts in lunar orbit with a specially prepared news brief:

"Among the large headlines concerning Apollo this morning there's one asking that you watch for a lovely girl with a big rabbit. An ancient legend says that a beautiful Chinese girl called Chango has been living there for four thousand years. It seems she was banished to the moon because she stole the pill of immortality from her husband. You might also look for her companion, a large Chinese rabbit, who is easy to spot since he is always standing on his hind feet in the shade of a cinnamon tree. The name of the rabbit is not recorded."

The half-asleep astronauts could not believe their ears. Since the wake-up call on Launch Day, they had flown nearly 240,000 miles from Earth on "the most hazardous and dangerous and greatest adventure on which man has ever embarked." In just a few short hours, they would attempt to make the climactic "giant leap" onto the surface of the moon. And now the voice of Mission Control, their only link to reason and reality in what otherwise seemed like a science fiction fantasy, was babbling about some mythical China doll and her big-eared bunny rabbit.

"Okay, we'll keep a close eye for the bunny girl," Aldrin promised dryly.

Armstrong typically made no comment, but Collins, who had a host of essential housekeeping chores and position checks to complete in the next hour, let Houston know in no uncertain terms that he was not amused.

"A three-ring circus!" Collins answered curtly, and then was heard muttering to himself, "I've got a fuel cell purge in progress, and am trying to set up cameras and brackets, watch an auto maneuver . . ."

The spacecraft slipped behind the moon for the tenth time at ninety-five hours twenty-five minutes GET, which was 8 A.M. in Houston. During this LOS period, Aldrin crawled through the access tunnel to the *Eagle,* which had been fully pressurized the day before, and powered up the main control panels. Then he floated back to the *Columbia,* where he and Armstrong doffed their long johns and began putting on their "moon cocoons."

"I got as far as my liquid-cooled underwear," Aldrin recalls, "and left to make some initial checks in the LM while Neil took my place in the navigation bay to begin changing. There is only room for one man to change clothes, with another standing by to help pull the long crotch-to-shoulder zipper closed. Both Neil and I were determined to put on our suits as perfectly and comfortably as possible—we were going to be in them for quite some time."

When the Apollo 11 astronauts reacquired radio contact with the ground fifty-five minutes later, the viewing room at Mission Control in Houston had begun filling up with NASA brass and the ranking geniuses of the U.S. manned space program. Dr. Thomas O. Paine, the agency's chief administrator, and deputy administrator Robert C. Seamans, Jr., arrived with George Low, George Mueller, and MSC director Dr. Robert

R. Gilruth, all members of the original NASA brain trust. Mission Control operations director Christopher Columbus Kraft, Jr., was there, as was burly Rocco Petrone, director of launch operations at the Cape, and the venerable Dr. Wernher von Braun.

Chief astronaut Donald K. "Deke" Slayton led a contingent of former and future spacemen standing by in the Control Room to share their firsthand insights and experiences if called upon. The group included John Glenn, the first American to orbit the earth, Apollo 8 crewman Jim Lovell, who had flown on man's first voyage around the moon and would later command the ill-fated Apollo 13 flight; Apollo 9 commander Jim McDivitt, who had made the first lunar module rendezvous in Earth orbit: and Apollo 10 lunar module pilot Gene Cernan, who had flown on the "dress rehearsal" for Apollo 11 and would later command the last lunar landing mission.

There was also a fresh new face at the CAPCOM console: future Apollo 16 astronaut Charley Duke. Then age thirty-three, Duke was a six-foot-tall, softly drawling North Carolinian who had graduated from the U.S. Naval Academy, then switched over to the air force for pilot training and won admission to the fifth NASA astronaut class in the spring of 1966. Though Duke had not yet flown in space or even been assigned to a crew, he had been one of the CAPCOMs for Apollo 10, and had so impressed Neil Armstrong that the Apollo 11 commander personally requested that he serve as CAPCOM during the landing phase of the mission. "I want you to do it for me," Armstrong had told him, "because you probably have a better knowledge of this than anybody who's not flying."

Duke wished with all his heart that he could have been one of the astronauts bound for the moon, and in less than three years' time, he would be. But in July of 1969, he was proud enough to have the vital role of CAPCOM. As Duke told an interviewer, "We're supposed to know how the crew responds, and whether this is good from an operational standpoint or not . . . To be a good CAPCOM, you've got to know the procedures used by the crew and also the software—in other words, the operational details of how they're flying the spacecraft and also the flow of information to the crew."

Duke got the first hint of how severely his knowledge would be tested at acquisition of signal (AOS) on Apollo 11's tenth revolution around the moon.

CAPCOM (DUKE): "Hello, *Columbia,* this is Houston. Do you read? Over."

Columbia did not respond immediately. Instead, there was a barely audible squawk that sounded like it might have come from the LEM.

CAPCOM (DUKE): "*Eagle,* did you call me? Over."

Eagle (ALDRIN): "Roger. How do you read? Over."

CAPCOM (DUKE): "Roger . . . a lot of noise on the [radio] loop. We think it's coming in from *Columbia,* but we can't tell. We're unable to raise voice with him. Would he please go to high gain [antenna]? Over."

Aldrin relayed the request to Collins in the command module, but the switch from the S-Band antenna to the high-gain antenna seemed only to solve half the problem.

Columbia (COLLINS): "Houston, *Columbia.* Reading you loud and clear. How do you read me?"

CAPCOM (DUKE): "Roger, about three by [numerical shorthand for medium clarity and volume] . . . Mike, we've got a lot of noise in the background . . ."

Duke and the Apollo 11 crewmen would soon realize to their mutual consternation that this wayward antenna was merely the beginning, not the end, of their communications problems.

In the meantime, the spacecraft completed its pass across the front side of the moon in one hour and twelve minutes, and "went over the hill" of the western horizon to begin lunar orbit number eleven. It was now a little past 10 A.M. in Houston, ninety-seven hours thirty-two minutes (GET). According to the flight plan, Armstrong and Aldrin were scheduled to touch down in the Sea of Tranquility in less than six hours.

The anxiety and activity levels inside *Eagle* and *Columbia* rose in tandem as the astronauts prepared for the undocking maneuver during orbit number twelve. After pressurizing their suits, Armstrong and Aldrin cross-checked the LEM's guidance platform, the rendezvous radar, the descent propulsion system (which was actually a single low pressure with 96,000 pounds of thrust), and the reaction control system (RCS) thrusters used for roll, pitch, and yaw maneuvers.

Although Collins would remain at the helm of the *Columbia* during his comrades' descent to the lunar surface, he, too, had to get back into helmet and suit in case of a sudden loss of cabin pressure. He also had to verify the deployment of the LEM's landing gear (before launch Collins had made a special trip to the Grumman plant in Bethpage, N.Y., just to see what the legs looked like when fully extended), as well as prepare the command module for solo lunar orbit.

"Apollo 11. Houston," called CAPCOM Duke as the spacecraft reemerged on the front side of the moon. "We're GO for undocking."

The tension in Mission Control was now building to the same anxious level reached before the lunar orbit insertion the day before. Like the LOI burn, the undocking maneuver had to take place on the back side of the moon. If something went awry, the men in the trench could not come to the rescue, a fact not lost on the increasingly harried Apollo 11 crew.

"There will be no television of the undocking," Collins informed Houston as the spacecraft approached the moon's western horizon. "I'm busy with other things . . ."

"Roger. We concur," replied CAPCOM Duke.

Apollo 11 lost contact with Earth for the twelfth time at ninety-nine hours thirty-one minutes GET, which was high noon back in Houston. Armstrong and Aldrin sealed the hatch of the *Eagle,* and started severing the umbilical cords that connected the LEM to the mothership. Inside the *Columbia,* Collins commenced the final countdown to the undocking, carefully making sure that he held the flexible joints between the two spacecraft in perfect alignment as his fellow crewmen calibrated their independent guidance and control instruments.

"I have five minutes and fifteen seconds since we started," Collins reported. "Attitude is holding very well."

"Roger, Mike. Just hold it a little bit longer," Aldrin requested.

"No sweat, I can hold it all day," Collins assured. "Just take your sweet time." Then he added: "How's the Czar over there? He's so quiet."

He was referring to Armstrong, who had been dubbed the "Czar of Apollo 11" by the Soviet news agency Pravda in reports of the trans-lunar voyage.

"Just hanging on," Armstrong replied, "and punching . . ."

Armstrong's punching was directed at the six hundred separate switches, buttons, dials, and warning lights festooning the cockpit of the *Eagle.*

"All I can say is, 'Beware the revolution!' " Collins punned.

Armstrong apparently did not appreciate the humor, for he failed to respond.

"You guys take it easy on the lunar surface," Collins admonished. "If I hear you huffing and puffing, I'm going to start bitching at you."

Collins threw the release switch, and a set of explosive bolts blasted the LEM away from the command module.

The *Eagle* did a slow pirouette in the silence of the lunar void, extending her four spindly legs and lumpy-headed body like a giant insect bursting from its cocoon.

A few minutes later, the *Columbia* and her now detached sibling sailed around the front side of the moon, where Houston impatiently demanded to know the results of the undocking maneuver.

"The *Eagle* has wings!" Collins cried triumphantly.

The *Eagle,* however, was still sixty miles above the lunar surface and only a few yards away from the *Columbia,* orbiting in tandem at a velocity of close to 5,500 feet per second or roughly 3,600 statute m.p.h. In order to descend to the Sea of Tranquility, the astronauts had to execute three equally crucial and difficult maneuvers before, during, and immediately after their next pass around the back side.

First, Collins initiated the front side "separation burn" by firing the thrusters of the service module for a few quick seconds. That pushed the *Columbia* far enough away from the *Eagle* for Armstrong and Aldrin to begin the two-step descent procedure. It also provided Collins with his first good view of the lunar module in all her awkward majesty.

"I think you've got a fine-looking flying machine there, *Eagle,*" he observed, "despite the fact that you're upside down."

"Somebody's upside down," Armstrong rejoined.

At this point, the Apollo 11 "Czar" and his fellow traveler were "lying" on their backs facing up at *Columbia* and the stars beyond. In the course of their thirteenth pass around the back side of the moon, they would flip around, slow down, and descend to a 60-mile-by-8.5-mile elliptical orbit by executing a braking maneuver called the "descent orbit insertion" (DOI) burn. Then, upon reemerging on the front side, they would roll over on their backs once again, and if all went well, make the "powered descent insertion" (PDI) burn, gradually making the *Eagle* "stand up" so her thrusters pointed down toward the lunar surface.

Contrary to popular misconception, the success of Apollo 11's historic lunar landing attempt depended much more on the skill of the two men inside the *Eagle* than on the giant computers back at Mission Control. The men in the trench could not assist with DOI because the astronauts had to make the burn while they were out of radio contact. Nor could the ground help guide the power descent due to the six-second round-trip lag in radio transmissions between them and the spacecraft.

Likewise, Armstrong and Aldrin knew they could not simply trust their fate to the *Eagle*'s automatic guidance system. Although the astronauts had to rely on the on-board computer to calculate the rapidly changing altitude, attitude, velocity, and fuel consumption figures in the initial stages of the power descent, Armstrong would have to assume full or partial manual control at five hundred feet above the lunar surface, perhaps even sooner. A twenty-four-inch by twelve-inch by six-inch box identical to the DSKYs in the command module, the LEM's computer was one thousand times smaller than the IBM 360/75s at Mission Control; the vocabulary devoted to unforeseen problem solving totaled only two thousand words. And as Aldrin had noted before lift-off with unconscious foresight, "The computer isn't going to dodge boulders for you."

Collins noted the approach of LOS period number thirteen. According to the official flight plan, the touchdown attempt was now less than three hours and three-quarters of a revolution away.

"Going right down U.S. 1, Mike," Armstrong assured in a cryptic reference to the nickname the Apollo 10 astronauts had given the orbital approach to the Sea of Tranquility.

"Okay, *Eagle,* one minute," Collins cautioned. "You guys take care . . ."

"See you later," Armstrong promised.

The *Eagle* disappeared behind the moon's western horizon at 101 hours 28 minutes GET, approximately 2 P.M. Houston time. Then, at Armstrong's command, the spacecraft tilted her feet and exhaust nozzle forward, and fired her descent engine in a 28.9-second braking burn. The LEM slowed to a forward velocity of about 4,600 feet per second, roughly 3,000 statute m.p.h., and started plunging feet first toward the crater-pocked surface of the moon.

When Armstrong and Aldrin came around the front side, they were "hanging" from their shoulder harnesses with heads pointed "backward," looking down at the lunar surface from an altitude of about eighteen miles. But as the *Eagle* continued to dive toward the 8.5-mile mark where the astronauts planned to begin their final powered descent to the landing site, they suddenly realized that something was wrong with the high-gain antenna, their radio lifeline to Mission Control.

CAPCOM (DUKE): "*Eagle,* Houston. If you read, you're GO for powered descent. Over."

No answer.

COLLINS: "*Eagle,* this is *Columbia.* They just gave you a GO for powered descent."

CAPCOM (DUKE): "*Columbia,* Houston. We've lost them again on the high gain . . . We recommend they yaw right ten degrees and then try for high gain again."

Collins relayed the message via the S-Band frequency, and *Eagle* yawed right as instructed.

CAPCOM Charley Duke fretted in silence along with his cohorts back in Houston. As Duke recalls: "It always happens that when we have the critical revolution or the critical pass, we have lousy communications. It just seems like that's our luck . . . I said to myself, 'Oh, no, here we go again,' because we had a mission rule that said we had needed adequate communications and data from the spacecraft before we could commit to powered descent."

Then, amid a cacophony of radio static, the sound of Armstrong's voice once again crackled in the earphones of the men at Mission Control: "Houston, *Eagle.* How do you read us now? Over."

CAPCOM (DUKE): "*Eagle,* Houston. We read you now. You're GO for PDI . . . MARK 3:30 [three minutes and thirty seconds] until ignition."

ARMSTRONG: "Roger, understand . . . Elevate the GO circuit breaker."

Armstrong and Aldrin were now due to arrive at the Sea of Tranquility in just twelve minutes. But at ignition of PDI, they were still fifty thousand feet above the lunar surface with a forward velocity of three thousand statute m.p.h. And unbeknownst to them and their guardian angels on the ground, they were about to confront a series of last-minute crises that almost forced them to abort—and very nearly cost them their lives.

Four minutes into the powered descent, at forty thousand feet, the *Eagle* rolled over to lock in her landing radar, and the two astronauts once again found themselves "upside down." They would have to remain in that inverted position, totally dependent on the LEM's computer guidance system, until "high gate" at seven thousand feet, where Mission Control would make the penultimate GO/NO GO decision. Then the spacecraft would start to tilt forward and downward until they were "standing up" like trolley car conductors and could once again see where they were going.

"*Eagle,* Houston . . . You are GO to continue powered descent," CAPCOM Duke advised.

"Roger," Aldrin acknowledged. "Got the earth right out our front window."

But just before he lost sight of the lunar surface, Armstrong realized he had an even more serious problem than the annoying radio lapses. As the *Eagle* was descending, he had been charting their progress by looking for previously photo-mapped landmarks and comparing their actual pass-over times to those in the flight plan. When a large easterly crater called Maskelyne W appeared in his window two seconds sooner than expected, he knew something was wrong. At three thousand m.p.h., two seconds translated into two miles off course.

"Our position checks downrange show us a little long," Armstrong reported.

"Roger, we confirm," replied Charley Duke.

This deliberately matter-of-fact exchange belied the true gravity of the *Eagle*'s descent trajectory error. According to ground-based radar, the spacecraft was diving toward the lunar surface a good fifteen m.p.h. (twenty-three feet per second) faster than called for in the flight plan. Mission rules stated that if the velocity increased by another eight m.p.h., the astronauts would have to abort their landing attempt and rendezvous with the command module.

The responsibility for deciding whether or not Armstrong and Aldrin should continue their descent rested with one of the men in the trench back at Mission Control: thirty-five-year-old computer specialist Steve Bales, the guidance officer (GUIDO). Flight director Gene Kranz, who was already sweating through his "lucky" white vest, demanded an immediate answer.

"I think this is going to hold steady and we're going to make it," Bales reported, adding an ominously inconclusive, "I think."

Five minutes into the PDI burn, as the *Eagle* dove to within twenty thousand feet of the lunar surface, it looked like Bales was right. The spacecraft came uncomfortably close to but did not exceed the mandatory abort mark of twenty-three m.p.h. (thirty-five feet per second) over the projected optimum descent velocity.

Then the LEM's on-board computer flashed an unexpected warning signal.

"Program alarm," Aldrin rasped. "It's a 1202."

The computer program alarm caught Steve Bales by surprise. At first, he could not even recall what a 1202 was. For a good ten seconds, he fumbled around his work station in a frantic search for a code book that could help him identify the nature of this mysterious malfunction.

"Give me a reading on that alarm," Aldrin demanded after another thirty seconds had elapsed.

Then one of Bales's teammates in the trench advised that a 1202 program alarm signified an "Executive Overflow." That was the on-board guidance computer's cryptic way of saying it was being taxed beyond maximum capacity and would temporarily have to delay execution of certain commands until it could catch up with itself.

"The most dangerous items," as Collins had observed before lift-off from the Cape, "are the ones we've overlooked."

The cause of the 1202 program alarm was one of the few items Houston had overlooked. At first, it looked like the combined currents of the spacecraft's guidance, control, and life support systems were producing an unforeseen overload of electromagnetism. The electromagnetic interference was apparently creating "sneak circuits," which disrupted and confused the on-board computer's program execution process. Much to their chagrin, the men in the trench would later determine that the program alarm was caused by a much less esoteric electronic bugaboo. It turned out that the *Eagle*'s rendezvous radar, which could not communicate with the separately designed landing radar, was simultaneously transmitting a series of contradictory signals to the onboard guidance system. As a result, the computer was vainly attempting to serve two masters—one intent on locating a touchdown site in the Sea of Tranquility, the other preoccupied with finding a route back to the mothership in lunar orbit.

Fortunately, the men at Mission Control had not overlooked ways to cope with an on-board computer failure during Apollo 11's final descent to the lunar surface. An almost identical problem had been thrown at them in a disastrous simulation exercise about a month before the launch. Although flight director Kranz had aborted the simulated mission to the mutual embarrassment of all involved, the experience had prompted the White Team to "get smart" about the general nature of last-minute computer alarms and the procedures for avoiding an unnecessary abort during the real thing.

Bales, who had participated in the aborted simulation exercise, decided "on instinct as much as anything else" to recommend that Armstrong and Aldrin get the GO to proceed with their powered descent despite the 1202 program alarm. As he recalled in a recent interview: "All that we could tell at the time was that the computer was being overloaded. Why, we didn't know, couldn't tell. The only thing we could do was rely on the rules we'd established before the flight. If this particular kind of alarm didn't come up more than once every minute or so, and if the computer continued to steer the vehicle properly, we'd continue with the descent. I looked at it for about thirty seconds, and it looked good to me, and that was it. We didn't have time to delve into it any more than that."

Both astronauts later applauded Mission Control's grace under pressure. "It was not a serious program alarm," Buzz Aldrin would insist. "It just told us that for a brief instant the computer was reaching a point of . . . having too many jobs for it to do. Unfortunately, it came up when we did not want to be trying to solve these particular problems."

Aldrin claimed the decision to ignore the 1202 alarm was "what we wanted to hear." It was that kind of positive thinking, that kind of "GO if at all possible" attitude, that truly separated the real mission from all the simulated missions. Indeed, as Armstrong would note in a postflight press conference: "In the simulator we have a large number of failures, and are spring-loaded to the ABORT position . . . In the real flight, we are spring-loaded to the 'LAND' position."

A little over eight minutes into the powered descent, the *Eagle* reached "high gate" at seven thousand feet, and "stood up" on her landing legs. The braking thrust of her descent engine had reduced her velocity to six hundred feet per second, about four hundred m.p.h., and her exhaust nozzle now pointed down toward the lunar surface. But according to the mission rules, the astronauts had to touch down in the Sea of Tranquility in less than six minutes or hit the abort handle.

Flight director Gene Kranz commenced the last formal GO/NO GO roll call of the men at the key consoles in the Control Room: the retrofire officers, the flight dynamics officers, the guidance officers, the flight surgeon, the electrical and communications systems squad, the telemetry communications team, the flight operations chief, and the ground tracking network controllers.

"RETRO?"

"GO!"

"FIDO?"

"GO!"

"GUIDO?"

"GO?"

"SURGEON?"

"GO!"

"EECOM?"

"GO!"

"TELCOM?"

"GO!"

"OPS?"

"GO!"

"NETWORK?"

"We're GO, FLIGHT!"

Kranz ordered CAPCOM Charley Duke to relay the unanimous verdict to Armstrong and Aldrin.

"Eagle, Houston. You're GO for landing."

"Roger, understand," Armstrong answered. "GO for landing . . ."

Although the *Eagle*'s faltering computer was still expected to pilot the descent for another 6,500 feet, Armstrong had already begun preparing to take over manual control much sooner if need be. On his right, there was a joy stick with a bright red pistol grip, the Attitude Controller Assembly (ACA), which could make the LEM pitch, roll, or yaw, thereby changing her course and/or slowing her descent. On his left was a toggle switch called the Thrust Translator Controller, which would supposedly slow the LEM's descent velocity one foot per second each time he clicked it.

Then at 2,500 feet, *Eagle*'s on-board guidance systems registered a second program alarm.

"1201," Aldrin barked.

"1201," Armstrong repeated.

Houston copied the alarm, determined that it indicated a computer overload similar to the 1202 alarm, and gave the GO to proceed. But the 1201 warning raised the tension still higher out of fear that the astronauts might face catastrophic computer problems if they had to abort the landing attempt. As CAPCOM Duke recalls, "Here we are with a computer that seems saturated during descent, and my gosh, we might be asking it to perform a more complicated task during ascent."

Only then did Armstrong and Aldrin begin to realize the true implications of the computer alarms. The "Executive Overflows" and the excess velocity of their powered descent were causing the *Eagle* to overshoot their designated landing site in the Sea of Tranquility by at least four miles. If the trajectory error increased to six miles, they would have to affect a mandatory abort.

In the meantime, as the *Eagle* passed the one-thousand-foot mark, the astronauts appeared to be plumeting toward a large crater full of jagged boulders. Armstrong later admitted: "I was surprised by the size of those boulders; some of them were as big as small motor cars. And it seemed at the time that we were coming up on them pretty fast; of course, the clock runs at about triple speed in such a situation."

Aldrin kept calling out the latest altitude and velocity readings in a steady drone:

"Seven hundred feet, twenty-one [feet per second] down . . ."

"Six hundred feet, down at nineteen . . ."

"Five hundred forty feet, down at thirty—down at fifteen . . ."

When Armstrong took over semiautomatic manual control at five hundred feet, he could not decide what to do. He was "sure the ejecta [in the crater] would have been lunar bedrock, and as such, fascinating to the scientists." But he also knew that the boulders could wreck his spacecraft on impact.

"I was tempted to land," he confided afterward, "but my better judgment took over."

Armstrong made the LEM pitch over from her vertical or "standing up" posture to a horizontal attitude with her thrusters pointing "backward" so he could "skim over the top of the boulder field" and look for a less hazardous alternate landing site. As he would later explain to media critics back on Earth, "I was absolutely adamant about my right to be wishy-washy about where I was going to land, and the only way I could buy time was to slow down the descent rate . . ."

Aldrin calmly continued to monitor the instruments.

"Four hundred feet, down at nine . . ."

"Three hundred fifty feet, down at four . . ."

"Three hundred thirty feet, three and one half down . . . We're pegged on horizontal velocity."

But as Armstrong continued to exercise his "right to be wishy-washy," man's first real attempt to land on the moon suddenly looked like it might turn into a repeat of his disastrous practice session in the Lunar Landing Research Vehicle (LLRV) at Ellington Air Force Base in Houston the previous spring.

Although the Apollo 11 commander's impromptu pitch maneuver did slow the *Eagle*'s rate of descent, it actually increased her forward velocity by several more feet per second. It also extended her approach trajectory another two miles. The spacecraft was now at least four miles off course—just two miles from a mandatory abort—and she was rapidly running out of descent fuel.

Armstrong, who was ordinarily the most unflappable pilot in the entire astronaut corps, started to feel the heat. During lift-off from the Cape, his heart rate had never exceeded the relatively moderate level of 115 beats per minute. But following the mysterious computer malfunctions during his descent to the lunar surface, his pulse reached a peak of 156 beats per minute, more than fast enough for him to break into a cold sweat. Later, when informed of how high his pulse had elevated, Armstrong replied, "I'd be really disturbed with myself if it hadn't."

"I changed my mind a couple of times again, looking for a parking place," Armstrong recalls. "Something would look good, and then as we got closer, it really wasn't good. Finally we found an area ringed on one side by fairly good-sized craters and on the other side by a boulder field. It was not a particularly big area, only a couple of hundred square feet, about the size of a big house lot. But it looked satisfactory. And I was quite concerned about the fuel level. We had to get on the surface very soon or fire the ascent engine and abort."

At two hundred feet above the lunar surface, the same altitude at which Armstrong had bailed out of the LLRV before it crashed and burned on the tarmac at Ellington, he hooked back around in the direction from whence he came, hovered for a few seconds, and jerked the LEM back into the full upright position required for landing.

"Sixty seconds," warned CAPCOM Duke, indicating that only one minute's worth of descent fuel was left.

"Forty feet," Aldrin reported, "down two and one half . . ."

The *Eagle* drifted into the darkness of her own shadow, and quickly became engulfed in an enormous cloud of dust kicked up by the exhaust of her descent engine.

By this time, the men in the trench at Mission Control had started the countdown for an emergency abort. But the astronauts were already in the "dead man's zone." As Aldrin noted later, "In this zone, if anything had gone wrong—if, for example, the engine had failed—it would probably have been too late to do anything about it before we impacted with the moon."

Yet, Aldrin also claims, "I felt no apprehension at all during this short time. Rather, I felt a kind of arrogance—an arrogance inspired by knowing that so many people had worked on this landing, people possessing the greatest scientific talent in the world."

"Thirty seconds," CAPCOM Duke reported.

The *Eagle* was down to less than half a minute of descent fuel.

"Forward drift?" Armstrong wanted to know.

"Yes," Aldrin confirmed, adding a reassuring, "Okay . . ."

Duke started to call out another fuel consumption reading, but chief astronaut Deke Slayton, who had taken the adjacent seat at the CAPCOM's console, cut him off.

"Shut up, Charley," Slayton ordered, "and let 'em land."

Moments later, the *Eagle* plopped down in an unobstructed clearing amid an unnamed boulder field on the western edge of the Sea of Tranquility. According to Aldrin, "The landing was so smooth that I had to check the landing lights from the touchdown sensors to make sure the slight bump I felt was indeed the landing."

"CONTACT LIGHT!" Aldrin exhulted.

Then, in keeping with the official flight plan, he immediately proceeded to run through the first items on the post-touchdown checklist.

"Okay," Aldrin noted, "ENGINE STOP. ACA—out of DETENT."

"Out of DETENT," Armstrong confirmed.

"MODE CONTROL both AUTO," Aldrin continued. "DESCENT ENGINE COMMAND OVERRIDE—OFF. ENGINE ARM—OFF. 413 is in . . ."

"We copy you down, *Eagle,"* CAPCOM Duke reported a few seconds later.

There was a pregnant pause.

Buzz Aldrin would later claim that he had uttered the first words ever spoken from the surface of the moon when he cried, "CONTACT LIGHT!" But at approximately 3:18 P.M. on the afternoon of July 20, 1969, Neil Armstrong blurted what most historians would declare to be man's first intelligible message from another planet: "Houston, Tranquility Base here. The *Eagle* has landed."

The men in the trench at Mission Control spontaneously erupted into cheers and applause. So did the Dr. Thomas Paine, Wernher von Braun, former Mercury astronaut John Glenn, and all the other NASA brass and scientific wizards in the VIP room. Row after row of flight controllers, system engineers, supervisors, and official spectators broke out pocket-size American flags and fired up fat cigars, shaking hands, hugging, slapping backs, and beaming with triumph.

CAPCOM (DUKE): "Roger, Tranquility, we copy you on the ground. You've got a bunch of guys about to turn blue. We're breathing again. Thank you."

One of the most prominent nail-biters was flight operations director Chris Kraft, who had been agonizing over the *Eagle*'s descent and the fate of the two astronauts on board ever since the 1202 and 1201 computer alarms started flashing. As Kraft confided to Duke, "Boy, Charley, I thought we were gone when we had those things."

White-vested flight director Gene Kranz suddenly found himself speechless. He tried to regain his voice, couldn't, and tried again. Still nothing. Then he started banging his fist against the housing of his computer console like a frustrated child. Finally, his vocal cords responded.

"All right," Kranz croaked. "Everybody settle down, and let's get ready for a T-1 STAY/NO STAY."

Meanwhile, the first two men to land on the moon still did not know whether they could remain at Tranquility Base, or if they would have to take off immediately due to excessively soft surface conditions or some other unforeseen problem. Houston would give them the first official word at the one-minute mark with the T-1 STAY/NO STAY decision. A second STAY/NO STAY decision would come two minutes later. Then command module pilot Mike Collins would slip around the back side of the moon, and the next optimum abort/rendezvous time would not come until *Columbia* reemerged on the front side once again.

According to Aldrin, "We gave in to our excitement long enough to pat each other on the shoulder, then we plunged into frantic activity . . . If there was any emotional reaction to the lunar landing it was so quickly supressed that I have no recollection of it. We had so much to do and so little time in which to do it, that we no sooner landed than we were preparing to leave, in the event of an emergency. I'm surprised, in retrospect, that we even took time to slap each other on the shoulders."

Oddly enough, Armstrong later recounted a quite different version of what happened in the *Eagle* after touchdown, indicating that he and Aldrin silently shook hands instead of patting each other on the back. "If there was an emotional high point," he said at the tenth anniversary of the landing in 1979, "it was the point after touchdown when Buzz and I shook hands without saying a word. That still in my mind is the high point."

Even more surprising was the fact that the Apollo 11 astronauts had managed to land safely in the first place. Amid the joy of their triumphant success, virtually no one on Earth save the men in the trench at Mission Control appreciated just how close they had come to disaster. In the process of detouring over the crater full of boulders to find an alternate landing site, Armstrong had used up nearly all of the *Eagle*'s descent fuel. Though the LEM's ascent stage carried a separate fuel supply, it could only be tapped during a conventional lunar lift-off. Had Armstrong exhausted his descent fuel after entering the so-called "dead man's" zone about fifty feet above the surface where the spacecraft was too low to execute an abort maneuver, the *Eagle* would have crashed into the Sea of Tranquility. Even in one-sixth G, such a crash would have caused the LEM to tip over sideways, ripping her fuselage on the jagged edges of the surrounding bolders. Armstrong and Aldrin would either have been killed on impact or from exposure.

Exactly how much descent fuel remained when the *Eagle* touched down was never determined. According to Aldrin, the dials on his side of the spacecraft showed just ten seconds' worth of fuel left.

"My own instruments would have indicated less than thirty seconds, probably fifteen or twenty seconds," Armstrong later claimed at a post-mission press conference. He added with characteristic professional cool that a subsequent analysis by the ground suggested that the on-board fuel gauge was slightly off, and that the LEM "probably" had about "forty or forty-five" seconds of fuel left at touchdown.

"That sounds like a short time," Armstrong noted, "but it really is quite a lot."

Even so, as the astronauts waited to hear their first STAY on the lunar surface, Armstrong felt obligated to apologize for the extra time he took to find a "parking place." "Houston, that may have seemed like a very long final phase. The AUTO targeting [of the on-board computer guidance system] was taking us right into a football field–sized crater with a large number of big boulders and rocks for about one or two crater diameters around us."

Houston made clear that no apologizes were necessary.

CAPCOM (DUKE): "Roger, copy. Sounds good to us, Tranquility . . . Be advised that there are lots of smiling faces in this room and all over the world."

ALDRIN: "There are two of them up here."

Then Mike Collins interrupted for the first time since the astronauts began their powered descent.

"And don't forget one in the command module," he echoed.

Ironically, Collins was probably the only human being in the universe with a direct personal interest in the *Eagle*'s landing who did not watch live TV coverage of the great event. Still orbiting sixty miles above the lunar surface, he was also too high to observe the descent through the windows of the *Columbia.* He could try only to monitor the radio dialogue between the *Eagle* and the ground, and even he wound up with the mistaken impression that everything had gone fairly smoothly.

"It sure sounded great from up here," Collins declared. "You guys did a fantastic job."

"Thank you," Aldrin replied. "Just keep that orbiting base ready for us up there now."

"Will do," Collins promised, then pressed for more details on the descent and landing.

"It really was rough, Mike," Armstrong confided. "It was extremely rough over the targeted landing area—craters and large numbers of rocks that were larger than five or ten feet in size."

Collins cheerfully reminded his comrades of the test pilot's tried-and-true adage: "When in doubt," he chirped, "land long."

"So we did," Armstrong returned.

But because of being forced to "land long," the Apollo 11 astronauts could not determine exactly where they had touched down and neither could Mission Control.

ALDRIN: "The guys who said we wouldn't be able to tell precisely where we are, are the winners today. We were a little busy worrying about program alarms and things like that . . . I haven't been able to pick out the things on the horizon as a reference yet."

CAPCOM (DUKE): "Roger, Tranquility. No sweat. We'll figure it out . . ."

Houston soon discovered that figuring out the location of Tranquility Base was no easy matter. The extended trajectory of the *Eagle*'s descent suggested that the astronauts had landed about four miles beyond their original target on the eastern edge of the Sea of Tranquility. But the spacecraft's primary and abort guidance systems gave two conflicting sets of latitudinal and longitudinal coordinates, while the ground tracking stations came up with a third set.

Mission Control's special team of lunar mapping experts eventually came up with fourteen possible locations, all within a five-mile radius of each other. But no one could say for sure which of the fourteen was the actual landing site.

Shortly after Houston gave a second STAY approval, Collins went out of radio contact behind the moon, and Armstrong commenced man's first on-site description of the lunar surface as he could see it through the windows of the LEM.

"It's pretty much without color," he reported. "It's gray and it's a very white chalky gray as you look into the zero phase line . . . And it's considerably darker gray, more like ashen gray, as you look up ninety degrees to the sun . . . Some of the surface rocks in close here that have been fractured or disturbed by the rocket engine plume are coated with this light gray on the outside, but when they've been broken they display a dark, very dark, gray interior, and it looks like it could be country basalt . . ."

When the *Columbia* came back around the front side of the moon an hour later, Houston radioed the third STAY approval, and the two astronauts at Tranquility Base proceeded to power down the LEM's ascent systems in preparation for their walk on the lunar surface.

The Apollo 11 mission clock now showed a GET of over 105 hours 30 minutes, about 5 P.M. Houston time, which meant Armstrong and Aldrin had been on the moon for a little more than an hour and a half. According to the flight plan, they were supposed to take a four-hour nap before beginning their Extra-Vehicular Activity (EVA). But as Aldrin noted later, they were "too excited to sleep," and therefore requested permission to skip the rest period.

ARMSTRONG: "Our recommendation at this point is planning an EVA with your concurrence starting at about eight o'clock this evening, Houston time. That is about three hours from now."

CAPCOM (DUKE): "Stand by . . . We will support it. We're GO at that time . . . You guys are getting prime time TV there."

ARMSTRONG: "Hope that little TV set works, but we'll see."

As the two astronauts prepared to eat their last meal before venturing outside the LEM, Aldrin pulled out his personal preference kit (PPK), and unpacked a white wafer, a tiny gold chalice, and a thimble of red wine. Because of the controversy stirred by atheist Madalyn Murray O'Hair's lawsuit over the Apollo 8 crew's reading of Genesis, he had been admonished not to mention these sacred items or his intended use for them over the radio. Instead, he simply said, "This is the LM pilot. I'd like to take this opportunity to ask every person listening, whoever and wherever they may be, to pause for a moment and contemplate the events of the past few hours, and to give thanks in his or her own way."

Then Aldrin quietly celebrated the first Holy Communion on the moon.

"I would like to have observed just how the wine poured in that environment," he reported after his return to Earth, "but it wasn't pertinent at that particular time. It wasn't important how it got into the cup, it was only important to get it there. I offered some private prayers, but I found later that thoughts and feelings came into my memory rather than words. I was not so selfish as to include my family, or so spacious as to include the fate of the world. I was thinking more about our particular task, and the challenge and the opportunity that had been given me. It was my hope that people would keep the whole event in their minds and see, beyond minor details and technical achievements, a deeper meaning behind it all—a challenge, a quest, the human need to do these things."

Although the inscrutable Armstrong did not join in Aldrin's communion in the LEM, he observed the few moments of silent contemplation in his own personal fashion. He also found himself staring out the window at every available opportunity. He later shared his unique (and for him, rather surprising) observations with the crew's official biographers.

"The sky is black, you know," Armstrong would report. "It's a very dark sky. But it still seemed more like daylight than darkness as we looked out the window. It's a peculiar thing, but the surface looked very warm and inviting. It looked as if it would be a nice place to take a sunbath. It was the sort of situation in which you felt like going out there in nothing but a swimming suit to get a little sun. From the cockpit, the surface seemed to be tan. It's hard to account for that because later when I held this material in my hand, it wasn't tan at all. It was black, gray, and so on. It's some kind of lighting effect, but out the window the surface looks much more like light desert sand than black sand."

Of course, Armstrong and Aldrin had no intention of venturing onto the lunar surface in "nothing but a swimming suit." Both men knew their survival would depend on their Extravehicular Mobility Units (EMUs) which consisted of three protective layers—a liquid cooling garment, a pressure garment, and a thermal-micrometeoroid garment—along with an awkward backpack. The key element on each backpack was the portable life support system (PLSS), a 26-inch-by-17.8-inch-by-10.5-inch box containing three and a half hours of oxygen, a water cooling loop, and their primary and backup voice communications systems. Manufactured by the Hamilton Standard division of United Aircraft, the PLSS was the product of over 2.5 million man hours of design and production work, but it had been space-tested only once before, during Apollo 9 astronaut Rusty Schweikart's spacewalk in Earth orbit.

The backpacks also included a forty-pound box called the Oxygen Purge System (OPS), which fit on top of the portable life support system. The OPS carried a thirty-minute backup oxygen supply that could also be used in the event the astronauts had to make an EVA after they lifted off from the lunar surface and attempted to rendezvous with the mothership. The third major element in each backpack was a remote control unit worn on the chest of the pressure suit that housed a water pump, an oxygen fan, a communications selector switch and volume control, and a camera mount. On Earth, the EMU assembly weighed more than 180 pounds, which made walking around in it almost impossible. But thanks the moon's one-sixth G, its effective weight on the lunar surface would be a far more manageable thirty pounds.

Armstrong and Aldrin expected their EVA preparations to take about two hours, but they ended up taking twice that long, partly because the exhaust gases from the backpacks compounded the difficulty of depressurizing the cabin of the lunar module. As Armstrong noted afterward: "That's a part of the exercise we never get to duplicate in any tests or simulations . . . You have to be in a vacuum to do that. We had trained for this aspect in a [simulated vacuum] chamber and under somewhat different conditions."

While the astronauts struggled with their cabin depressurization problems, command module pilot Mike Collins tried in vain to fix the exact location of Tranquility Base with the help of his navigational sextant. The ground kept sending up new sets of latitudinal and longitudinal coordinates each time the *Columbia* came around the front side of the moon. But as it turned out, Collins was being told to look in all the wrong places. At 8 P.M. Houston time, nearly five hours after touchdown, newly arrived CAPCOM Bruce McCandless initiated the same disappointing exchange that had been repeated all afternoon.

CAPCOM (MCCANDLESS): *"Columbia,* this is Houston . . . Were you successful in spotting the LM on that pass? Over."

COLLINS: "Negative. I checked both locations and no joy."

At approximately 9:45 P.M. Houston time, nearly two hours later than anticipated, the *Eagle* finally bled off enough cabin pressure for the astronauts to open the hatch.

With Aldrin's help, Armstrong methodically climbed out the hatch of the LEM, and stepped onto the narrow metal porch atop the exit ladder. There were now just nine rungs between him and the lunar surface. But before descending further, he had to hook up the lunar equipment conveyor cord (LEC) and pull open the Modular Equipment Stowage Assembly (MESA) attached to the fuselage of the spacecraft.

Houston then reminded him to activate the remote controlled TV camera mounted inside the MESA so the worldwide audience back on Earth could watch him take man's first step on another planet.

ALDRIN: "Roger. TV circuit breaker's in. LMP reads loud and clear."

CAPCOM (MCCANDLESS): "And we're getting a picture on the TV."

The resolution of the TV pictures was shadowy to say the least, but the bulky white silhouette backing off the porch of the LEM was clearly visible against the black void behind it.

CAPCOM (MCCANDLESS): "Okay, Neil, we can see you coming down the ladder now."

Armstrong backed all the way to the bottom of the ladder, then hopped down to the round metal landing pad. Although the rungs of the ladder were spaced only twelve inches apart, the distance from the lowest bar to the landing pad was about three and a half feet. Just to make sure there would be no problem when it came time to return to the LEM, he jumped back up from the pad to the ninth rung of the exit ladder, and reported the results to his comrade inside the cabin:

"Okay, I just checked—getting back up to that first step, Buzz, it's not even collapsed too far, but it's adequate to get back up . . . It takes a good little jump."

Then Armstrong hopped back down to the landing pad.

"I'm at the foot of the ladder . . . The LM footpads are only depressed in the surface about one or two inches, although the surface appears to be very, very fine grained as you get close to it. It's almost like a powder. Now and then, it's very fine . . . I'm going to step off the LM now . . ."

Buzz Aldrin would later recount the mystery and controversy surrounding how Armstrong chose his first words from the lunar surface:

"There was, in fact, widespread discussion prior to our flight about what he should say. When it reached epidemic proportions, [public affairs officer] Julian Scheer wrote a terse memo to the NASA heads saying in effect: Did Queen Isabella tell Christopher Columbus what to say? There were also rumors—unconfirmed—that Neil was advised what to say by Simon Bourgin, a United States Information Agency official who was in frequent contact with the astronauts and . . . was also the rumored source of [Apollo 8 commander] Frank Borman's reading of Genesis. For his part, Neil declined to comment on any of these rumors to anyone, including Mike and myself."

Nor did Armstrong tell his fellow crewmen what he planned to say when they asked him about the matter during the voyage from Earth.

"I had thought about what I was going to say largely because so many people had asked me to think about it," Armstrong claimed later. "I thought about that a little bit on the way to the moon . . . [but] I hadn't actually decided what I wanted to say until just before we went out [onto the lunar surface]."

At 9:56 P.M. Houston time, the Apollo 11 commander lifted his Beta cloth-covered right boot off the landing pad, planted it in the colorless dust of the Sea of Tranquility, and blurted:

"THAT'S ONE SMALL STEP FOR MAN, ONE GIANT LEAP FOR MANKIND."

Although neither Armstrong nor anyone else realized it at the time, he had just blown what was automatically destined to become one of history's most famous lines. As critics back on Earth would soon point out, the statement, "That's one small step for man, one giant leap for mankind" sounded like a tautology, redundant to the point of being meaningless, as well as rather sexist. What Armstrong reportedly meant to say was, "That's one small step for *a* man"—that "man" being Armstrong himself—so as to make it clear that he was merely the lucky individual chosen to make "one giant leap" on behalf of the entire human race. But in the glorious heat of the moment, he inexplicably failed to insert the crucial article "a" before the word "man."

NASA spokesmen would disingenuously deny and/or try to explain away Armstrong's soon to be immortalized miscue. The most widely spread official line maintained that he had actually said "That's one small step for *a* man," but that radio transmission interference had made the "a" inaudible to his listening audience 240,000 miles away. Armstrong himself was more candid. Shortly after emerging from post-splashdown quarantine in Houston, he confessed: "I thought it [the article 'a'] had been included. Although it is technically possible that the VOX didn't pick it up and transmit it, my listening to the recording indicates it is more likely that it was just omitted."

Armstrong's second, third, and fourth steps were a series of short backward shuffles. Still gripping the railing of the exit ladder, he lifted his left boot off the landing pad, and planted it next to his right boot. Then, after pausing to steady himself, he started kicking at the lunar topsoil to test its depth and texture.

"As the surface is fine and powdery, I can . . . I can pick it up loosely with my toe . . . It does adhere in fine layers like powdered charcoal to the sole and sides of my boots . . . I can only go in a small fraction of an inch—maybe an eighth of an inch—but I can see the footprints of my boots and the treads in the fine sandy particles."

Armstrong let go of the ladder, slowly turned around, and started to take a few more carefully chosen small steps away from the LEM. He was both pleased and relieved to discover that contrary to the dire predictions of some NASA scientists and official worrywarts, he did not sink into a lunar quicksand or fall over backward from the weight of his portable life support system.

"There seems to be no difficulty in moving around as we expected . . . It's perhaps even easier than the simulations of one-sixth G that we performed in the simulations on the ground . . . It's actually no trouble to walk around."

Armstrong then made another reassuring—but also rather puzzling—observation. Although the *Eagle* had landed in between a jagged boulder field and a still unidentified crater formation, she appeared to be resting in an almost perfect upright position and in no danger of tipping over. And despite the fact that the LEM's exhaust had kicked up a thick cloud of dust during the powered descent, it also appeared that the impact of the touchdown had barely scratched the "fine and powdery" surface of the landing site.

"The descent engine did not leave a crater of any size. There's about one foot of clearance on the ground [beneath the exhaust nozzle]. We're essentially on a very level place here."

Armstrong asked Aldrin to lower the 70 mm Hasselblad still camera down the lunar equipment conveyor line so he could record what he saw for posterity.

"Looking up at the LM . . . I am standing directly in the shadow now looking up at Buzz in the window . . . And I can see everything quite clearly. The light is sufficiently bright—backlighted into the front of the LM—that everything is quite clearly visible."

Armstrong hooked the camera to the remote control unit on the chest of his pressure suit and moved several more feet away from the lunar module.

"I'll step out and take some of my first pictures here."

Houston reminded him to take a "contingency sample" of lunar rocks as soon as possible in case he had to terminate his EVA prematurely.

"Rog. I'm going to get to that just as soon as I finish these picture series."

Armstrong then retreated back to the LM, reached into the MESA cabinet attached to the near strut, and removed a long metal tube with a scoop on the end. Due to the weight of his backpack and the inflexibility of his "moon cocoon," he could not simply bend over and pick up a handful of rocks with his gloves because if he happened to fall down he might not be able to get back up. But he quickly discovered that scooping the contingency sample from the lunar surface into the pouch of his pressure suit was much more difficult than expected.

ARMSTRONG: "This is very interesting. It's a very soft surface, but here and there where I plug with the contingency sample collector, I run into a very hard surface . . . I'll try to get a rock in here . . . Here's a couple . . ."

ALDRIN: "That looks beautiful from here, Neil."

ARMSTRONG: "It has a stark beauty all its own. It's like much of the high desert of the United States. It's different, but it's very pretty out here."

Aldrin continued to watch him struggle with the contingency sample collection procedure from inside the LEM.

ARMSTRONG: "You can really throw things a long way up here . . . That pocket open, Buzz?"

ALDRIN: "Yes, it is, but it's not up against your suit, though . . . Hit it [the contingency sample] back once more . . . More toward the inside . . . Okay, that's good."

ARMSTRONG: "That in the pocket?"

ALDRIN: "Yes, push down . . . Got it? . . . No, it's not all the way in. Push it . . . There you go."

ARMSTRONG: "Contingency sample is in the pocket . . . Are you getting a TV picture now, Houston?"

CAPCOM (MCCANDLESS): "Neil, yes, we are getting a TV picture . . . You're not in it at the present time. We can see the bag on the LEC being moved by Buzz, though . . . Here you come into our field of view."

Armstrong had now been alone on the lunar surface for almost fifteen minutes, and it was time for Aldrin to climb out of the LEM and become the second man to walk on the moon.

ARMSTRONG: "Okay, you saw what difficulties I was having [getting through the hatch]. I'll try to watch your PLSS [Portable Life Support System] from underneath here . . . Okay, your PLSS is—looks like it's clearing okay . . . The shoes are about to come over the sill . . . Okay, now drop your PLSS down . . . There you go . . . About an inch of clearance on top of your PLSS . . ."

ALDRIN: "Okay, you need a little bit of arching of the back to come down. . . . How far are my feet from the . . ."

ARMSTRONG: "Okay, you're right on the edge of the porch . . . Looks good."

ALDRIN: "Now I want to back up and partially close the hatch, making sure not to lock it on my way out."

ARMSTRONG: "A particularly good thought."

ALDRIN: "That's our home for the next couple of hours and I want to take good care of it."

Though Aldrin was still miffed about not getting the chance to be the first man to walk on the moon, he later reported that he felt "buoyant" and "full of goose pimples" as he hopped the rest of the way down the exit ladder and made his own "giant leap" to the lunar surface. But his first words hinted at his deep-seated ambivalence as he surveyed the "beautiful view" from Tranquility Base.

ARMSTRONG: "Isn't that something? Magnificent sight out here."

ALDRIN: "Magnificent desolation."

"I immediately looked down at my feet," Aldrin would report in his post-mission memoir, "and became intrigued with the peculiar properties of the lunar dust. If one kicks sand on a beach, it scatters in numerous directions with some grains traveling farther than others. On the moon the dust travels exactly and precisely as it goes in various directions, and every grain of it lands very nearly the same distance away."

Aldrin then tried to exorcise his preoccupation with being "only the second man" to walk on the moon with a bizarre act of physical catharsis. As he recalls:

"The second thing I did was record my own first on the moon, a first that became known only to a select few. My kidneys, which have never been of the strongest, sent me a message of distress. Neil might have been the first man to step on the moon, but I was the first to pee in his pants on the moon. I was, of course, linked up with the urine collection device, but it was a unique feeling. The whole world was watching, but I was the only one who knew what they were really witnessing."

After secretly wetting his pants, Aldrin commenced a running commentary on the joys and pains of trying to walk on the moon. As he ventured away from the LEM, he reported:

"I say, the rocks are rather slippery . . . Very powdery surface. When the sun hits, the powder fills up all the very little fine porouses. My boot tends to slide over it rather easily . . . About to lose my balance in one direction and recovery is quite natural and very easy. And moving arms around . . . doesn't lift your feet off the surface. We're not quite that light-footed . . . Got to be careful that you are leaning in the direction you want to go, otherwise you . . . (static) . . . In other words, you have to cross your foot over to stay underneath where your center of mass is . . ."

Aldrin then turned to his official first priority on the moon—helping

Armstrong collect additional rock specimens. According to the checklist, the astronauts were supposed to gather about fifty pounds of geologically valuable material from the lunar surface. Their "bulk sample" container consisted of a single bag designed to hold about twenty pounds of lunar topsoil. Their "core sample" container was equipped with fifteen smaller bags that they intended to fill with rocks from deeper subsurface layers.

But like most other Apollo astronauts, the first two men to walk on the moon regarded their scientifically important lunar prospecting as a tedious and annoying act of forced labor. When asked at a prelaunch press conference what sorts of unique specimens might be bagged on the lunar surface, Aldrin had sarcastically predicted they would find some "purple rocks." Shortly after setting foot on Tranquility Base, he discovered to his great delight and amusement that many of the local stones and pebbles actually did look purple through the tinted visors of his space helmet.

ALDRIN: "Say, Neil, didn't I say we might see some purple rocks?"

ARMSTRONG: "Find the purple rocks?"

ALDRIN: "Yes, they are small, sparkly . . . I would make a first guess, some sort of biotite [commonly known as mica]. We'll leave that to the lunar analysts . . ."

Before anyone at Mission Control could challenge the ruse, Armstrong momentarily interrupted the rock-collecting project to take care of some equally important ceremonial duties. The first item on the agenda was unveiling the commemorative plaque attached to one of the struts on the lunar module. For the benefit of those who had not seen the plaque, he described its design and read the brief inscription:

"First, there's two hemispheres . . . showing each of the two hemispheres of the earth . . . Underneath it says, 'Here Man from the planet Earth first set foot upon the Moon, July 1969 A.D. We came in peace for all mankind.' It has the crew members' signatures and the signature of the president of the United States."

A few moments later, the two astronauts attempted to plant the Stars and Stripes on the moon. But they had never practiced the flag ceremony in training, and this seemingly simple task of jamming the pole into the lunar surface proved to be much tougher than either had anticipated.

"It took both of us to set it up, and it was nearly a disaster," Aldrin recalled afterward. "Public Relations obviously needs practice just as everything else does. A small telescoping arm was attached to the flagpole to keep the flag extended and perpendicular in the still lunar atmosphere. We locked the arm in its ninety-degree position, but as hard as we tried, the telescope wouldn't fully extend. Thus, the flag, which should have been flat, had its own unique permanent wave. Then, to our dismay, the staff of the pole at first wouldn't go far enough into the lunar surface to support itself in an upright position. After much struggling, we finally coaxed it to remain upright, but in a most precarious position. I dreaded the possibility of the American flag collapsing into the lunar dust in front of the television camera."

Armstrong, who still had the Hasselblad camera mounted on the chest of his pressure suit, took a picture of Aldrin saluting the flag. Then, just as the astronauts were about to switch positions for a second flag salute picture, the ground announced that President Richard Nixon wanted to speak to them via a special communications link between Washington and Houston.

ARMSTRONG: "That would be an honor."

CAPCOM: "Go ahead, Mr. President. This is Houston. Over."

NIXON: "Hello, Neil and Buzz. I am talking to you by telephone from the Oval Room at the White House, and this certainly has to be the most historic telephone call ever made. I just can't tell you how proud we all are of what you have done. For every American, this has to be the proudest day of our lives. And for people all over the world, I am sure that they, too, join with Americans in recognizing what an immense feat this is. Because of what you have done, the heavens have become a part of man's world . . . And as you talk to us from the Sea of Tranquility, it inspires us to redouble our efforts to bring peace and tranquility to Earth. For one priceless moment in the whole history of man, all the people on this earth are truly one—one in their pride in what you have done. And one in their prayers that you will return safely to Earth."

ARMSTRONG: "Thank you, Mr. President. It's a great honor and privilege for us to be here representing not only the United States but men of peace of all nations—and with interest and a curiosity, and men with a vision for the future. It's an honor to be able to participate here today."

NIXON: "And thank you very much. And I look forward—all of us look forward—to seeing you on the [recovery ship] *Hornet* next Thursday."

ARMSTRONG: "Thank you."

Aldrin later claimed that Nixon's phone call caught him off guard. Although the typically tight-lipped Armstrong had known about the possibility of a presidential teleconference well in advance of lift-off, he had not informed his fellow crewmen of the planned p.r. gambit. Now even more miffed about being left in the dark and excluded from this historic conversation, the disgruntled lunar module pilot couldn't resist putting in his two cents.

ALDRIN: "I look forward to that very much, sir."

Before indulging in further ceremonial ado, Armstrong and Aldrin tackled the second most important scientific chore on their lunar checklist—setting up the Early Apollo Scientific Experiments Package (EASEP). The three main components were: a sixteen-pound seismic meter designed to measure "moon quakes" and/or any other significant geological movements; a honeycombed laser reflector, which would enable astronomers to measure the exact distance between the earth and the moon; and a "solar wind detector," which consisted of a rectangular sheet of aluminum foil designed to trap ambient gases emitted from the sun.

But as the astronauts assembled the EASEP equipment, they found themselves becoming ever more fascinated—and more than a little unnerved—by the alien properties of the lunar environment. In keeping with the alleged NASA mandate to confirm that the moon was indeed a "friendly" place, Armstrong, who would claim that the lunar surface looked like "a nice place to take a sunbath," later elaborated on the pleasantly surprising ease of movement and relative stability of one-sixth G: "After landing, we felt very comfortable in the lunar gravity. It was, in fact, in our view preferable both to weightlessness and Earth gravity."

"There's one sight I'll never forget," Armstrong declared years later. "As I stood on the Sea of Tranquility and looked up at the earth, my impression was of the importance of that small, fragile, remote blue planet. People everywhere have, by television and photographs, shared that perspective and shared our concern for the security of our globe."

The Apollo 11 commander would add that his only regret was not being able to spend more time on the moon: "We had the problem of the five-year-old boy in the candy store. There are just too many interesting things to do."

Aldrin, however, would later confess to experiencing some diametrically opposite impressions. As he reported in his post-mission memoir: "I quickly discovered that I felt balanced—comfortably upright—only when I was tilted slightly forward. I also felt a bit disoriented: on the earth when one looks at the horizon, it appears flat; on the moon, so much smaller than the earth and quite without high terrain (at least in the Sea of Tranquility), the horizon in all directions visibly curved down away from us."

Aldrin added that attempting to run or jog on the moon was also a disorienting experience that required special precautions to avert a disastrous accident: "The exercise gave me an odd sensation and looked even more odd when I later saw films of it. With the bulky suits on, we appeared to be moving in slow motion. I noticed immediately that my inertia seemed much greater on the moon . . . Earthbound, I would have stopped my run in just one step—an abrupt halt. I immediately sensed that if I did this on the moon, I'd be face down in the lunar dust."

Shortly after the stroke of midnight in Houston, the astronauts were ordered to wrap up their EASEP assignments and return to the LEM. Armstrong had been on the lunar surface about two hours and fourteen minutes. Aldrin's total EVA time was approximately twenty minutes less. Both astronauts still had a good two hours of oxygen and water left in the PLSS backpacks, but they had also been working for over eighteen hours since their morning wake-up call in lunar orbit.

Houston notified their solitary colleague orbiting in the mothership of their safe return from the lunar surface.

CAPCOM (MCCANDLESS): *"Columbia, Columbia,* this is Houston. Over."

COLLINS: "Roger . . . How do you read?"

CAPCOM (MCCANDLESS): Roger, *Columbia.* This is Houston. We're reading you loud and clear . . . The crew of Tranquility Base is back inside their base, repressurized, and they're in the process of doffing the PLSSs. Everything went beautifully. Over."

COLLINS: "Hallelujah!"

Armstrong and Aldrin later reported that the first thing they noticed when they got back inside the *Eagle* and removed their helmets was a peculiar odor. According to Aldrin, "there was a distinct smell to the lunar material—pungent, like gunpowder or spent cap pistols. We carted a fair amount of lunar dust back inside the vehicle with us, either on our suits and boots or on the conveyor system we used to get boxes and equipment back inside."

The astronauts had to spend the next three hours inside the LEM attending to housekeeping chores and answering various technical questions about their problems during the powered descent. Most of the questions were posed by Owen Garriott, the CAPCOM of the Maroon Team who came on to relieve Black Team counterpart Bruce McCandless. It was already past 3 A.M. Houston time on the morning of Monday, July 21, 1969, when the ground made one last vain attempt to determine the exact location of Tranquility Base.

CAPCOM (GARRIOTT): "You commented, Neil, that on your approach to the landing spot you passed over a 'football field–sized crater' containing rather large blocks of rock perhaps ten to fifteen feet in size. Can you estimate the distance to it from your present position? Over."

ARMSTRONG: "I thought we'd be close enough so that when we got outside we could see its rim back there, but I couldn't. But I don't think we're more than half a mile beyond it. That is, half a mile west of it."

CAPCOM (GARRIOTT): "Okay, well . . . unless you have something else, that will be all from us for the evening. Over."

But after a brief interval of radio static, Houston came back with one more question.

CAPCOM (GARRIOTT): "A few more verifications here. Can you—will you verify that the disc with the messages was placed on the surface as planned, and also that the items that are listed on the flight plan, all of those listed there, were jettisoned? Over."

The disc in question was inscribed with messages from President Nixon and various heads of state from the countries of the Free World. Included in the same commemorative packet were the patch selected by the late crew of Apollo 1, two medallions memorializing the Soviet cosmonauts who had lost their lives in space, and a miniature gold olive branch identical to that carried by the eagle on the Apollo 11 patch. The items that the astronauts were supposed to have jettisoned upon returning to the LEM included their PLSS backpacks and the boots they had worn on the moon.

ARMSTRONG: "All that's verified."

CAPCOM (GARRIOTT): "Roger. Thank you, and I hope this will be a final good night."

Out of sight but by no means out of mind, command module pilot Mike Collins could not decide how he should spend his last night alone in lunar orbit. He would later report that following the undocking and separation maneuvers, "I looked forward to a chance to relax for a little while and look out the window—get some assessment of what it's all about."

But when the crew of Tranquility Base finally signed off for the night, Collins found himself torn between his own need to rest up for the critical rendezvous maneuver in lunar orbit later that same morning and his responsibility to watch over the mothership that all three astronauts depended on to fly safely back to Earth.

"There is a problem," Collins had told the crew's official biographers prior to lift-off from the Cape, "in that when you go around the back side [of the moon] there is a period of forty-five minutes or fifty minutes when the ground is unable to give you any telemetry, so you're completely in the dark, literally and figuratively, and with nobody awake to watch the candy store. This sort of rubs you the wrong way, and I'm not sure whether I'll come to some mental accommodation with that and sleep like a log or whether I won't . . ."

Collins eventually reached a "mental accommodation" by taking intermittent naps during each pass across the front side of the moon. But his exhausted colleagues on the lunar surface had no such luck.

Aldrin tried to curl up on the floor of the LEM, only to discover that he was too "elated" and also too "cold" to sleep during the astronauts' scheduled seven-hour rest period before lunar lift-off. As he reported afterward, "The thing which really kept us awake was the temperature. It was very chilly in there. After about three hours it became unbearable. We had the liquid cooling system in operation in our suits, of course, and we tried to get comfortable by turning the water circulation down to minimum. That didn't help much. We turned the temperature control on our oxygen system up to maximum. That didn't have much effect either. We could have raised the window shades and let the light in to warm us, but that would have destroyed any remaining possibility of sleeping."

Armstrong suffered an even more fitful sleep. Since there was not enough room for both astronauts to lie on the floor of the LEM, he tried to rest in a hammock slung between the ascent engine cover and an adjacent support bar. But his gerry-rigged berth forced him to assume an awkward semi-vertical posture, which might have been commodious enough in zero G but proved to be decidedly discomforting in the one-sixth gravity of the moon.

The Apollo 11 commander's insomnia was further exacerbated by the fact that his head was facing up toward the unshuttered portal at the top of the cabin. As fate and historical circumstances would have it, the earth happened to be suspended directly above the window, staring down at him like "a big blue eyeball."

And while Neil Alden Armstrong, the first man to walk on the moon, tried in vain to snatch a few sorely needed hours of uninterrupted rest, his unique view of the fragile but bountiful planet he called home served as a real and symbolic reminder of the extraordinary feat he had just accomplished for all mankind. As he would later declare in a typically succinct and matter-of-fact summation: "It's a beginning of a new era."

10

SNOOPY AND THE SURVEYOR

Man's Second Moon Landing

Apollo 12: November 18–20, 1969

Nobody ever remembers what the second person to do something does.
APOLLO 12 COMMANDER CHARLES "PETE" CONRAD, JR.

On July 21, 1969, *The New York Times* announced Apollo 11's triumphant success with the biggest headline in the paper's entire history of publication. The one-inch-high block letters bannered across the front page summarized the most extraordinary technological feat of the twentieth century in four short, simple, but unforgettable words:

"MEN WALK ON MOON."

Man's first lunar landing was one of the few epochal events in human history that personally affected the witnesses as much if not more than the actual participants. Regardless of whether they reacted with speechless wonder or passionate outrage, the estimated one billion people in the worldwide TV and radio audience would always remember exactly where they were and how they felt when Neil Armstrong made his "giant leap for mankind."

But as Apollo 12's Charles "Pete" Conrad, Jr., wryly observed in a recent interview: "Nobody ever remembers what the second person to do something does."

Conrad speaks from unique personal experience. In November of 1969, less than four months after Apollo 11's heroic return to Earth, he commanded man's second lunar landing mission.

The Apollo 12 astronauts had, by definition, the most anticlimactic assignment in the annals of modern space exploration. Barring the unlikely prospect of encountering intelligent life on or en route to the moon, nothing they hoped to accomplish could possibly equal—much less upstage—the immortal deeds of their predecessors.

But contrary to widespread public misconception, Apollo 12 was not merely a replay of Apollo 11. Neither were the five subsequent lunar landing missions, especially not unlucky Apollo 13. Each trio of astronauts was targeted for a different region of the lunar surface with different scientific objectives in mind and with different operational demands to meet. And although the uninformed public grew blasé about space travel, each crew faced incrementally greater dangers, for the statistical probability of catastrophe increased in the wake of every success.

Apollo 12 commander Pete Conrad and lunar module pilot Al Bean were bound for the Ocean of Storms ("Oceanus Procellarum"), a rolling crater field 995 miles due west of the Sea of Tranquility on the far "left" of the moon's front side. One of their primary objectives was to retrieve a piece of the Surveyor III, an unmanned photo reconnaissance probe that soft-landed in the area in April of 1967. But in order to do so, they would have to make man's first pinpoint lunar landing, a far trickier and more hazardous maneuver than Apollo 11's hit or miss landing procedure.

Last-minute computer program overloads and guidance system malfunctions had caused Armstrong and Aldrin to overshoot their targeted landing site by some four miles, nearly dashing them against a boulder field. Conrad and Bean had to touch down within walking distance of the Surveyor, which meant they could not err by more than a quarter of a mile. And simply locating the probe even with the help of voluminous photo reconnaissance maps and the computers at Mission Control was like finding a needle in a haystack 240,000 miles away. As Conrad recalled: "I wasn't worried about landing—I was worried about landing in the right place."

At one of his last preflight press conferences, the Apollo 12 commander reckoned he had no better than a 50-50 chance of landing close enough to the Surveyor, assuming the probe was actually where the ground tracking team calculated it to be. But in keeping with the astronaut code, Conrad did not mention that he and Bean might have less than a 50-50 chance of making it back alive if they had to abort their landing attempt at the last minute.

In hopes of getting a better look at the Ocean of Storms before attempting a touchdown, Conrad and Bean planned to pitch their LM into an upward attitude at about seven thousand feet above the lunar surface, thereby abandoning their "free return" trajectory. Had Armstrong and Aldrin aborted at the last minute, their on-board guidance system would have automatically calculated the proper trajectory for returning to lunar orbit. But the Apollo 12 astronauts would have to steer themselves back into orbit with a complex series of manually controlled engine burns in the event of an abort, and hope that an inadvertent human error did not hurl them into the far reaches of trans-lunar space.

Even if Conrad and Bean did land in the right place, they would face a whole new set of dangers their predecessors did not risk. The flight plan called for them to spend thirty-two hours on the lunar surface, ten hours more than the Apollo 11 crew, and to take two moon walks instead of one. They were also expected to do roughly twice as much work, which in turn required them to pack and unpack twice as much equipment. And each additional task, each extra minute of EVA exposure, increased the chances for a fatal accident.

The unique hazards of the Apollo 12 mission were compounded by an item not included in Apollo 11's equipment kit: a nuclear generator. The various scientific monitoring devices Armstrong and Aldrin deployed on the moon had shut down a few days after their departure. NASA engineers hoped that Apollo 12's portable nuke plant would power their ALSEP (Apollo Lunar Scientific Experiments Package) devices for a full year. But lunar module pilot Al Bean had to activate the generator by inserting a tube of plutonium 238. Any slipup in this delicate procedure could expose both astronauts to radiation and/or heat of up to 900 degrees Fahrenheit.

Despite the redoubled burdens of being number two, the Apollo 12 astronauts always maintained a refreshingly irreverent sense of humor, in part because they were the most congenial crew that ever launched for the moon. Because all three were navy men, they patriotically dubbed their command module *Yankee Clipper* and their LM *Intrepid.* But in keeping with their self-effacing attitude, they chose the cartoon character Snoopy as their quality control symbol, and ate their final prelaunch breakfast in the company of a stuffed gorilla, symbolically mocking the "astronaut as trained ape" stereotype.

Conrad, who at five feet six inches was the shortest of the Apollo astronauts, took a similarly impish delight in composing the first words he would say if and when he first set foot on the moon. Like many other skeptics in the general public, one of Conrad's close friends believed Neil Armstrong's immortal first words had been written by someone else under orders from NASA officials. Conrad insisted that was not the case, and offered to put money on what would come out of his own mouth.

"You can say anything you want when you get to the bottom of the ladder," Conrad declared. "I'll bet you five hundred bucks that when I get to the bottom of the ladder, I'm going to say, 'It may have been a small step for Neil, but it's a big step for a little fellow like me.' "

"No way you're going to do that," his friend replied, and agreed to take Conrad's bet.

"If you were going to make a movie about an astronaut," former Apollo 7 crewman Walt Cunningham wrote later, "Pete is the guy you would pick to play the lead."

Already semi-infamous for racing his Corvette on the Houston freeways, the thirty-nine-year-old Conrad was an amiably balding Princeton graduate and former navy flight instructor who had flown on both Gemini 5 and Gemini 11. Many of his peers had considered him the most likely candidate for man's first lunar landing attempt, in part because he weighed a good fifteen pounds less than average and would therefore lighten the load on the LM. But as it turned out, fate and NASA's decision to accelerate the timetable conspired to give Apollo 11 first crack at making the "giant leap" to the surface of the moon.

Apollo 12 command module pilot Richard F. "Dick" Gordon, Jr., had been one of Conrad's star pupils back at Naval Test Pilot School. He had also been Conrad's co-pilot on Gemini 11. Though one year older and nearly as short, Gordon had a full head of brunette hair, an even more outgoing macho personality, and, according to Apollo 7's Cunningham, "a raw sex appeal that overshadowed most of the rest of us." He also had a chip on his shoulder that would never get the chance to heal or be knocked off.

Gordon realized the honor implicit in being chosen as a command module pilot. Since the CMP had to fly the mothership all by himself in lunar orbit, space agency officials decreed that only the best pilots in the astronaut corps could qualify. But the CMP was the only member of the crew who would not get the chance to land on the moon. And though Gordon genuinely prided himself in the importance of his supporting role, he longed to play the lead so he, too, could explore the lunar surface.

"At the time," he recalls, "it looked like I was going to command my own mission somewhere down the line . . . But the [Apollo] program ended too damn soon, and I never got the chance."

Apollo 12 lunar module pilot Al Bean was the youngest and most artistic member of the crew. He was also the only rookie. Balding but still physically fit, the thirty-seven-year-old former college gymnast had been born in a small town near Ft. Worth and educated at the University of Texas. Though most of his peers had pursued advanced engineering degrees, he had attended various art classes, and secretly dreamed of becoming a great painter.

But even though Bean happened to be one of the most creative and sensitive of the twenty-four men who flew to the moon, he was no wimp.

Like fellow crewman and former classmate Dick Gordon, he had also been a pupil of Pete Conrad's at Naval Test Pilot School. And despite the fact that he had never flown in space, he pressed personal recommendations with the authority of an old hand. As Conrad once chided: "When you think you have the door closed on one of Bean's ideas, you had better keep running around to all the side doors to make sure that he's not slipping it under one of them."

Apollo 12's nearly catastrophic lift-off on the stormy morning of November 14, 1969, strengthened the bond between the three crewmen. Seconds after clearing the launch tower, their Saturn V booster was struck by lightning. Although the lightning bolts temporarily knocked out the spacecraft's primary electrical systems, Bean knew and threw the right switches to activate the backup systems. His grace under pressure boosted the crew's morale to a level of invincibility. Even so, their brush with disaster had an ever more sobering effect as the astronauts got closer to the moon.

"Before we went down to land," Bean recalls, "I was thinking, 'When I get back from here—if I get back—I'm really gonna try to live my life like I want to. I'm taking a big risk in going to the moon, and life is short.' "

Apollo 12's moment of truth came in the wee morning hours of November 18, 1969, when Conrad and Bean undocked from the mothership and began their descent to the Ocean of Storms. The separation maneuver was particularly poignant for CMP Dick Gordon, who felt the excruciating ambivalence of having to remain in the command module. As he recalled in a recent interview:

"I wanted to go with them so bad I could taste it. As far as I was concerned, that was what it was all about. Not only going to the moon, but going down on the surface and walking. Matter of fact, Pete [Conrad] made the comment that he wished the damn thing [the LEM] could hold three people so we could all three go down and do that."

At the same time, Gordon claims he was "glad" when his comrades departed for the lunar surface: "I was very delighted to be alone in the command module. You rationalize your own position, you rationalize your own importance so that it really doesn't bother you. But from an egotistical, selfish, or professional point of view, I would have rather gone and landed on the moon, there's no question about that."

Shortly before 1 A.M., Conrad and Bean rounded the back side of the moon, and made an almost perfect Descent Orbit Insertion (DOI) burn that lowered the spacecraft to fifty thousand feet. A few minutes later, the two astronauts reestablished radio contact with Houston on the front side, and began their powered descent toward the Ocean of Storms. But according to Bean, the only way he could cope with the overpowering reality of the descent was by reverting to make-believe.

"I found it was most convenient not to look out the window," he reports, "because when I did I was kind of amazed and it was also slightly frightening. I found if I looked inside [the cockpit] and concentrated on the displays and the computer, it was more like the simulator and I could perform in a more normal mode. When I looked out for a second and saw what was happening, that was about all I could stand, and I'd look back in and try to concentrate on the job again."

At seven thousand feet, Conrad pitched the *Intrepid* into the upward attitude that would improve his view of the lunar surface. At five hundred feet, he took over manual control, gripping throttle and thruster controls in either hand while scanning the horizon for the so-called Snowman craters near which the Surveyor was supposed to have landed. That's when the astronauts realized they were in trouble.

Conrad and Bean had laboriously studied all the available photographs of the Ocean of Storms, many of which had been relayed back to Earth by the Surveyor III. They were also carrying a thick atlas of detailed lunar maps. But the totally alien character of the lunar landscape and the thick dust clouds stirred up by the descent engine of the LM defeated all their preparations.

"I obviously didn't recognize a thing after studying all those photographs," Conrad recalls. "Nothing looked right. There was nothing familiar to judge distances with, no trees, houses, roads, and so forth. It was a little hard to tell where we were going with all that dust blowing."

Conrad eventually managed to locate the Snowman crater formations, only to discover that the photographs taken by the Surveyor were dangerously misleading. According to the photos, the desired landing spot appeared to be at the four o'clock position in the middle of the Snowman formation. But instead of being relatively smooth and flat, the area proved to be forbiddingly rough and jagged.

The Apollo 12 astronauts now faced the very predicament they had hoped to avoid—a last-minute fuel consumption crisis like Apollo 11's. Rather than heading straight for his designated landing site, Conrad had to fly around it, desperately hoping to find a more suitable parking place before the LM ran out of descent fuel.

"There it is! There it is!" Conrad barked as he honed in on a crater in the "body" of the Snowman formation. "Son of a gun, right down the middle of the road."

"Outstanding, Pete!" Bean exclaimed.

Moments later, the Intrepid touched down next to the belly of the Snowman. But as Conrad later admitted, he came awfully close to landing on the rim of the crater, which could have caused the LM to tilt over and crash.

"We landed at about the two o'clock position," he recalls, "twenty feet or less from the side of the crater. I couldn't see the crater anymore through the narrow, small windows. I didn't realize we were quite that close."

"When you land, you're not really congratulating yourself," Bean reports, "because you don't know whether you can stay. Your first concern is, 'How are the systems? Are we going to have to abort, and take off real soon?' So you go through these procedures that get you ready to launch again in the event that you can't stay. You're waiting for information from Houston, and your heart's beating very fast. Time is passing, and you want to be ready."

Fortunately, it didn't take long for Houston to give Apollo 12 official permission to stay on the moon. Then the astronauts enjoyed a brief celebration neither is able to recount in detail. "I forget what we did," says Bean, "but we were real glad to be there."

Bean, however, candidly admits that his initial euphoria quickly gave way to what he describes as "a slight sinking feeling." Having barely managed to land, he couldn't help wondering if they would also be able to lift off from the moon when the time came. "One of the feelings I always had when I was going fast on this mission, which was most of the time, was that we would keep going fast and keep progressing to our goal. When the lunar module came to a stop on the moon, there was kind of a letdown . . . I remember thinking to myself, 'You know, we're going to have to get it going fast again to get back home.' "

The Apollo 12 astronauts nevertheless proceeded to record a unique and yet not totally unexpected "first." Though Conrad and Bean were the third and fourth human beings to land on the moon, they were the first men to have fun there.

By 6:30 A.M., about five hours after touchdown, Conrad was ready to open the hatch of the *Intrepid* to begin the first of his two scheduled moon walks. At 6:45 A.M., he reached the bottom of the ladder, hopped down onto the surface, and said just what he had wagered he would say: "Whoopie! Man, that may have been a small step for Neil, but that's a long one for me."

Conrad ventured a few more steps away from the LM, steadied himself in the powdery surface dust, and started scanning the surrounding moonscape. He promptly spied the half-sunken metallic object he was looking for.

"Boy, you'll never believe it! Guess what I see sitting on the side of the crater? The old Surveyor!"

"The old Surveyor!" Bean echoed. "Yes, sir!"

"Does that look neat!" Conrad declared, laughing with glee. "It can't be any further than six hundred feet from here. How about that?"

"Well planned, Pete," complimented the voice of Mission Control.

"Hey," Conrad replied, "you don't know how happy I am."

The Apollo 12 commander's impish enthusiasm was in such sharp contrast to the stilted commentaries of Apollo 11 astronauts Armstrong and Aldrin that many people back on Earth suspected that he might be high on something besides the intoxicating thrill of walking on the moon.

"I was always accused of being euphoric or drunk on the moon and doing a lot of giggling," Conrad recalls, "but I had a lot of reason to do it, and I didn't really care to discuss why I was so happy on the moon. Obviously, the reason was that we got where we were supposed to be. The Surveyor was there. And once I ascertained that, it was downhill. We were obviously going to have a successful flight, so it was a very comfortable feeling."

Conrad continued on his merry way for the duration of his first moon walk, and even sang his own merry tune. "I could work out here all day," he declared, and began humming aloud, "Dum de dum dum . . . Dum de dum dum dum . . ."

Unbeknownst to his earth audience, Conrad's giddiness was also inspired by a series of mischievous practical jokes secretly perpetrated by the ground crew back at the Cape. The operative metaphor involved Snoopy, the Apollo 12 quality control symbol, and the surprise appearance of an unclad female sex symbol in the checklist attached to his pressure suit.

"When I turned to the first page of the checklist on my cuff," he remembers, "why there was [a cartoon of] Snoopy, and it had an inside joke on it. I came to find out that on every page there was another Snoopy with another inside joke. After I got about three pages into [the checklist],

I flipped over to the next page, and lo and behold, there was the Playmate of the Month."

Forty-five minutes after Conrad set foot on the moon, Bean emerged from the LM and joined him on the lunar surface.

"You look great," Conrad told his partner. "Welcome aboard."

Bean immediately noticed that he couldn't move about as quickly as he had expected. But once he found his bearings, he began to enjoy his first moon walk almost as much as Conrad.

"Hey, it's real nice moving around up here," Bean confirmed. "You don't seem to get tired. You really hop like a bunny."

The astronauts' sense of ease was not an illusion. The monitoring devices attached to their bodies showed that Conrad and Bean had average heart rates of 105 and 121 beats per minute respectively. Similarly, Conrad used up only sixty percent of his portable air supply, while Bean consumed slightly less than seventy percent of his oxygen.

But the two "bunny hoppers" soon encountered the first of several increasingly serious problems when they tried to set up the TV camera designed to broadcast the first live color pictures from the lunar surface. Although Bean said he could feel the motion of the wheels running inside the camera housing, Houston complained that it wasn't receiving a clear transmission. Then the picture suddenly snapped into focus.

"I hit it on the top with my hammer," Bean reported. "I figured we didn't have a thing to lose."

"Skillful fix, Al," Conrad teased.

Unfortunately, Bean's hammering didn't quite do the trick, and the camera again went on the blink.

The astronauts were by now falling badly behind schedule, so Houston told them to give up on the camera and tackle the next item on their checklists, planting the second American flag brought to the surface of the moon.

"O.K., we have the flag up," Conrad reported. "Hope everybody down there is as proud of it as we are to put it up."

"Affirmative, Pete," Houston replied. "We're proud of what you're doing."

Conrad and Bean then ventured about a thousand feet to the west of the LEM—five times as far as the Apollo 11 crewmen had gone on their moon walk—to set up the ALSEP equipment. Along the way, they collected about fifty pounds of lunar rock samples, including several of the glittering glasslike beads that Armstrong and Aldrin had found littering the floor of the Sea of Tranquility.

The two astronauts became so carried away with the wonders of the moon that they got careless as they traversed the rims of the Snowman formation. At one point, Bean nearly stumbled backward into the mouth of the "head" crater. Only an alert warning cry from Conrad saved him from what could easily have been a fatal tumble.

A few minutes later, Bean flirted with yet another potential disaster when he tried to deploy the ALSEP's nuclear generator, and found that the plutonium fuel canister was stuck in its casing. As he yanked away with a frustrated grimace, his otherwise easygoing partner became alarmed that the radioactive canister might rupture.

"Don't touch that," Conrad warned. "If you touch that, that's all."

Then, almost miraculously, Bean managed to dislodge the canister, and carefully inserted the plutonium tube into the generator. Both he and Conrad let out sighs of relief.

The rest of Apollo 12's first EVA on the lunar surface was, in Conrad's words, "a piece of cake." While polishing off their official chores, the astronauts kept trying to snatch a few private moments to look around and reflect on where they were, only to find themselves overwhelmed by the fact that they had actually made the "impossible dream" come true.

"It seemed very unreal to me to be there," says Bean. "The thing I remember most visually is looking back at the earth, and thinking how far, far away it was. Frequently on the lunar surface, I said to myself, 'This is the moon. That is the earth. I am really here. I am really here.' "

It was this otherworldly view of the distant Earth—even more than what he saw on the lunar surface—that made Bean's moon walk seem truly magical and mystical. "On the moon, all the stars circle around you every twenty-eight days instead of every twenty-four hours, and the sun moves around you the same way. Yet, the earth stays right in the same spot. I remember thinking if ancient man had been born on the moon instead of the earth, he would have had much more difficulty determining what was going on because things would have been moving in slow motion. But I felt pretty sure ancient cultures would have worshiped the earth and thought it was an eye because it would change from blue to white and you could see something moving up there that did look like a colored eye. No doubt they would think that was a god up there watching them. There's no telling the virgins they would have sacrificed to this thing."

Bean adds that he felt "a little guilt" whenever he caught himself indulging in such "daydreaming," and would silently remind himself, "I shouldn't be doing this. I've got work to do, and every minute counts. If I've got any spare minutes, I ought to be picking up rocks, making observations. I shouldn't be trying to think about what this might mean. There will be plenty of time to do that when we get back."

The Apollo 12 crewmen returned to the LM shortly before 11 A.M. Their first EVA had lasted four hours and one minute, and except for failing to get the TV camera to operate, they had accomplished all they set out to do. They were now scheduled to have a ten-hour rest period before beginning moon walk number two. But neither astronaut managed to get much shut-eye.

Conrad complained that the medical sensors attached to his chest were rubbing itchy, running sores that were "driving me buggy." He suffered additional and even greater discomfort from his poorly fitting pressure suit, whose leg sections were each a quarter of an inch too short. "Being a quarter of an inch too short doesn't sound like much," he says, "but believe me, it was like being in a vise, and it was just driving my shoulder bananas. It was really painful. After I'd had about four hours' worth of sleep, there was no way I could sleep anymore because of that."

Conrad's tossing and turning aroused Bean, who obligingly got up in the middle of the rest period and tried to adjust his partner's boot laces. Though Bean's suit fit better, he says that he too suffered considerable discomfort. "You were sleeping on your neck ring. The next flight, they quit doing that because it made you sleep poorly. They put the seats on the side, and took off their clothes."

But Bean claims that pressure suit problems were not the sole cause of the astronauts' insomnia. "We were sleeping at the times the earth was awake, and vice versa, so we hadn't shifted our circadian rhythms around. We tried to, but we never did make it."

He also blames their inability to sleep on various distractions emanating from the bowels of the *Intrepid:* "The hydraulic pumps woke both of us up, and we wondered what the hell was going on. It was like being on an airliner. You hear the noise, and when the noise changed, it got your attention."

Bean adds that his mind kept racing over the events of the day, mulling the various success and failures of EVA number one. He says he was particularly concerned about burning out the TV camera, and kept wondering if there was something he could do to get the camera working again. He had also developed a runny nose during his trek across the lunar surface, and worried if the mild decongestant pill he had taken upon returning to the LM would dry up his sinuses in time for the second EVA.

"If I had to do it over again, I'd definitely take a sleeping pill. Although we had sleeping pills aboard, we didn't do that mostly because of the kind of guys we are. We don't take a lot of medicine. However, I'd rethink it, and look at it as more of a tool in order to get the kind of rest I needed so the next day I'd perform better. Although the performance was adequate, I think it could have been better if I'd had a good night's sleep."

Meanwhile, Apollo 12 command module pilot Dick Gordon continued to orbit the moon, alone and half-forgotten but seldom idle. Along with watching over the mothership, Gordon's duties included shooting what was to be the most extensive and accurate photographic contour map man had ever made of the lunar surface. The checklist called for him to take 150 snapshots at twenty-second intervals using a special multispectral still camera designed to highlight contrasts in topography with alternating red, blue, green, and infrared lenses.

Gordon knew that the trickiest part of his photo-mapping assignment would be keeping the *Yankee Clipper* on the proper course and maintaining a steady attitude as he zoomed around the front side of the moon. But much to his chagrin, he found these chores even harder to accomplish because of all the happy talk coming over the radio from his comrades on the lunar surface.

"Do you have those [mapping] coordinates?" Houston asked as the *Yankee Clipper* made yet another orbital pass over the Ocean of Storms.

"No," Gordon grumbled. "I can't hear because those guys keep yacking."

Gordon managed to complete his photo-mapping assignment only after Houston admonished Conrad and Bean to keep their mouths shut for a few minutes. Then, just as the two moon walkers climbed back inside the LM, he shuttered the windows of the command module, and settled down for what proved to be a surprisingly restful sleep.

Conrad and Bean stepped back out onto the lunar surface at 10:01 P.M. that same night. The two astronauts regarded their second EVA, which was scheduled to include their long-awaited expedition to the Surveyor III, as considerably more important than the first. But despite their lack of sleep and the lingering annoyances of their minor physical maladies, they tackled their preliminary assignments with the same lighthearted enthusiasm as before.

"Do you know what I feel like, Al?" Conrad giggled as the two men left the LM and began bounding across the lunar landscape. "Did you ever see those pictures of giraffes running in slow motion? That's exactly what I feel like."

"Say," interjected the amused but unfailingly conscientious voice of Mission Control, "would you giraffes give us some comment on your boot penetration as you move across there?"

After dutifully obliging Houston's request, Conrad and Bean retraced the westward route of walk number one to check out the scientific equipment they had planted that morning. Then they ventured south toward Sharp Crater, periodically stopping to collect more rock samples.

"Oh, boy, I want that rock," Conrad blurted at one point. "It's a dandy grapefruit-size goodie."

Conrad went after the "grapefruit-size goodie" as if it were a chunk of gold, and gleefully pouched the specimen in his sample bag.

"Man, have I got the grapefruit rock of all grapefruit rocks," he boasted. "This has got to come home in the spacecraft. It'll never fit in the rock box."

Typically, Conrad and Bean couldn't resist the chance to play a practical joke on Houston during their rock-gathering expedition. The two astronauts breathlessly informed Mission Control that they had sighted a pile of "purple rocks" in the path of their moon walk.

"Gee whiz," Conrad chortled. "That's neat-o."

Conrad later admitted that their report was a ruse, and that the "purple rocks" were merely concoctions of their own imaginations.

The astronauts turned back to the east, and made their way past Bench Crater and Halo Crater in the general direction of the Surveyor III. They had now been on the moon for nearly twelve full hours, but both men kept marveling at how comfortable they felt in the lunar environment. As Bean noted later, their sense of security was partly real and partly make-believe: "I always had the feeling that people were just over the hill, just over the horizon because the [radio] communications [with Earth], even though they had a delay, were so clear and calm, like you are close to the [ground] station."

As the Apollo 12 crewmen continued their meandering trek toward the Surveyor III probe, they noticed the texture of the lunar surface abruptly change from the "rich brown color . . . [of] a good plowed field" to a rocky gray mosaic of cracks and crevices. A few minutes later, they made an even more startling discovery—they were lost!

"I can't believe we're at the right place," Bean muttered.

"I'm not sure that we're at the right place, either," Conrad agreed. "Let me look at the top of this hill here . . ."

Upon scaling the "hill," Conrad realized that he was actually standing on the rim of a crater. And nestled on the floor of the crater less than a hundred yards below was the inanimate hunk of metal they had been looking for.

"There's the Surveyor!" he cried with joy.

Although Houston shared his joy, the men at Mission Control were now worried that Conrad and Bean might overexert themselves in a hasty dash down the inner slope of the crater. The astronauts promised to take a break and "case the joint" before starting their descent, but they were simply too excited to keep their word. Instead, they committed man's first act of mutiny on the moon, and forged on toward the Surveyor.

"Don't worry about it, Houston, because it's really no strain," Conrad assured. "I'm about two hundred feet away from it [the Surveyor]. The ground is firm and I can go right back up the way I came down with no strain at all."

"Now we can see which way he [the Surveyor] came in," Bean observed. "See the way the pads dug in over there . . . That'll make a good shot. We're not supposed to take pictures of that leg . . . we'll have to do it, though."

Bean reported that the half-buried hunk of metal on the crater's floor, which had been painted white back on Earth, appeared to have turned brown in the course of its voyage to the moon, but was otherwise in remarkably good shape.

"Houston, not a bit of this glass is cracked," Bean assured. "One little piece down here looks like it no longer reflects. But other than that, it's in perfect condition."

Bean took out a pair of wire cutters and snipped off three of the Surveyor's body parts: a strand of wire cable, a section of aluminum tubing, and the probe's clawlike soil scoop, which had performed the first remote-controlled test of the lunar topsoil two and a half year before.

"Bit of extra sample for you, Houston," Bean chuckled. "The scoop's got dirt in it."

Because the return trek back to the LM was only about six hundred feet, the Apollo 12 astronauts deliberately took their time. Unbeknownst to Houston, they were also preparing to play one more practical joke on the NASA public relations department by taking a seemingly impossible photograph—a self-portrait of the two of them standing side by side on the surface of the moon.

"We got the idea to smuggle a timer on board the spacecraft," Conrad recalls. "What Al and I were going to do was run over to the Surveyor after we set this little deal up on the [camera] stand, and give it our best pose standing next to the Surveyor . . . That would obviously be the picture that the p.r. people would grab and send out to the world . . . until some smart guy asked, 'Who took the picture?' "

So just before leaving the floor of the Surveyor crater, Conrad and Bean took an unscheduled break, and started communicating with each other in sign language.

"You guys resting?" Houston inquired.

"Yeah," Conrad fibbed, "we're resting."

In truth, Conrad was fumbling frantically for the camera timer he had hidden in the bottom of his rock sample collection bag. By the time he finally dredged up the dust-covered camera attachment, Houston impatiently ordered the astronauts to march back to the LM right away. But even though he could not take the "impossible" photograph. Conrad managed to get one last laugh by literally chucking the idea's key device.

"I didn't want to take [the timer] back in my rock box," he says, "so I gave it a good heave. You know on the moon it's going to go a long distance, so it whistled away somewhere . . . Two million years from now our future archeologists and historians who are tromping around on the moon at the original sites of the lunar landings are going to find this thing they can't account for—that was the joke that didn't work."

Conrad and Bean finally climbed back inside the *Intrepid* and sealed the hatch at 1:43 A.M. on November 18, 1969. Their second EVA had laster three hours and thirty-two minutes, about twenty minutes longer than the first. During that time, they had traversed a total linear distance of six thousand feet, three times that of EVA number one and six times the ground covered by the crew of Apollo 11.

More important, by touching down within walking distance of the Surveyor III, the Apollo 12 astronauts had proven that man could land at a specified location on the moon with the same pinpoint accuracy he could land an airplane on the earth. But as fate would have it, the seemingly effortless success of man's second lunar landing also set the stage for the most dramatic crisis in the history of the U.S. manned space program.

11

UNLUCKY THIRTEEN

Explosions in Cislunar Space

Apollo 13: April 11–17, 1970

Apollo 13 did something that's never happened before in the history of man: for a brief instant of time, the whole world joined together.

—APOLLO 13 COMMAND MODULE PILOT
JOHN L. "JACK" SWIGERT

Apollo 13 was scheduled to make man's third lunar landing in the tumultuous spring of 1970. But prior to lift-off, as the war in Vietnam raged on and the nation's college campuses erupted in violent protests, many otherwise patriotic Americans were beginning to question the value of another voyage to the moon. What would it prove? What, if any, important new knowledge would be gained? Was the Apollo 13 mission really necessary?

The even more cynical attitude of America's youthful "New Left" was graphically illustrated by a T-shirt on sale in New York's Times Square. Emblazoned beneath a silk-screened photo of an anonymous astronaut standing on the lunar surface was a rhetorical caption expressing the frustrations and disillusionment of an entire generation: "So What?"

New York Times science reporter Walter Sullivan, the unofficial apologist for the U.S. manned space program, gallantly tried to counter the growing public apathy and antipathy by underscoring the scientific importance of the Apollo 13 mission.

"It may turn out, in fact," Sullivan declared the day after the Apollo 13 launch on April 11, 1970, "that centuries from now, as historians look back on early explorations of the moon, they will decide that the current flight . . . was the critical mission."

The best way to grasp the significance of Apollo 13, Sullivan contended, was by comparing it to a "hypothetical examination of the earth by scientists from a distant planet." If alien visitors made only two landings in basically similar areas like the prairies of the western U.S. and the steppes of the Soviet Union, they would get "a very incomplete picture" of the earth. Likewise, man had so far gotten a very incomplete picture of the moon since both Apollo 11 and Apollo 12 had landed in basically similar lunar lowlands.

Apollo 13, on the other hand, was bound for the Fra Mauro region, a rugged lunar highland named after a famous medieval cartographer. Located about 110 miles west of Apollo 11's landing site in the Sea of Tranquility, the Fra Mauro region rested on an uncharted plateau creased by deep canyons and narrow valleys that could only be reached by navigating the lunar module between the peaks of an adjacent mountain range.

One of the key scientific questions Apollo 13 hoped to answer was whether the moon was geologically "dead" or "alive." In the course of exploring the Fra Mauro highlands, the astronauts planned to set up seismic devices to detect evidence of ongoing "moonquakes." They also planned to conduct "heat flow" experiments to determine if the lunar interior contained a molten core that might still be prone to periodic volcanic eruptions capable of resculpturing the surface—and/or imperiling human exploration and colonization—with massive lava flows.

"While the current mission may seem 'old hat' to some, its hazards are no less than those of previous flights," Sullivan noted. "In fact, the planned landing in a comparatively hilly area inevitably makes for increased risk. And the last-minute change in team makeup introduces a further element of uncertainty. The mission can by no means be described as routine."

Ironically, Apollo 13 commander James A. Lovell, Jr., disagreed with that assessment even after the above-mentioned "change in team makeup" shortly before lift-off. Recalling his own private prognosis in a recent interview, Lovell maintained: "It was a routine flight. The only difference between 13 and 11 and 12 was the fact that we were going to explore the highlands—the hilly areas—whereas 11 and 12 went into the [relatively flat lunar] seas where we were a bit more sure of the landing spots."

Even so, Jim "Shakey" Lovell would not have traded the chance to command man's third lunar landing mission for all the gold on Earth. The most experienced member of the NASA astronaut corps with over 572 hours in space, he had orbited the earth on Gemini 7 and Gemini 12, then orbited the moon on the December 1968 Apollo 8 mission. Now age forty-two, Lovell, a gregarious navy captain with a wife and four children, still had what he described as "an addiction to space flight." He planned to retire after the Apollo 13 mission, and was anxious to make the most of his second and final lunar voyage.

"I felt as if I were looking back in history," Lovell had reported after getting his first close-up look at the moon on the Apollo 8 flight fifteen months before, adding that he sensed at the time that "if we could only get that scant sixty miles closer—really down there—then we'd have a chance to pry open some of the secrets of creation. The lunar surface was so clear. It beckoned."

The Apollo 13 commander and his crew, however, encountered the first of several unprecedented strokes of bad luck before the mission ever got off the ground. In late February of 1970, less than two months before the scheduled launch date, Charley Duke came down with a case of German measles. Although Duke was not a member of the Apollo 13 crew, he had inadvertently exposed all his fellow astronauts before his symptoms became apparent. That, in turn, prompted a controversial last-minute change in the Apollo 13 roster.

Medical records showed that Lovell and fellow crewman Fred W. Haise, Jr., had already survived bouts with the German measles during their boyhoods. They had also been exposed to the disease by their own children in years past, and were therefore presumed to be immune. But NASA doctors predicted that command module pilot Ken Mattingly, a bachelor who had never contracted the measles as a youth, would probably break out in a rash about the time the three astronauts were scheduled to begin their return trip from the moon back to Earth.

"A big argument ensued," Lovell recalls. "NASA wanted to replace Mattingly with his backup, Jack Swigert. I argued against that from the beginning. Not because I didn't think Jack was qualified; I did. But I didn't think Ken was going to come down with the measles. And even if he did come down with the measles on the way home from the moon, what difference did it make? I can't think of a more comfortable environment than floating around the lower equipment bay of the command module eating aspirin and chicken broth. The only real factor [of concern] would be the reentry to Earth, which all three of us could make. If Ken comes down with the measles, then Fred and I will take over the spacecraft. If he doesn't, then everything will be fine."

For better or worse, Lovell lost out and so did Mattingly. Two days before lift-off, it was announced that the Apollo 13 command module pilot would be backup crewman John L. "Jack" Swigert, Jr.

NASA administrator Dr. Thomas Paine publicly admitted that the last-minute substitution was a "gamble." Like the luckless Mattingly, the thirty-eight-year-old Swigert was a bachelor reputed to have "a girl in every airport." He was also a veteran military and commercial pilot who had earned his license at the age of sixteen. Born in Denver and educated at the University of Colorado where his party-loving fraternity brothers nicknamed him "Big Swig," he had flown combat missions for the air force during the Korean War, then picked up a masters in aerospace science from Rensselaer Polytechnic Institute and a masters in business administration from the University of Hartford.

Even though Swigert was an accomplished understudy, there was no way he or anyone else could possibly master all of the scientific experiments and photo-mapping assignments Mattingly had planned to perform within twenty-four hours of lift-off. But by this time, the Apollo 13 launch was already a month behind schedule, and Paine contended that scratching a few items on the checklist was vastly preferable to delaying the mission for another month at an estimated cost of $800,000.

Like the outspoken Lovell, Apollo 13 lunar module pilot Fred Haise greeted the decision with profound ambivalence. Having already suffered the disappointment of being bumped from the CMP slot on Apollo 11, the thirty-six-year-old former marine corp fighter jock from Biloxi, Mississippi, could sympathize with Mattingly's plight more than any other astronaut. But unlike Lovell, he also planned to continue his astronaut career until "they force me to retire," and he had no desire to anger his NASA superiors or forfeit his once in a lifetime opportunity to walk on the moon.

The same budgetary concerns that prompted Apollo 13's last-minute crew change caused NASA to minimize and mishandle a much more serious last-minute problem with the Apollo 13 spacecraft. Following the final prelaunch "countdown demonstration" at the Cape, the pad techs were supposed to drain the liquid oxygen tanks supplying breathing air to the command module with a series of high-pressure oxygen gas injections. But instead of flushing out LOX, one of the tanks merely recirculated the gas injections back out its drainage pipes.

Rather than postponing the Apollo 13 launch, the space agency's top brass assigned a special team of engineers and technicians to correct the mysterious LOX tank malfunction in less than seventy-two hours so the astronauts could lift off on schedule. According to Lovell, "They went into the history of the tank just to see what the story was, and they found out that it was originally scheduled for Apollo 10, but that it had been dropped at the factory, so it had been recycled, refurbished, and set up for Apollo 13. They looked at the schematics and saw that there was a tube that guides this gaseous oxygen in, and that if the tube was broken or moved away somehow, it would not guide the gases down to force the liquid out, but it would bypass the liquid and just let the gas go out the vent line.

"Well, the engineers all sat around to philosophize on what to do. They could order a new tank, or take one out of another vehicle down the line. But by the time they did all that, several weeks would go by, and they'd have to slip the launch . . . The tank worked perfectly for all the flight aspects—it fit all the systems, it pressurized the spacecraft, it fit the fuel cells, it was good for breathing . . . The only thing that didn't work was the fact that we couldn't get the doggone oxygen out of it, which in a normal flight we would never do. In other words, that was something that was strictly a ground test device."

Under the pressure of their hasty and ill-conceived official mandate to get Apollo 13 launched on schedule, the engineers then proposed what seemed like an ingenious ad hoc solution to the LOX tank's drainage problem. "There was a heater system," Lovell explains, "a long tubelike affair with regular wires in it submerged in the liquid oxygen. And they said, 'Why don't we turn on the heater system and boil the oxygen out?' They took a poll, and everybody said, 'Gee, that's a good idea, didn't think of that.' So they turned on the switch for about eight hours, and by gosh, they were absolutely right. All the oxygen boiled out. The tank was absolutely dry. Everything was in good shape. The tank was loaded again a day or so before the launch . . . then we took off."

The Apollo 13 astronauts left the launchpad at 1:13 P.M. Houston time on the afternoon of April 11, 1970, and settled back for what they still hoped would be "a routine flight" to the moon. Still skeptical about the decision to ground Mattingly, Lovell devised a secret code by which to query Mission Control about the latest news on his former comrade's health status.

"Are the flowers blooming in Houston?" the Apollo 13 commander would ask.

"No," Houston answered each time, indicating that Mattingly had not yet broken out in red spots.

As Lovell had predicted, the flowers never bloomed in Houston. At least not in the metaphorical sense. Mattingly did not come down with measles before, during, or even after the completion of the Apollo 13 mission. But Lovell was gravely mistaken in predicting that the flight would be routine.

Shortly before 9 P.M. on the night of April 13, just as the astronauts were nearly halfway into Day Three of their trans-lunar voyage, an explosion rocked the fuselage of the Apollo 13 spacecraft.

The audible alarm system in the command module erupted in an ear-piercing cacophony of staccato beep-beeps, and the instrument panel ignited a blaze of flashing yellow warning lights. All three crewmen froze in their respective places, staring at each other in stunned disbelief.

"I was in the back of the spacecraft when the explosion occurred," says Lovell. "It was a sharp bang . . . I really didn't know what happened, but I couldn't imagine why it was so loud. I looked up at Fred Haise, and I knew Fred didn't know what the story was. Then I looked over at Jack Swigert, and Jack's eyes were as big as saucers."

Swigert, who was in the front of the cabin, happened to be farther from the epicenter of the explosion than Lovell. As a result, he says, the sound was muffled. But since Swigert also happened to be touching part of the command module's metal frame at the time, he could feel the aftershocks as well as hear them. Even so, he had no idea what caused the blast. "Our first thoughts were that we had been hit by a meteorite. But we didn't know whether the explosion had occurred in the lunar module or the command module."

Before the astronauts could locate the center of the explosion, another mysterious blast of approximately the same magnitude knocked out the command module's entire electrical system. This second explosion also knocked out two of the spacecraft's three main fuel cells.

"That really put a perturbation through the crew," says Lovell. "One fuel cell was sufficient to get us to the moon or back to the earth. But the mission rules stated that you needed all three fuel cells to leave the command module to go to land [on the moon]. Losing two meant that we couldn't land . . . so we were very much disappointed."

Lovell made an even more unsettling discovery when he floated over to the center of the cabin to check the instruments monitoring the spacecraft's twin oxygen tanks. One gauge read zero, and the other was going down fast. "I kept looking out the window on the left side of the spacecraft, and I could see a gaseous substance venting at a very high rate of speed. It didn't take much intelligence for me to realize that the gas escaping from the back end of my spacecraft and the gauge on the second [oxygen supply] tank were one and the same, and that we would soon be completely out of oxygen in the command module."

To make matters worse, the Apollo 13 spacecraft was now more than 200,000 miles from home. It was also headed in what had suddenly and ironically become "the wrong direction." Unlike a conventional airplane, it did not have the maneuverability to make a quick U-turn. The only way the astronauts could reverse their trans-lunar trajectory was by swinging around the moon, and using the momentum to "slingshot" back to Earth.

It was at this point that the Apollo 13 astronauts first realized the life-or-death nature of their plight. As Lovell later confessed: "Our fears changed from 'I wonder what this is going to do to the lunar landing,' to 'I wonder if we're going to get back home.' "

Although the astronauts had dealt with all sorts of imaginary catastrophes in preflight simulations, they simply weren't prepared to cope with the sheer number of problems that resulted from the unexpected and unexplained series of explosions that had disrupted their otherwise routine voyage to the moon. At least not all at once.

"We never trained for multiple failures," Lovell explains. "We relied on redundancy and the reliability of the equipment. We knew that something was going to fail, but we had two of everything. There was such an array of possible double failures that you couldn't possibly train for all of them anyway."

The men in the trench were also caught with their pants down. At first, Mission Control didn't know what was going on aboard the Apollo 13 spacecraft. Although Swigert attempted to report the available details almost immediately, the blasts had destroyed the spacecraft's high-gain radio antenna, so his words were lost in outer space.

Houston eventually reestablished voice contact on a low-gain frequency, and determined that the mysterious explosions had occurred somewhere in the service module. Under the generalship of Gene Kranz, the crew-cut veteran flight director with the distinctive white vest, the men in the trench then spent a frantic forty-five minutes trying to find a way to revive the crippled power, electrical, and oxygen supply systems of the command module. But nothing they suggested seemed to work.

Former Apollo 7 crewman Walter Cunningham later ridiculed the ground crew's efforts as an unconscionable waste of time. Cunningham claims that he and fellow astronauts Dave Scott and Rusty Schweickert rushed to the Manned Spacecraft Center viewing room within fifteen minutes of the first report of the Apollo 13 mishap, only to find that Mission Control was at least a step or two behind the embattled Apollo 13 crew in dealing with the crisis. According to the account in his autobiography *The All American Boys,* "When we arrived, the flight controllers were still huddling, going through the usual crisis format: What happened? Where are we? What are we going to do about it. As we monitored the air-to-ground [communications] loop, it was obvious that Lovell, Swigert, and Haise . . . [were] standing by for the official go-ahead on a decision they had already begun to implement."

Shortly before 11 P.M., Mission Control officially authorized the Apollo 13 astronauts to do what they were already in the process of doing on their own volition—evacuating the command module *Odyssey* and stuffing themselves into the cramped confines of the lunar module *Aquarius* in hopes of using the LEM as their "lifeboat."

All concerned recognized the enormous risks involved in that allegedly obvious decision. The two-seat lunar module was never intended to serve as a lifeboat. Its sole purpose was to ferry Lovell and Haise from lunar orbit down to the lunar surface and back up to a preordained orbital rendezvous with the mothership. Mission Control didn't know whether or not the LM had enough fuel and oxygen to fly all three Apollo 13 crewmen safely back to Earth, but it offered the only means of saving Lovell, Haise, and Swigert from suffering a slow and agonizing death in outer space.

The first and most critical hurdle the astronauts had to clear was converting the *Aquarius* from a passive capsule to an active spacecraft. Like the mothership, the LEM had to know where it was and where it was heading. That required Lovell to transfer the essential computer guidance data known as the "reference" from the navigational console of the command module to the navigational console of the LEM. And because of the dwindling oxygen supplies in the command module and Mission Control's delay in deciding on the appropriate evacuation procedure, Lovell had less than fifteen minutes to complete this mind-numbing computational task.

"This was the day before the electronic calculator," he notes, "so we didn't have any of those things aboard. I had a piece of paper and I'm writing down these angles and transferring them, and I'm saying. 'How about double-checking my arithmetic? Is two and two really four? Ten minutes to go . . . is two and three really five?' "

Lovell's arithmetic checked out okay, but it did not ensure that the Apollo 13 astronauts would make it home alive. After relocating to the LEM, Lovell and his crew had to figure out how to steer their crippled spacecraft—command module, service module, and all—with a system that wasn't designed to do such a thing. According to Lovell, "We had to continue on with about four hundred thousand pounds of unburned fuel plus all the mass it had otherwise. We had to reinterpret [the throttle controls] . . . Pushing down really means it's going to go left . . . Going left really means it's going to be pushed up . . . And so forth."

By this time, news of the Apollo 13 crisis was spreading to virtually every corner of the globe, prompting an unprecedented international reaction of sympathy and concern. Once blasé and apathetic American citizens became glued to their TV sets. Offers of assistance poured in from Argentina, Brazil, Burundi, England, France, Greece, Italy, Pakistan, the Malagassy Republic, and the Soviet Union. Pope Paul VI offered prayers, as did holy men and devotees of most of the world's major religions.

Vito Bellafore, mayor of a small Sicilian town recently devastated by an earthquake, summed up the sudden change in public attitude toward Apollo 13 in an interview with *The New York Times:* "We know human suffering here. After the first Apollo success people did lose interest. It seemed that it was only a mechanical exercise. Now we see again the human drama, and all our worry is for those three lonely men."

Back on board the beleaguered Apollo 13 spacecraft, "those three lonely men" remained remarkably calm in the face of their predicament, partly because all three were former test pilots who realized it would do them no good to be anything but calm.

"People often ask me, 'Did you panic?' " Lovell recalls. "I say, 'No, why panic?' If you panic you can bounce off the walls for ten minutes . . . and you're right back where you started . . . In an airplane accident, if you're flying along and the engine falls off, something is going to happen. You can panic all the way down to the ground and crash. There you're already falling. But we were under the clutches of the three gravitational bodies [the sun, moon, and Earth], and we're just flying to the moon. If you don't panic, nothing is going to happen unless there's a wide gap in the spacecraft or something like that."

Adds command module pilot Jack Swigert: "I liken it to the analogy of having a flat tire in the middle of the desert when it's a hundred twenty degrees outside. You don't run around in circles and shout and panic. You say, 'Hey, there's a series of steps I have to do in order to get back on the road. I've got to go out and open the trunk, and take out the jack and the spare tire, and I'm going to jack up the car and I'm going to take off the tire and put on the spare tire and lower the car, put everything back into the trunk, and then I'm back on the road.'

"We had to do a whole series of steps in order to get back home. We had to power up the lunar module, transfer guidance, get back under free return trajectory, make course corrections, set up a whole new reentry procedure, power up the command module, jettison the service module and the lunar module, and reenter. The only difference between that and the flat tire is that the time between some of the steps is twenty-four hours."

Yet for all their outward cool, each of the Apollo 13 astronauts silently bemoaned their misfortunes in his own personal way. Lovell, for example, mused about the fact that he had originally been scheduled to fly on Apollo 14. Apollo 13 was supposed to have been Alan Shepard's mission. But in the summer of 1969, NASA officials decided that Shepard, who had been grounded for several years because of an ear problem, needed more training, and asked Lovell's crew to take Apollo 13. As he recalls:

"I thought, 'Yeah, that's a great idea. Let's do it.' But after the explosion occurred, I thought, 'My gosh, why me? Why now? Why here?' "

LMP Fred Haise felt much the same way, only more so. Having been bumped from Apollo 11, his fate was double-crossing him again. Now he was going to lose the chance to walk on the moon—maybe even lose his life.

Jack Swigert, the obliging last-minute replacement, suffered an equally cruel twist of fate. If Ken Mattingly, the original Apollo 13 CMP, had been immune to the German measles, Swigert would never have been on board in the first place, and probably would have flown on Apollo 15, the first moon rover mission. And that, Swigert admits, made the irony of the Apollo 13 explosion even more painful: "I thought, 'Lord, just because I had the measles when I was a kid, here I am.' I thought I was going to have the shortest-lived time as a prime crew member in existence."

Lovell later confessed that he, too, felt that all three of them might very well die in space. But he added he and his fellow crewmen were also determined to get themselves as close to home as possible even if they died trying. As he later told a congressional investigating committee: "Had we not done anything, we would have been in a permanent orbit [around the moon and the earth] with an apogee of two hundred thirty thousand miles and a perigee of one thousand miles. I felt it would be much better to get the spacecraft into the [Earth's] atmosphere than not get back at all."

The astronauts realized that getting back alive—or at all—depended on successfully circumnavigating the moon. That, in turn, depended on pulling off two key engine burns. The first burn would slingshot the spacecraft around the back side. The second and even more critical burn was the Trans-Earth Injection or TEI, which would blast the spacecraft toward home. As Lovell noted afterward, "We didn't know whether we would have enough fuel to get home until we made that second burn."

The moment of truth came on the night of April 14 as the Apollo 13 spacecraft veered around the left side of the moon. The first burn went off without a hitch, and thrust the astronauts into a thousand-mile elliptical curve. The astronauts had to make the second burn, their TEI, using the lunar module's descent engine, and they had to do it on the back side of the moon while temporarily out of radio contact with Earth. The rocket firing began at approximately 9:41 P.M., and it, too, went off without a hitch.

"That was a good burn," Mission Control noted with obvious relief as the astronauts scored back around the right side of the moon and angled their spacecraft on course for the Earth.

The men in the trench now rated the astronauts' chance for a safe return as "excellent," but they also knew that the crisis they termed "the most critical situation" in the history of the U.S. manned space program was far from over. Ever since the initial explosion, a special team of astronauts and engineers—which included the grounded Ken Mattingly—had been running a round-the-clock series of simulations to determine if the Apollo 13 spacecraft had enough consumables remaining for the rest of the journey, and if so, what sort of reentry and splashdown procedures ought to be used.

Mission Control calculated that it should take Apollo 13 about sixty-five hours to fly the 240,000 miles between the moon and the earth. Assuming that the astronauts followed strict conservation guidelines, they should have at least twelve hours worth of fuel and oxygen remaining by the time they reached the earth's atmosphere. But as Lovell recounts, the Apollo 13 crewman soon encountered a series of unanticipated mini-crises, any one of which could have cost them their lives:

"We had the situation where our speed was so great on the way home that if we missed the earth we would go into orbit around the sun. Then we had a problem where one of the batteries vented gas. If it had exploded or quit, we would never have had enough electrical power to get home. Then we had the problem of being poisoned by carbon dioxide."

Mission Control corrected the speed problem by ordering the astronauts to execute a fifteen-second mid-course engine burn on Day Four. That slowed the spacecraft's velocity by about 7.6 feet per second, just enough to reduce the chance of hurtling past the earth without using up too much of Apollo 13's precious fuel supply.

Nothing could be done about the faulty battery except to hope for the best. Some later suggested that the world's collective prayers must have been heard by the god and/or gods to whom they were directed, for the venting gas did not explode as the astronauts feared it might. However, it was the uniquely human inventiveness of the men in the trench—what Lovell described as a classic display of "Yankee ingenuity"—that kept the astronauts from succumbing to carbon dioxide poisoning.

Under ordinary conditions, the air in the command module was purified by an assemblage of lithium hydroxide canisters. But the command module had been powered down to conserve the backup electrical system not damaged by the mysterious explosions, and the supply of lithium hydroxide canisters in the lunar module was not sufficient for all three astronauts. By the time the Apollo 13 spacecraft rounded the moon, the carbon dioxide level inside the LEM was twice the desirable level, and climbing fast.

Houston's solution was elegant in its primitive simplicity. The ground simply instructed the astronauts on how to devise a homemade filter system using friction tape and air hoses borrowed from their "moon cocoons." In essence, this jerry-rigged contraption consisted of a network of tubing that diverted the fan in the lunar module so that it blew air over the lithium hydroxide canisters in the command module.

The water supply situation created additional problems for the Apollo 13 astronauts. They needed water to drink, and the spacecraft needed water to operate its essential life support systems. Lovell and his crew agreed with Mission Control that the needs of the spacecraft's life support systems had to come first. As a result, they abstained from drinking more than a couple of ounces of water each day. Instead they tried to quench their thirsts by such unsavory indirect means as chomping on hot dogs in order to extract the few drops of moisture inside the skins, a diet that may have contributed to a mild urinary infection and low-grade fever Haise developed on Day Five.

However, the astronauts' greatest physical discomfort was sheer exhaustion resulting from their inability to get to sleep. Their insomnia resulted in large part from the loss of their primary electrical system. Although they spent their waking hours in the lunar module, they spent their rest periods in the darkness of the command module. With the power shut down, the temperature inside the mothership dropped to thirty-eight degrees. The astronauts tried to put the spacecraft into a thermal roll, but the maneuver, which turned out to be more of a wobble than a roll, failed to warm up the interior of the command module more than a few degrees. Appropriately, they dubbed the mothership "the refrigerator."

Then, almost by accident, Lovell and his crew made what was literally a heartwarming discovery—a way to create their own body heat "cocoons." As Lovell explains: "The interesting thing about zero gravity is that there is no heat convection because there is no difference in the density of the air, so if you're very quiet, and don't move around, and don't move the air back and forth, your body heat will heat up the air right next to your body. It will stay there and form a cocoon of hot air around your body. Therefore, you could be very quiet and still your body would stay warm because you are protecting yourself against the outside cold because your body is throwing off this heat."

The discovery of the body heat "cocoon" proved to be a good omen for the Apollo 13 astronauts. Early on the afternoon of April 16, the beginning of Day Six on the mission time clock, the astronauts passed the halfway point between the moon and the earth. They were now less than twenty-four hours away from their projected splashdown time.

"*Aquarius* is coming home," Lovell radioed from his seat in the lunar module cabin.

From that point forward, Apollo 13's problems seemed to dissipate, and the spirits of the three crewmen lifted accordingly. Lovell even managed to find a humorous silver lining in their failure to make a lunar landing—at least he, Haise, and Swigert would not have to endure the standard three weeks of quarantine in the Lunar Receiving Laboratory (LRL) that NASA doctors had required of the Apollo 11 and 12 crewmen to make sure they had not been contaminated by their exposure to the lunar surface.

"Would you tell the people at the LRL to turn the thing off?" Lovell jokingly requested of Mission Control.

"Oh, no," replied the deadpan voice of the CAPCOM. "We're going to do the whole bit."

The astronauts then tackled the first step in the checklist for their anxiously awaited reentry to Earth the following afternoon.

In the interests of conserving power and electricity, Mission Control wanted them to remain in the lunar module until the last possible moment. But just before attempting to dive into the earth's atmosphere, the astronauts would have to return to the command module, and jettison both the LEM and the service module. So that night, approximately fifteen hours before the projected separation deadline, the crewmen began powering up the dormant systems of their mothership.

At 5:10 A.M. Houston time on the morning of April 17, CMP Jack Swigert crawled through the hatch between the lunar module and the command module, and commenced final preparations for reentry and splashdown.

"It's going to be interesting today," Lovell observed.

Then some 44,000 miles from Earth, the Apollo 13 spacecraft was zooming along at over seven thousand knots, and picking up speed every second thanks to the steadily increasing gravitational pull.

"I'm looking out the window," Swigert reported, "and that Earth is whistling in like a high-speed freight train."

Once again, the astronauts were in danger of whistling on past the earth, and hurtling into a fatal orbit around the sun. In order to prevent such a catastrophe, Mission Control ordered one more course correction burn. The engine firing began shortly before 7 A.M., and lasted for twenty-three seconds, slowing the spacecraft's velocity to a much safer and more manageable speed.

About fifteen minutes later, Swigert threw the explosive bolts that jettisoned the service module. Lovell, who was still in the lunar module cabin, then executed a series of short rocket blasts that separated the LEM and the attached command module a safe distance away. At this point, the Apollo 13 astronauts got their first look at the damage the dual explosions had wreaked on the service module.

"There's one whole side of that spacecraft missing," Lovell reported, noting, "Right by the high-gain antenna, the whole panel is blown out almost from the base to the engine . . . It's really a mess . . ."

Shortly before noon, when the astronauts were about ten thousand miles from Earth, Lovell and Haise joined Swigert in the command module. Then the astronauts closed the hatch to their "lifeboat" for the last time, and Swigert threw a second set of explosive bolts that jettisoned the lunar module. Thanks to the oxygen trapped in the connecting tunnel, the *Aquarius* shot away from the *Odyssey* like a ruptured balloon.

"LM jettisoned," Swigert informed Mission Control a few seconds later.

"O.K.," replied CAPCOM Joe Kerwin, hastening to add, "Farewell, *Aquarius,* and we thank you."

At approximately 11:53 A.M., the astronauts completed a brief pirouette that pointed the blunt end of their spacecraft toward the earth, and fired their engines. The command module plunged through the outer edges of the earth's atmosphere, seared by temperatures of up to five thousand degrees Farenheit.

The control room in Houston suddenly fell silent. As anticipated, the buildup of electrically charged particles around the fuselage of the command module temporarily disrupted radio contact with the Apollo 13 astronauts. But the blackout lasted no more than six minutes. At 11:59 A.M. the men in the trench heard Lovell's voice rasp a terse response to an overture from the pilot of one of the rescue craft: "Okay, Joe."

Three minutes later, two sixteen-foot nylon drogue parachutes popped out of the cone of the plummeting spacecraft. The astronauts were then about 23,000 feet above the Pacific Ocean, and about 610 nautical miles southeast of the Samoan Islands. At 12:03 P.M. the spacecraft's three main chutes were deployed with a brilliant flourish of orange and white, and the astronauts floated gently down toward the splashdown site.

Less than fifteen minutes later, the *Odyssey* hit the water with the blunt end down just eight hundred yards from its designated target, and remained almost perfectly upright, floating casually in rhythm with the waves.

The first frogman reached the *Odyssey* within ten minutes of splashdown. The astronauts opened the hatch, releasing a blast of frigid air, then scrambled out of their bobbing spacecraft, and climbed aboard the rescue helicopter.

When Lovell, Haise, and Swigert landed on the aircraft carrier *Iwo Jima,* the ship's band struck up "The Age of Aquarius," and a mob of cheering sailors rushed out on the flight deck to greet them. The welcoming ceremony was wisely limited to a two-sentence prayer offered by Commander Philip E. Jerauld, the ship's Protestant chaplain:

"O Lord, we joyfully welcome back to Earth astronauts Lovell, Haise, and Swigert, who by Your good grace, and their skill and the skill of many men, survived the dangers encountered in their mission and returned to us safe and whole. We offer our humble thanksgiving for this successful recovery."

When the prayer ended, the astronauts mingled briefly with Captain Leland E. Kirkemo and other high-ranking ship's officers.

"Man, it sure is nice to be warm," exhulted LMP Fred Haise.

"It was so damn cold," chimed in CMP Jack Swigert.

"We've been cold for six days," Commander Lovell confirmed.

Then, without further ado, the Apollo 13 crewmen made a beeline for the mess hall, where they devoured their first full-course meals in nearly a week, thirstily washing down the food with tumblers full of orange juice and water. The next day, the trio flew to Honolulu, where President Richard Nixon awarded them the Medal of Freedom.

"Greatness does not simply come in triumph but in adversity," Nixon proclaimed. "It has been said that adversity introduces a man to himself."

Nixon went on to assure Lovell, Haise, and Swigert that even though they did not land on the moon, he officially regarded their ill-fated voyage as "a successful mission."

The Apollo 13 astronauts felt basically the same way. Though the disappointment of failing to reach the lunar surface still rankled in their hearts, they were proud of their success in overcoming such unprecedented outer space adversity, and took genuine as well as philosophical pride in their accomplishment of returning safely to Earth. As Lovell noted later: "We could've been assured a catastrophe. But the dedication and knowledge of the ground and the flight crews were such that we were able to make it a successful failure."

CMP Jack Swigert went even further. Apollo 13, he claimed in a recent interview, was more than just a "successful failure." It was, as he saw it, a unique and unparalleled triumph for all mankind: "Apollo 13 did something that's never happened before in the history of man: For a brief instant of time, the whole world joined together."

Be that as it may, the repercussions from the Apollo 13 mission also cast a shadow over the future of the U.S. manned spacecraft program. A two-month-long official inquiry led by Edgar M. Cortright, head of NASA's Langley Research Center in Hampton, Virginia, placed the ultimate blame for the calamity on space agency officials and on two prominent private contractors, Beech Aircraft Corporation and North American Rockwell Corporation.

"The accident was not the result of a chance malfunction in a statistical sense, but rather resulted from an unusual combination of mistakes, coupled with a somewhat deficient design," the Cortright report concluded.

More specifically, the review board determined that the explosions in the Apollo 13 service module were caused by a preventable malfunction in the oxygen tank that failed to drain properly following the countdown demonstration, a finding that foreboded the fatal oxygen tank explosion on the space shuttle *Challenger* sixteen years later. The heating system made by Beech Aircraft was designed to carry only twenty-eight volts of electrical power rather than the officially specified sixty-five volts. When the ground crew used the heater to boil out the oxygen after the prelaunch dress rehearsal, the thermostats melted and were rendered "inoperative," thereby permitting the temperature inside the tank to rise up to a thousand degrees Farenheit.

"From that time on . . . oxygen tank No. 2 was in a hazardous condition when filled with oxygen and electrically powered," Cortright and his colleagues contended, adding that if the first explosion had occurred on the launchpad, which it easily could have, "the entire launch vehicle might perhaps have been destroyed."

Luckily, Apollo 13's faulty heater system in oxygen tank No. 2 had ignited when the spacecraft was fifty-six hours into the frictionless vacuum of outer space, which reduced the impact and the related fire hazards and probably saved the astronauts' lives. That first explosion had, in turn, ignited the second explosion in oxygen tank Number 1. But according to the review board, neither explosion would have occurred if space agency officials and NASA contractors had taken the proper precautionary steps before lift-off.

"The thermostatic switch discrepancy was not detected by NASA, North American Rockwell [the prime Apollo contractor], or Beech in their review of documentation," the Cortright report noted. "It was a serious oversight which all parties shared."

The question then became what to do next? Did the Apollo 13 accident mean that the U.S. should terminate plans to send any more men to the moon? Or was it merely a sobering close call that underscored the need to proceed more slowly and with significantly more care and oversight?

NASA officials predictably took the latter view. Instead of canceling plans for all future lunar exploration missions, they called time out, and proceeded to make the changes recommended by the review board. These changes, which cost an estimated $10–15 million, included the addition of a third oxygen supply tank and the redesign of the electrical wiring in the heater systems. But the net delay incurred for the Apollo program amounted to less than four months.

In January of 1971, NASA announced that the Apollo 14 mission, originally scheduled for the previous October, was ready for launch. Man's first lunar exploration program would resume. Three more astronauts were GO for the moon.

12

THE MOON SHOT NEVER TO BE FORGOT

Man's Third Landing

Apollo 14: February 5–6, 1971

I'm going to try a little sand trap shot here.
APOLLO 14 COMMANDER ALAN B. SHEPARD

No one suspected that Apollo 14 was destined to make man's most memorable moon shot when commander Alan Shepard and his crew launched from the Cape on January 31, 1971. Having all but forgotten the Apollo 13 crisis, most of the world considered Apollo 14 a non-event. The Soviets didn't even bother to provide their citizens with live TV coverage. And although an estimated forty-five million people watched the U.S. network broadcasts, that was less than half the audience for Apollo 11.

Apollo 14's dismal Nielsen ratings reflected the bitterly divisive political and social changes that had ravaged America since man's first lunar landing in the summer of 1969. During these post-Woodstock but pre-Watergate years, the war in Vietnam had claimed the lives of over fifty thousand U.S. servicemen. College campuses and urban ghettos were in a bloody uproar, and the ballooning military budget had put the nation's economy into a tailspin.

According to an informal *New York Times* survey, a profound apathy toward the space program in general—and a growing antipathy toward Project Apollo in particular—prevailed all across the heartland.

"If you've seen one moon dock, you've seen 'em all," opined a tobacco-chawing general contractor in Maryville, Kansas, adding, "They could cure cancer almost tomorrow if they would stop spending all this money on space research and put some of it into medical research."

"We feel they've gotten there, and now we'd like to save a little money," said a Garden City, New Jersey, housewife. "We feel that perhaps they are spending too much money on repeated trips to the moon. When money is so tight and with inflation, there are too many areas where the money could be spent more advantageously—such as unemployment and for the poor."

Ironically, Apollo 14 would fuel the flames of public discontent with a physical stunt the world would never forget. This was the mission on which Shepard would whack three golf balls across the "sand trap" of the lunar surface. Though this display of lunar linksmanship would make a more dramatic impression on many TV viewers than Neil Armstrong's "giant leap," NASA's critics would charge that it confirmed the frivolity of sending men to the moon.

Apollo 14 also featured an equally controversial psychic stunt. This was the mission on which lunar module pilot Ed Mitchell secretly conducted an unauthorized experiment in extrasensory perception (ESP) during the voyage to and from the moon.

The Apollo 14 astronauts also played a very serious and vitally important role in salvaging the nation's embattled manned space program. Shepard and Mitchell made the first successful lunar landing in the wake of the Apollo 13 calamity. By so doing, they completed the last "small steps" and performed the crucial "giant leap" needed to get Project Apollo back on track.

NASA officials claimed that alone justified the extra cost of landing two more men on the moon. They also insisted that Apollo 14 was actually a bargain for the American taxpayer. The estimated $400 million allocated to the mission included at least $10 million worth of technical modifications called for by the Apollo 13 investigative review board. Not counting the bill for these admittedly essential but expensive changes, the Apollo 14 budget was approximately the same as the budget for each of the three previous moon shots.

And contrary to popular misconceptions, man's third successful lunar landing mission was not all fun and games. Like their predecessors, the Apollo 14 astronauts faced a statistical probability of over 5,600 technical failures, even if all their spacecraft systems functioned with 99.9 percent efficiency, as well as several unique risks and hazards that increased the negative odds.

Shepard and Mitchell had the unenviable task of landing in the Fra Mauro region, the rugged lunar highland originally assigned to the crew of Apollo 13. They also had to perform yet another newly revised descent procedure intended to eliminate the last-minute fuel consumption problems of Apollo 11. And like their predecessors, they, too, encountered a series of unanticipated crises that nearly forced them to abort their mission before they could even attempt to touch down on the moon.

All three Apollo 14 crewmen, however, were accustomed to coping with adversity and defying the odds, especially mission commander Al Shepard. Born in East Derry, New Hampshire, and educated at Annapolis, Shepard was a forty-seven-year-old navy captain with crew-cut brown hair and a stern demeanor. He had earned the immortal distinction of becoming the first American in space by taking a historic seventeen-minute roller coaster ride in the Freedom 7 capsule on May 5, 1961. But since that time, the second most famous member of the Original Seven (next to John Glenn, the first American to orbit the earth) had been grounded by an inner ear affliction known as Meniere's disease, which caused vertigo and hearing loss.

Although Shepard further enhanced his power and prestige by staying on to help Deke Slayton run the NASA astronaut office, he would never have qualified to command a lunar landing mission without a space agency–financed ear operation in August of 1968. Dr. William F. House, a Los Angeles ear surgeon, managed to relieve Shepard's symptoms by inserting an inch-long silicone rubber drainage tube behind his left ear.

Shepard then determined to use his considerable inside stroke to get himself on a flight to the moon. Though he had no reservations about pulling rank, he realized that it was already too late to reshuffle the crews assigned to Apollo 11 and Apollo 12, so he convinced Slayton that he ought to get the honor of commanding man's third lunar landing mission. But NASA's top brass insisted that Shepard needed more time in training, and assigned Lovell's crew to Apollo 13, a controversial decision for which Shepard now thanked his lucky stars.

Apollo 14 lunar module pilot Edgar D. "Stud" Mitchell's personality was diametrically opposite from Shepard's. Like Shepard, he was a navy man, but he was an independent thinker, not a company man. Then age forty, he was in the throes of a midlife crisis, and planned to leave NASA and divorce his wife shortly after completing his lunar mission. The tapes he chose to listen to on his way to the moon were hit soundtracks of *Camelot* and *My Fair Lady,* but his new consuming interest was "noetics," a branch of a parapsychology devoted to more than mental telepathy and by no means as far out as it might seem.

"In noetics," Mitchell explains, "it is the experiencing of the mental perception that is important, that's the reality. The matter world that we live in, that we've been trained in, that's the matter world only because we perceive it that way. That's why the work in the psychic area has been important. It has taught us there is no objective reality. As a matter of fact, the true physicists, not the technicians who call themselves scientists, but the true physicists, are the ones closest to understanding what it is we're talking about."

Apollo 14 command module pilot Stuart A. Roosa was a redheaded former Forest Service smoke jumper from Oklahoma with what one colleague described as "a boyishness and earnestness about him that would do justice to Huck Finn." Like Mitchell, he was a space rookie who belonged to the fifth generation of astronauts selected in the spring of 1965, too late to qualify for one of the Gemini missions. A former air force fighter pilot with a degree in aeronautical engineering from the University of Colorado, Roosa, age thirty-six, had a reputation for being a "straight arrow." But like most of his fellow CMPs, he was blessed with exceptional aviating talent, keen powers of observation, and a special romanticism that would inspire him to play Sonny James's version of "How Great Thou Art" as the spacecraft drifted around the far side of the moon.

Although the media billed Apollo 14 as "Shepard's flight," it was Roosa who took charge of the mission's first major crisis on Day One. The trouble began shortly after the trans-lunar injection on the night of January 31, 1971, as Roosa attempted the transposition and docking maneuver. After separating the command module *Kitty Hawk* from the lunar module *Antares,* he gently steered the probe of the mothership into the drogue of the LEM according to the standard "soft" docking procedure, but the capture latches inexplicably refused to lock. Roosa repeated the maneuver four more times to no avail. Finally, after a frustrating two-hour struggle, he improvised a "hard" docking procedure similar to the slam-bang coupling of two railroad cars, which popped the capture latches into place.

"It was a period of high anxiety," Roosa admitted later, noting that despite his ultimate success, the difficulty of the docking cast a shadow over the rest of the trans-lunar voyage.

Shepard and Mitchell anticipated no major problems undocking the LEM from the mothership when they reached lunar orbit, but they now had reason to worry about how they would get back inside the CM when they finished exploring the lunar surface. If the probe and drogue did not lock properly at rendezvous time, they would have to attempt a "space-walk" back to the mothership or they would die inside the LEM when their oxygen supply ran out.

Rather than dwelling on such grim thoughts, Mitchell occupied the spare moments of his in-flight rest periods with an ESP experiment. "It was a laboratory-type experiment that had been conducted in Dr. J. B. Rhine's Foundation for Research on the Nature of Man [in Durham, North Carolina) and dozens of other laboratories for the last thirty years. My motivation was to repeat that type of experiment to see if the space environment had any effect. Contrary to popular opinion, there were no cards carried on the flight; it was simply a blank page on my knee board and a table of random numbers that I put symbols against. I thought about the numbers during my rest periods, and allowed four people in the United States to try to guess what the orderings were. I would concentrate on each number for about fifteen seconds, and there were twenty-five numbers in each set, so it would take me about fifteen minutes or so to do the experiment. I was not trying to 'send' anything, I was just holding the picture of a number in my mind, seeing the image."

Mitchell later reported that two of his four associates back on Earth correctly guessed fifty-one of two hundred number sequences, where random chance would have resulted in only forty correct answers. He claimed that the results were "far exceeding anything expected," but were still only "moderately significant."

"This is an acceptable significant result in the other sciences," Mitchell told an Associated Press interviewer, "but parapsychology is more conservative and considers such odds as only suggestive or extra chance performance. We're much too uninformed, unknowledgeable in this mechanism of telepathy or ESP to project its uses, but I think once we start to understand what the mechanism is, then we can start talking about uses."

The Apollo 14 crewmen encountered their second major crisis on Day Five, at 12:30 A.M. Houston time on February 5, 1971, just as Shepard and Mitchell prepared to make their final descent to the lunar surface. Instead of undocking in a seventy-mile-high lunar orbit as their predecessors had done, they had first put the spacecraft into a seventy-mile-by-ten-mile elliptical orbit, then separated from the mothership at the fifty-thousand-foot level in hopes of conserving the LEM's descent fuel. But as the *Antares* completed the second of three scheduled orbits preliminary to the landing attempt, the control panel suddenly started flashing the ABORT light signaling a critical problem in the powered descent, a malfunction Mitchell describes as "a failure we just couldn't live with."

Houston quickly surmised that because the astronauts had not yet begun their powered descent, the ABORT signal was a false alarm triggered by an unidentified computer bug. But unless the error could be corrected before Shepard and Mitchell actually did start their powered descent, the false alarm would trigger a real abort, hurtling the astronauts right back into lunar orbit.

Although Shepard and Mitchell unceremoniously shut off the ABORT light by banging their fists on the control panel, correcting the underlying computer error required considerable brain power rather than old-fashioned brawn. The astronauts had enough fuel and oxygen to postpone their landing attempt for another three or four more orbits, but such a delay would surely wreak havoc on their already churning nervous systems. And if the mission fell that far behind schedule, the likelihood of running into all sorts of other unanticipated problems increased exponentially.

Houston had only one viable way to save the landing attempt. "They had to break out the M.I.T. people," Mitchell recalls, "and reprogram the computer in less than two hours while we were on the back side of the moon."

The brain trust Mitchell referred to as "the M.I.T. people" was a seven-man team at the Charles S. Draper Laboratory in Cambridge, Massachusetts. Only three of them were on round-the-clock duty; the other four had to be roused from their beds in the middle of the night. The crisis also demanded the aid of seventeen men from the Grumman Aerospace division in Bethpage, New York, who had to rush to their work stations over icy roads.

The NASA computer wizards had no idea whether the fault lay with the hardware, the software, or both. As a result, they were forced to improvise a complex set of reprogramming instructions that unavoidably created several new problems, not the least of which involved the time factor. As Mitchell notes: "After we came back around on the front side of the moon, we had a little under ten minutes from reacquisition of communications until we had to ignite the descent engine. All that reprogramming had to be accomplished in that period of time."

In order to feed in the reprogramming instructions, the astronauts had to deactivate part of the LEM's on-board computer guidance system so that Shepard could temporarily assume manual control of the spacecraft. Then Mitchell had to punch sixty keys on his computer terminal in rapid succession and with flawless accuracy. If the spurious ABORT signal had flashed again—even once—during his keypunching, Mitchell would have triggered a real abort.

Shepard and Mitchell managed to complete their reprogramming feat on schedule, but getting past that hurdle opened the door to yet another danger. Having instructed their computer to ignore the ABORT signal, the astronauts now had only one fail-safe option available during their descent. If a bona fide emergency occurred on their way down to the lunar surface, they would have to rely on the abort mechanism of their backup guidance system to get them safely back up into lunar orbit.

Mitchell later acknowledged that the ABORT light problem remained "a cause of great consternation" as the Apollo 14 astronauts ignited their descent engine. But Shepard, appropriately nicknamed the "Icy Commander" by his fellow astronauts, remained cool and upbeat.

"It's a beautiful day in the land of Fra Mauro," Shepard exulted as the LEM dove toward the moon.

Apollo 14's target was in a narrow valley between a pair of crater clusters dubbed Doublet and Triplet. Like the Apollo 12 astronauts, the Apollo 14 astronauts were expected to make a pinpoint lunar landing within walking distance of their primary objective: a geologically intriguing 250-foot-deep hole known as the Cone Crater. But the jagged terrain of the Fra Mauro region made Apollo 14's landing considerably more hazardous. Shepard literally had to thread the needle between the eight-thousand-foot-high peaks along his approach path, then take extra care to avoid the many sharp crater rims surrounding his landing site so the LEM would not tip over and crash after touchdown.

According to Mitchell, the uncontaminated clarity of the moon's airless environment compounded the difficulty of judging distances and visually cross-checking instrument panel data on the LEM's flight path and position. As he points out: "Here on Earth, people in the mountains or out in the desert on a clear day totally misjudge distances because it's so clear. You have no atmosphere on the moon, so it's ultra clear . . . You really don't have any familiar sights and familiar objects to gauge with. You don't know how big those boulders are."

Nevertheless, the LEM's reprogrammed computer guidance system managed to keep the Apollo 14 astronauts on a beeline for the land of Fra Mauro. At five hundred feet, Shepard reassumed manual control, and as

Mitchell droned out the latest altitude and velocity readings, calmly guided the spacecraft down through its treacherous approach. A few short but anxious minutes later, the two crewmen spotted the Doublet and Triplet crater clusters guarding their designated landing site.

"There it is—right on target!" Mitchell rejoiced. "Beautiful . . . right out the window . . . just like you said it would be."

Antares touched down at 3:18 A.M. Houston time on February 5, 1971, only two minutes behind schedule despite the heart-stopping disruption caused by the faulty ABORT light. The LEM came to rest in the valley between Doublet and Triplet, tilting slightly but in no danger of tumbling over. Mitchell says that he and Shepard breathed "a sigh of relief." Only after their return to Earth did they learn that the bug illuminating the ABORT light was a loose solder ball in the wiring.

"We're on the surface . . . we made a good landing," Shepard informed the men back at Mission Control. "That was a beautiful one. We landed on the slope. But other than that we're in great shape—right on the landing site."

Of course, Shepard and Mitchell had a long way to go and much to do to make their mission a true success. They had proven that NASA could still get a crew to the moon in the wake of the Apollo 13 failure. But so far, they had merely repeated the feats of Apollo 11 and 12. Now, in the course of the two moon walks they were scheduled to make during their thirty-three-and-a-half-hour stay on the lunar surface, they had to do something new and different, something to prove that landing men on the moon was—as opposed to unmanned probes—well worth the multimillion-dollar price tag.

The Apollo 14 astronauts knew that their multifaceted assignments dovetailed to a single, overall goal—unlocking the secrets of the universe. They hoped to find clues not only to the origins and physical composition of the moon, but also to the origins and evolution of life itself. Their objective wasn't just showing that men could land on the moon and walk on its surface. Rather, as *New York Times* space reporter Walter Sullivan had observed during Apollo 14's trans-lunar voyage, "For the first time, men have made scientific exploration the primary objective for landing on the moon."

Consequently, when the Apollo 14 crewmen began their first moon walk at 8:49 A.M. that morning, they had to unload the most numerous and complex array of scientific equipment ever brought to the lunar surface. Their ALSEP package contained more than a dozen separate instruments, including: a seismometer; a "solar wind" detector; a cold cathode ion gauge for recording possible subatomic traces of a lunar atmosphere; a charged particle detector for measuring any proton and electron volleys impinging on the moon from outer space; a laser reflector for tracking the exact and cyclically shifting distances between the moon and the earth; and a nuclear-powered generator similar to Apollo 12's.

In order to conduct a risky new experiment unique to their mission, Shepard and Mitchell also unloaded an arsenal of grenade launchers and a packet of explosive charges. NASA scientists hoped that by blowing a series of holes in the lunar surface, the astronauts would be able to probe the underlying rock layers, and help solve the mysteries of the moon's geological structure.

That was also the motive for what Shepard and Mitchell regarded as the most important specific objective of their mission. On their second moon walk, they planned to climb down into and explore the uncharted bowl of the Cone Crater. With a reported depth of some 250 feet, the floor of the Cone Crater was believed to be one of the lowest points on the lunar surface. NASA scientists therefore believed that it would yield rock specimens from some of the deepest and oldest layers of the moon.

Fortunately for the Apollo 14 crewmen, their LEM cargo included the first two-wheeled vehicle ever brought to the moon, the Modularized Equipment Transporter (MET), a metal-ribbed rickshaw that they referred to as their "golf cart." The astronauts happily informed Houston that they could both see the Cone Crater nearby, and that they would have "no trouble" pushing the instrument-loaded MET from the landing site to the rim of the crater when the time came. At that point, says Mitchell: "We started feeling comfortable, that we were safe, and we could experiment and explore the lunar environment."

Their sense of security increased as they ventured farther away from the LEM.

"I think they put champagne instead of iodine in the drinking water this time," Mitchell bubbled, as the astronauts giggled about the way they were "hopping around like kangaroos."

"There was that real sense of euphoria that you were exploring another planet, being out on a surface people had never been on before," Mitchell recalls. "There were just so many things to command your attention, so many things to look at and experience."

Foremost among these "aha!" experiences, says Mitchell, was "looking up and seeing the earth directly overhead," then turning to see the

LEM off in the distance "looking a little bigger than a cigarette lighter. It made you realize that you were standing out in a wilderness all by yourself."

But Shepard and Mitchell spent much more time working than wondering at the marvels of Fra Mauro. In addition to blasting deep rock samples with their grenade launchers, they set up most of the experiments in their ALSEP kit, including the seismometer, which immediately demonstrated its ultrasensitivity. The first transmissions the device sent back to Earth were recordings of their own footsteps, sounds too faint for them to hear with their own ears.

The astronauts crawled back inside the LEM at 1:31 P.M. after an EVA of four hours and forty-one minutes. They were now scheduled to have a ten-hour rest period before commencing preparations for moon walk number two. Like their predecessors on Apollo 12, both men had trouble getting to sleep, primarily because their spacecraft kept emitting disturbing creaks and groans. But Mitchell reports that he was surprised at how quickly they adapted to their alien environment. "When you were taking a break in the LEM and took off your helmet, you had a chance to relax and reflect a little bit."

Meanwhile, up in the *Kitty Hawk,* command module pilot Stu Roosa spent much of his solitary time in lunar orbit carrying out the most extensive photo-mapping assignment to date. Though hardly as thrilling or as hazardous as exploring the lunar surface, his project proved to be more scientifically valuable than all the film and snapshots taken by the six men who had so far walked on the moon. In addition to photographing the Apollo 14 landing site, he captured the first detailed images of the "zodiacal light," believed to be a cloud of reflective dust near the sun, and the mysterious solar glow known as the "gegenschein."

"That dim light photography was very complicated," Roosa remembers, "because you had to do it in total blackness, the blackest you can ever put a human being in without closing him in an absolute black room. You have no Earth light, you have no sunlight, you have no reflected light bending the corners anywhere. It is black-black. Because of the problems of light coming in and the high-speed film, you want to take your pictures as close to the moment of sunrise as you can, so you get your timer going and you count by listening to the click every second. You knew when sunrise was supposed to be, it was noted right there in your flight plan.

But boy, at the time, the sunrise was just instantaneous—POW! There's the sun right in your window."

Roosa's nights in the command module were less hectic but also considerably less enjoyable than his days. Like his comrades down in the LEM, he, too, found it hard to relax, and slept fitfully, dozing off only for a few short, unrefreshing, and frequently interrupted naps. Their insomnias differed only in cause. Well accustomed to the mechanical noises of the *Kitty Hawk* and relatively comfortable in his flight suit, Roosa says his overnight discomfort resulted mainly from the subtle drop in cabin temperature during passes around the far side of the moon.

"It never bothered me during the day," he says, "but during my rest period, I do recall that I got a little chilly . . . It's dark . . . the darkness is something you feel rather than see . . . You feel the cold clamminess . . . The humidity inside the spacecraft is a hundred percent, and as you come out of the sunlight and start in through the dark, the temperature doesn't have to drop but a degree or so, and you've got moisture condensing on the side of the spacecraft . . . When the water would condense, you could feel that clamminess in there."

But he also insists that the sheer adventure of flying alone in lunar orbit more than made up for his discomforts, especially the thrill of seeing the sun rise every two hours as he emerged from each pass around the dark side of the moon. "The feeling you got was just so good—like you're just born to a new day every time."

Shepard and Mitchell began their new day by stepping back out onto the lunar surface at 2:20 A.M. on the morning of February 6, and proceeded on what they thought was a direct route to Cone Crater with their equipment-packed MET "golf cart." But the undulating terrain and airless clarity of Fra Mauro kept causing them to misjudge distances and elevations.

"I thought we were in a low spot with the LEM," Mitchell informed Houston, "but it turns out we're really in the lowest spot around. You can sure be deceived by the slopes here. The sun angle is very deceiving."

As the trek wore on, Shepard and Mitchell repeatedly spied what they believed to be the rim of Cone Crater a few yards away, only to traverse the intervening incline and discover they were actually on some other unidentified hill or crater rim. Finally, after hiking for over two full hours, the astronauts admitted they were lost.

"Our positions are all in doubt now," Shepard reported. "I'd say that the rim [of Cone Crater] is at least thirty minutes away . . . We're approaching the edge of a boulder field here from the south flanks, and what I'm proposing is perhaps that we use that as the turnaround point. It seems to me that we would spend a lot more time in traverse if we don't."

Still intent on reaching the Cone Crater, Mitchell wanted to keep going to at least the far edge of the boulder field, but Shepard did not think there was enough time to do that and still make it safely back to the LEM.

"Oh, let's give it a whirl," Mitchell insisted. "Gee whiz, we can't stop without looking into the Cone Crater."

"I think that we're looking at it right here," Shepard replied, "this boulder field, Ed, is the stuff that is ejected from the Cone."

"But not the lowermost part," Mitchell countered, "which is what we're interested in."

With Houston's approval, the astronauts decided to extend their search for the Cone Crater thirty more minutes. But as the clock kept ticking, the moon walkers grew wearier, and their heart rates kept rising rapidly. By the end of the alloted extra half hour, the Cone Crater was still nowhere in sight. The men at Mission Control decided to call it quits.

"We've already eaten in our thirty-minute extension and we've passed that now," announced CAPCOM Fred Haise. "I think we'd better proceed with the [rock] sampling . . ."

Shepard seemed to greet the order to turn back with bittersweet resignation. But according to Mitchell, both astronauts were crushed by their failure to reach Cone Crater. "We felt that we had failed in some sense to achieve the mission."

The Apollo 14 astronauts tried to take solace in the fact that at least they were not returning to the LEM empty-handed. With the help of their MET "golf cart," they had managed to collect over 93 pounds of moon rocks, far more than Apollo 11's 47.5 pounds and Apollo 12's 75.4 pounds. And before they got back inside their spacecraft for the last time, they still had one final surprise for their TV audience on Earth.

Upon reaching the landing site, Shepard produced a small metal flange from one of the pouches of his "moon cocoon" and attached it to the long aluminum handle of his rock sample collector. Then he hopped several yards away from the LEM, stopped, and turned back toward Mitchell and the TV camera.

"Houston," rasped the Apollo 14 commander, "you might recognize what I have in my hand is the handle for the contingency sample return. It just so happens to have a genuine 6-iron on the bottom of it . . ."

Shepard reached back into one of the pouches of his pressure suit and pulled out a shiny new golf ball.

"In my left hand, I have a little white pellet that's familiar to millions of Americans," he announced. "I'll drop it . . ."

When the ball plopped to rest beneath his feet, Shepard tried to assume a conventional golf stance only to find that the girth of his "moon cocoon" made it impossible for him to grip his makeshift 6-iron with both gloves at the same time. He had no choice but to attempt man's first extraterrestrial golf shot one-handed.

"I'm going to try a little sand trap shot here," Shepard declared, then jerked his club back and through.

The ball popped almost straight up in a cloudy divot of moon dust, and seemed to hang in mid-flight as if suspended on a string. Then it tailed off to the right, and fell back down to the lunar surface less than a hundred yards away.

"That looked like a slice to me, Al," teased CAPCOM Fred Haise.

Shepard twinged with embarrassment. Any weekend golfer could do better than that, even one-handed, in the one-sixth gravity of the moon. He pulled out another golf ball, and determined to try again.

"There we go," he grunted. "One more . . ."

Shepard's second shot went over 500 yards, rising in a majestic arc, then soaring away clean out of sight.

"Miles and miles and miles," Shepard exclaimed triumphantly.

"Very good, Al," complimented CAPCOM Haise.

Shepard put away his now invaluable 6-iron, and helped Mitchell finish packing up the rest of their gear. At 6:38 A.M. Houston time, having spent over four and a half hours on the lunar surface during their second EVA, the astronauts prepared to climb back inside the LEM.

"Okay, Houston, the crew of *Antares* is leaving Fra Mauro Base," Shepard reported.

"Roger," replied CAPCOM Haise, who would have explored Fra Mauro the year before had the Apollo 13 service module not exploded on the way to the moon. "You and Ed did a great job. I don't think I could have done any better myself."

"That's debatable," Mitchell retorted, still rankling over their failure to reach the Cone Crater, "isn't it, Fredo?"

"Well, I guess not now, Ed," Haise replied, and quickly commenced a play-by-play commentary of Shepard's departure from the lunar surface.

"Al Shepard preparing to start up the ladder now . . . Moving forward . . ." Haise continued until Shepard at last shut the hatch. "And with that closeout, Al Shepard—now two score and seven years—becomes the undisputed leader in time spent walking and working on the moon, more than nine hours . . . A close second is his partner, Ed Mitchell."

Despite their failure to reach the Cone Crater, the Apollo 14 astronauts realized during and after their return to Earth that they had done much better than they thought. As it turned out, Mitchell had been right about the location of the Cone Crater. After comparing data on their moon walk route and Fra Mauro photo maps, NASA scientists estimated that the astronauts had stopped within fifty feet of the Cone Crater's rim.

But Shepard had also been right about the value of the rocks they collected at their boulder field turnaround point. They proved to be some of the oldest and deepest samples ever gathered, with geological histories of over 4.5 billion years. In short, the Apollo 14 astronauts had effectively accomplished their primary scientific objective after all, prompting to Shepard to brag that the mission was "a smashing success."

True or not, Apollo 14 proved to be a major turning point in the history of manned lunar exploration. Shepard and Mitchell were the last of the six original moon walkers. The next six men who landed on the lunar surface would arrive with four-wheel vehicles that enabled them to roam farther away from their LEMs and explore more territory per EVA than all their predecessors combined. But the three "moon rover" missions to come would also signal the beginning of the end of Project Apollo.

13

THE MOON ROVERS

Man's Last Three Lunar Landings

Apollo 15, 16, and 17: 1971–1972

I felt the presence of the Lord up there on the moon.
APOLLO 15 LUNAR MODULE PILOT JAMES B. IRWIN

On the morning of July 31, 1971, Apollo 15 astronauts David R. Scott and James B. Irwin emerged from the lunar module *Falcon,* and silently surveyed the alien terrain around their landing site in the Hadley-Apennine region. They were the seventh and eighth men to set foot on the lunar surface, but they were as filled with awe and exhilaration as all six of their predecessors combined.

"As I stand out here in the wonders of the unknown at Hadley, I try to realize there is a fundamental truth to our nature," Scott proclaimed for everyone listening back on Earth to hear. "Man must explore. And this is exploration at its greatest."

A few minutes later, the Apollo 15 astronauts opened a new age in lunar exploration by driving across the boulder fields in a four-wheel vehicle appropriately dubbed the "Moon Rover."

"We're on the way!" Scott cried with glee. "Oh, boy, is this traveling! It's great sport, I tell you . . ."

"Man, oh, man," Irwin chimed. "What a grand prix this is!"

A topless two-seater with titanium-coated solid core tires and an inverted umbrella antenna dish sticking up in front, the so-called Lunar Roving Vehicle (LRV) looked like a cross between a golf buggy and a dune buggy, but it represented a "giant leap" in extraterrestrial transportation. Designed by a Boeing–General Motors joint development team, the rover could cruise at speeds of up to ten m.p.h. for at least thirty-six linear miles, enabling the astronauts to explore ten times the area covered by previous moon walkers.

In addition to making automotive history, the Apollo 15 astronauts managed to give the nation's embattled manned space program a sorely need p.r. boost by documenting their Moon Rover rides with vastly improved ground-controlled color TV cameras. Having planned only minimal coverage of the mission due to declining audience ratings, the major U.S. networks preempted their regularly scheduled programs to broadcast the dramatic live footage beamed back from Hadley Base.

"That added a whole new dimension to our visit," Irwin recalls. "Houston panned and zoomed the camera lenses so they could focus attention on what we were doing and pick up the sense of excitement of exploring another world. People on the earth could feel like they were actually up there exploring the mountains and seeing the moon firsthand."

Of course, the Moon Rover show wasn't cheap, a fact that continued to stir debate among American taxpayers. The cost of the vehicle's development program, originally projected at $19 million, had doubled to some $38 million. As a GM spokesman pointed out in defense of the expenditures, it cost at least $70 million to develop a new automobile. The difference was that GM could sell millions of units at about $5,000 each (in 1971 dollars). NASA would send only three Moon Rovers to the lunar surface. Space agency officials placed the average construction cost per vehicle at $2 million, but when total development costs were factored in, the real price per rover was over $12 million.

In any event, the Apollo 15 astronauts saw the Moon Rover as a relatively small part of their mission's overall contribution to the advancement of human knowledge. As Irwin points out: "Ours was labeled the first 'extended' scientific mission to the moon, and that's exactly what it was. It was science. It was extended operation. Extended radius. Extended time. We were there about twice as long as the previous missions, and we had a very exciting area to explore."

Scott and Irwin were the first men to land in a lunar mountain range, and the first to strike what they accurately described as a major geological "gold mine." Along with making unprecedented firsthand observations that would help scientists determine the exact origins and history of the moon, they bagged the oldest and most awe-inspiring specimen ever recovered from the lunar surface—the "genesis rock."

Irwin would later report making another unique and even more extraordinary "discovery" of an intangible nature during his expeditions across the lunar surface. According to the devout, "born-again" Christian: "I felt the presence of the Lord up there on the moon."

Not surprisingly, the Apollo 15 lunar module pilot's incredible claim made international headlines following his return to Earth. But much to the dismay of all three crewmen, news of Irwin's religious epiphany was quickly overshadowed by revelations of an Apollo "commercialization" scheme involving first-edition stamps and commemorative coins that eventually scandalized the entire NASA astronaut corps.

For these and other reasons, most of the world did not appreciate the true importance—or the true dangers—of the Apollo 15 mission, and probably never would. That unfortunate fact was already becoming abundantly clear before lift-off on July 27, 1971, when a reporter asked the astronauts to justify the cost and potential benefits of "going back" to the moon.

Scott, a six-foot-tall, thirty-nine-year-old air force colonel with a "kinda corny" old-fashioned American patriotism, fairly bridled at the question. Born on an airfield in San Antonio, he was the son of an influential brigadier general, a West Point graduate with a masters in planetary science from M.I.T., and an avid swimmer whose triceps threatened to burst the seams of his custom-tailored pressure suit. Married to a San Antonio sweetheart who had borne them two daughters, he had also flown on the nearly disastrous Gemini 8 mission with Neil Armstrong, and on the crucial Apollo 9 mission that made the first LEM-CSM rendezvous in Earth orbit.

"I don't think it's a matter of 'going back,' " the Apollo 15 commander retorted. "It's a matter of going to a new place for new data. We're visiting the same planet, but my goodness, did they say they were 'going back' after Columbus had gone to North America?"

Jim Irwin, a forty-year-old fellow air force officer who had suffered more than his share of hard knocks just to qualify for Project Apollo, saw no need to elaborate. Born in Pittsburgh and raised in Salt Lake City, the dark-haired and handsome LMP was the son of a plumber, and a married man with four children. After graduating from Annapolis, Irwin had applied for air force flight school only to be rejected and assigned to a desk job. Although he successfully reapplied a few months later, he had subsequently received two more NASA rejection notices before being accepted for astronaut training.

Besides suffering these official insults, Irwin had also suffered some serious physical injuries. A near-fatal plane crash back in 1961 had resulted in a broken jaw, two broken legs, and a brain concussion that caused severe amnesia. As he confided in a prelaunch interview ten years later,

"I had to go through a lot of psychiatric help and treatment with truth serum and hypnosis to bring back that lost memory." And though there was no doubt that he had fully regained his senses, the experience had obviously fortified his religious faith and hardened his resolve to fly to the moon.

Apollo 15 command module pilot Alfred M. Worden, who prided himself on being a member of the first all air force crew bound for the moon, was even more gung-ho. Recognized as the swinger of the threesome, the thirty-nine-year-old son of a movie projectionist had once aspired to being a jazz musician, but after a year of undergraduate studies at the University of Michigan, he had transferred to West Point, then gone on to air force flight school. Recently divorced, he lived across the street from his ex-wife and their two daughters, but led an "active social life," dressing in fancy Western-cut suits and dating flashy Manhattan models.

At the same time, Al Worden had his serious side. When asked what he would be doing if not in the astronaut corps, he replied, "I would be volunteering for Vietnam. Not because I think the war is right or just or anything, but because almost all of my buddies are over there . . . It's unfair for them to be carrying all the burden by themselves." Yet, in light of the June 1971 Soyuz 11 tragedy that claimed the lives of three Soviet cosmonauts, he felt that he and his fellow astronauts were no less courageous than his buddies in Vietnam. As he put it, "We're not fighting a war, but damn if our risks aren't as great as the ones they take."

Apollo 15's first major in-flight crisis served as a startling reminder of those risks. On Day Three of the mission, when the astronauts were about two-thirds of the way to the moon, the command module *Endeavor* sprang a water leak that threatened to flood the entire cabin. Scott, Irwin, and Worden realized that a plumbing emergency in zero G could turn into a terrible nightmare, for there was no gravity to help them bail out the ship. Luckily, the astronauts managed to stop the leak in less than half an hour thanks to step-by-step instructions from Houston, otherwise all three would have drowned in space.

Early on the morning of July 31, 1971, the day of the landing attempt, Scott and Irwin had to confront the special dangers posed by the Hadley-Apennine region, whose rugged topography resembled the southern Rockies of the U.S. Huddled between the Mare Imbrium ("Sea of Rains") and the Mare Serenitatis ("Sea of Serenity") that formed the left and right "eyes" of the "Man in the Moon," the Apennine range curved like a giant amphitheater around the Palus Putredis ("Putrid Swamp"), bristling with jagged peaks up to twelve thousand feet high. Hadley Rille, the lunar counterpart of the Grand Canyon, was a winding crevice about a mile wide and some 1,200 feet deep. Both the mountains and the rille could dash the LEM to pieces.

The precariousness of Apollo 15's descent through the lunar mountains was not lost on either astronaut. Irwin found that the only way he could cope with the overwhelming reality outside the windows of the LEM was by retreating to make-believe. "I figured that if I really thought that I was landing on the moon, I might get so excited that I might forget something. So all during the powered descent I kept telling myself, 'Jim, this is not for real. Jim, this is not for real. You're back in the simulator. Just remember that. This is not for real.' "

Apollo 15's second major crisis occurred at the worst of all possible times—touchdown. Due to newly added payload items like the Moon Rover and related design changes, the lunar module *Falcon* weighed 2,470 pounds more than previous LEMs. Although the landing gear was equipped with heavy-duty struts and shock absorbers, Scott and Irwin had to use the same primitive touchdown procedures as their predecessors, which was basically a calculated crash landing.

At about eight feet above the lunar surface, the *Falcon*'s contact lights beamed on, and the astronauts cut their descent engines and simply let the LEM fall the rest of the way down, relying on "blind faith" to survive the bone-jarring impact that followed.

"Man, we hit hard," Irwin recalls. "Then we started pitching and rolling to the side. My first thought—and I'm sure Dave's, too—was, 'Man, are we going to get into an unstable position? Are we going to have to abort?' I thought surely we had ruptured something, that something might be leaking, and we were going to have to leave right away. Finally, the vehicle stopped rolling, and came to rest on the side of a crater, and we just held our breath for about ten seconds."

Then word came back from Houston: "You have a STAY."

The astronauts did not yet realize just how close they had come to disaster. Several hours later when they crawled out to begin their first EVA, they would see that the *Falcon* had landed less than fifteen yards from the rim of a crater big enough to swallow the LEM and/or dash it to pieces. But upon receiving initial approval to stay put, they reveled in an ignorant bliss.

"At that point," says Irwin, "Dave and I just hugged each other, pounded each other on the back like we had just made a winning touchdown in a football game. Then we started looking out the window and making remarks about what an unusual place we had just landed at. We got so carried away with the thrill of being on the moon that we forgot we had a time line to follow, and Houston called and reminded us, 'Hey, guys, there is some work to be done.' "

But instead of proceeding directly onto the lunar surface, Scott and Irwin started to get ready for bed. The astronauts planned to stay on the moon for nearly three full days, and make three separate excursions in the Moon Rover, a schedule that demanded that they do a great deal more work and deploy a great deal more equipment than their predecessors. But the astronauts could not retire without taking one quick look at the moonscape around Hadley Base.

"Before we went to sleep," Irwin recalls, "we opened the top hatch, and Dave climbed up in that area and took a panorama view around and described what he saw, which just whetted the appetite—our appetites and the appetites of the geologists and scientists on the earth . . . Then he came back down, and we closed the hatch and we fixed our hammocks and we went to sleep."

Irwin claims that he and Scott got "the best night's sleep of the entire mission" during that first rest period, adding with a mischievous grin: "Maybe we slept well just to confound the doctors."

The Apollo 15 astronauts also enjoyed a new freedom that obviously enhanced their long and sweet repose. Unlike their predecessors, they were allowed to remove their pressure suits and sleep in the nude. Free from the confinement of their "moon cocoons," Scott and Irwin could toss and turn in their hammocks without scraping against collar rings and tight knee joints. Likewise and perhaps even more important, they could get up and go to the bathroom in the middle of the night instead of discharging their bodily wastes into a disposable diaper. Says Irwin: "I can't imagine keeping the suit on. No way I'd want to use that diaper."

Scott emerged from the LEM and hopped down to the lunar surface at 8:26 A.M. that same morning. Irwin followed less than ten minutes later, a new exit record for an Apollo lunar module pilot that testified to man's increasing confidence about being on the moon.

However, the two astronauts found the task of unloading their Moon Rover much tougher and more time-consuming than either expected. In order to fit the rover's ten-foot-long frame inside the LEM, the Boeing-GM team had hinged the chassis so the front and rear wheels could fold over like the legs of a lawn chair. But that space-saving design also required the astronauts to engage in a two-step tug of war to get the vehicle out.

First, Scott had to yank on a long white lanyard that opened the cargo bay and allowed the Rover's front wheels to unfold while slowly lowering them to the surface. Then he had to hustle around to the opposite end of the vehicle and pull a second lanyard that unfolded and lowered the rear wheels. But even with Irwin's assistance, it took nearly an hour to complete the procedure, and several more minutes of grunting and groaning to detach the lanyards and carry the Rover a safe distance away from the landing pads.

Upon climbing into the driver's seat for a preliminary test run, Scott discovered a serious glitch in the steering mechanism. The rover was equipped to make "gentle" cornering maneuvers by simultaneously turning all four wheels. If the front wheels turned to the right, the rear wheels were supposed to turn to the left, and if the front wheels turned left, the rear wheels were supposed to turn right, thereby steadying and breaking the directional drift. But for some unknown reason, the front wheels refused to turn at all.

"Any suggestions?" the Apollo 15 commander asked the men back at Mission Control.

"Cycle over the forward steering circuit breaker, please," advised CAPCOM Joe Allen.

Scott did as instructed, only to report that the front wheels still failed to unlock. The astronauts now began to fear that they would be forced to curtail severely or even scratch all three of their scheduled long-range excursions in the rover. But because preflight ground tests had shown that the vehicle could turn reasonably well with only rear-wheel steering, Houston decided to take a chance, and radioed the words they wanted to hear.

"Press on," ordered CAPCOM Allen.

Scott and Irwin set out on man's first moon ride at 10:20 A.M., and drove a total of five linear miles over the next five hours. Their meandering route first took them south to the so-called Elbow Crater, where they stopped to collect rock samples and marvel at the sharply curved rim of nearby Hadley Rille. Then they veered off to the west to inspect the gaping mouth of the St. George, collected more rocks, and hooked back east along the northern edge of the Apennine Front.

"The rover handles quite well," Scott informed Mission Control. "We're moving, I guess, at an average of about eight kilometers an hour.

It negotiates small craters quite well, although there's a lot of roll . . ." Then, turning to Irwin, he added: "It feels like it needs seat belts, doesn't it, Jim?"

Irwin, who was assigned to provide a running commentary of the ride, couldn't resist a little "backseat" driving.

"You better watch the road, Dave," he cautioned as they approached a steep slope.

"You keep talking," Scott snapped. "Let me drive."

A few moments later, the Moon Rover skidded down the slope and went into a 180-degree spin.

"We just did a christie," Scott reported.

Although the Apollo 15 crewmen ventured more than three times farther away from their landing area than previous astronauts, they expressed no qualms about losing sight of the LEM nor did they worry about breaking down in some remote crater beyond walking distance from Hadley Base. Rather, they gained increasing confidence in themselves and their vehicle with each passing mile. "The rover made us feel more at home," says Irwin, "like we were on the earth where we could just get in our car and drive around wherever we wanted to go."

Even so, Irwin admits that the lunar landscape was so starkly lit and so alien in both form and content that he had a hard time believing his own eyes, especially after encountering a startling variety of rock and soil colorations.

"It looked unreal." he says, "Going around the moon looking down from lunar orbit, it was brown gray. Right at the terminator line between lightness and darkness, it was gun-metal gray, and as you travel from that point closer to the area under the sun, the surface gets lighter going from gray to brown to light tan, almost white. Down the surface, there were these shades of gray and brown, and then you find these rocks of all colors, if you could imagine, in miniature, much like walking along a sandy beach . . . We found black rocks, we found white rocks, we found green rocks with crystals, all conceivable colors, but the overall effect is just brown."

At the same time, Irwin claims that he also felt surprisingly at ease on the lunar surface. "I felt like an alien when I traveled through space, but when I got to the moon, I didn't feel that at all. I felt at home there even though the earth was a long ways away, and we could see it directly above, about the size of a marble . . . I felt like I was at the end of a thin cord that could be cut at any time . . . Yet I felt comfortable, and would have liked to have stayed there a longer time."

Irwin adds that the similarities between Apollo 15's mountainous landing site and the mountainous regions of the southwestern U.S. enhanced his sense of security: "I've always loved the mountains of the earth, done quite a bit of hiking and climbing in the mountains, and I always feel very inspired and elated when I am in the beauty of the mountains. Here to be camped in the shadow of the Apennine Mountains . . . to be able to drive that little automobile on the slope of the mountains . . . I felt very comfortable, very protected . . . I felt quite at home."

Irwin says that his confidence also came partly from a religious feeling, a sense "that someone was there that I cannot see."

"The feeling was so strong," he reports, "that several times I looked over my shoulder to see if someone was there. Perhaps it was because so many people on the earth were focusing their attention on us, they were sending us signals somehow . . . I guess we were very receptive to these things . . . I just sense that a spiritual presence was there. I refer to it as the presence of the Lord because I don't know any other words to describe it . . ."

The astronauts returned to the LEM at 2:40 P.M. that afternoon, after spending six hours and forty-three minutes on the lunar surface. Before closing the hatch, they set up an ALSEP package of scientific instruments powered by a nuclear generator, and drilled a sixty-four-inch-deep hole to test the underlying rock layers.

Having already felt "the presence of the Lord," Irwin says he now felt a seemingly contradictory sense of aloneness as he surveyed their landing site. "I guess it was similar to the feeling Adam and Eve had as they were standing on the earth and they realized they were all alone, there was no one else on the earth, I realized that Dave Scott and I were all alone on this vast planet, another world. We were the only two there. We felt very, very special, and yet we felt an unseen law, so we were not alone."

That night, the astronauts did not sleep well. At first, they thought the LEM had sprung an oxygen leak. The problem turned out to be caused by a urine dump valve they had neglected to close. But the crisis and the events of the preceding day seemed to have put both crewmen on edge, and they managed only five hours of real sleep during their fifteen-hour rest period.

Scott and Irwin began their second moon ride at 6:47 A.M. Houston time on the morning of August 1, 1971, and it proved to be the most scientifically productive of all. The astronauts once again drove all the way to the base of the Apennine Front, stopping to visit the South, Front, and Secondary craters along the way. But it was the Spur Crater that proved to be, in Scott's words, the real "gold mine." There, sitting up on a pedastal-shaped boulder, was the oldest moon rock ever found by man—a specimen that dated back almost 4.5 billion years.

"Oh, boy, guess what we found?" Scott exclaimed. "I think we found what we were looking for!"

Irwin felt even more elated. "That was remarkable to see that little white rock sitting on that little pedestal almost free from dust. Most of the rocks up there are covered with dust, dust that has been moved about the surface of the moon for millions and millions of years. Here was that rock clearly displayed for us . . . like it was just there for us to pick up . . . There was great elation in our souls and in the souls of the people back on Earth . . . Someone in the press room immediately labeled it 'genesis' because it seemed to be a modern-day revelation."

The Apollo 15 crewmen capped off their second EVA by using a special battery-powered drill to excavate a narrow trench from which they hoped to obtain valuable "deep" rock and soil samples, thereby becoming, as Irwin later noted with a self-effacing grin, "the first ditchdiggers on the moon." They would eventually amass over 170 pounds of samples, nearly twice the amount brought back by the Apollo 14 astronauts. Then, after spending a record seven hours and thirteen minutes on the lunar surface, the exhausted astronauts climbed inside the LEM, and tried to go to sleep.

Later that afternoon at a specially convened press conference in Houston, veteran flight director Gerald D. Griffin jubilantly informed the world media; "We have witnessed the greatest day of scientific exploration that we've ever seen in the space program."

At 4 A.M. the following morning, Scott and Irwin got back into the Moon Rover, and took off on a third and final expedition to the edge of Hadley Rille, where they made the most significant geological discovery of the entire mission.

In the course of describing the fantastically sculptured features of Hadley Rille, Scott reported that he could see "maybe ten well-defined layers" in the canyon walls. His seemingly mundane observation had enormous implications for the ongoing debate about the origins and history of the moon, for it was the first eyewitness evidence that indicated that the nearby Sea of Rains and other great lunar maria were not formed by meteorite impact or by a single volcanic eruption but by a series of "episodic" lava flows over many millions of years. According to CAPCOM Joe Allen, "That one sentence alone was perhaps worth the trip."

The Apollo 15 commander had one more surprise for Houston and his worldwide TV audience when the astronauts returned to their landing site. After parking the Rover on a relatively level plateau a few yards away from the LEM, Scott retrieved a hammer from one of the equipment kits and pulled a feather out of one of the pouches of his pressure suit. Then, as Irwin obligingly refocused the remote control camera, he announced: "In my left hand, I have a feather; in my right hand, a hammer. I guess one of the reasons we got here today was because of a gentleman named Galileo a long time ago who made a rather significant discovery about falling objects in gravity field, and we thought where would be a better place to confirm his findings than on the moon . . . So we thought we'd try it here for you . . . The feather happens to be, appropriately, a falcon feather for our *Falcon,* and I'll drop the two of them here, and hopefully, they'll hit the ground at the same time."

Just as Scott and Galileo predicted, the feather and the hammer did land on the lunar surface at the same time.

"How about that?" Scott exclaimed. "That proves Mr. Galileo was correct about his findings."

Before climbing back into the LEM, the astronauts set up the cameras on the Moon Rover to film the first live footage of a lunar lift-off, the lift-off they planned to make that afternoon. Scott also stopped to cancel a series of newly issued commemorative postage stamps, an act that would later become a matter of public controversy back on Earth. In the meantime, Irwin stopped to enjoy one last religious moment on the lunar surface.

"I am reminded of my favorite biblical passage from Psalms," Irwin declared. "I'll look unto the hills from whence cometh my help . . ."

Then he quickly added: "But of course, we get quite a bit from Houston, too."

What Irwin failed to note at the time was that he was looking at a scene neither God nor man had ever witnessed before. The lunar hills unto which he looked were no longer unmarked solitary dunes. For the first time in history, they bore tire tracks made by man. And henceforth and forever more, those hills would be the resting place of the odd-looking four-wheel vehicle Scott and Irwin were leaving behind—the first Moon Rover.

APOLLO 16
APRIL 20–23, 1972

Charley, you said you were going to see some other tracks on the moon.
—Apollo 16 commander JOHN W. YOUNG

Apollo 16 lunar module pilot Charles M. Duke had a very strange dream one night in the spring of 1972 about a week before launching for the moon. Though he claims it was "not a nightmare situation," Duke decided not to share the dream with his worry-prone wife, Dotty. But he did tell Apollo 16 commander John W. Young, for the dream involved both of them and their forthcoming roles in man's second Moon Rover mission.

"In my dream, John and I were driving the Rover up to the north, up to one of our stops," Duke recalls, "and as we came across this ridge, there was this set of tracks out there in front of us. We asked Houston if we could follow the tracks, and they said yes. So we turned and followed the tracks, went about an hour or so, and we found this vehicle that looked just like the Rover, and it had two people in that looked like me and John.

"I didn't feel like this was a premonition that we were gonna die up there on the moon or anything like that. It didn't upset me. I felt comfortable that these two other astronauts looked familiar, looked just like us, and I took off one of the parts from this other vehicle so that we could show the people in Houston. And I knew that we made it home okay, because in my dream I can remember handing the part to one of the lab technicians in Houston after we got back."

Ironically, the Apollo 16 astronauts claim that landing on the moon was no big deal—at least not for them. Although they were only the ninth and tenth men to accomplish a feat that was still an "impossible dream" less than four years earlier, they felt as if they were merely giving an umpteenth repeat performance of a traveling road show.

"It didn't overwhelm me," says Duke. "It was all technical. I had to make sure things were running right inside the LEM, so I didn't look out the window that much during our descent. But from the best I could tell, it was just like another simulation in the simulator. By the time John and I got out on the lunar surface, it seemed to be anticlimactic because it was so much like training."

Most of the world public also seemed to regard the Apollo 16 mission as anticlimactic because it did not feature any dramatic historical firsts. But as *New York Times* science reporter Walter Sullivan pointed out at the time, Apollo 16 was "the longest, most ambitious, and, in many respects, most productive" of the first five Apollo lunar landing missions.

"Astronauts John W. Young and Charles M. Duke spent more time on the moon . . . than anyone to date," Sullivan noted. "They traveled farther and faster over the lunar surface. They brought home more specimens . . . And they conducted several novel observations that could be of major scientific importance."

In fact, two of Apollo 16's "novel observations"—the discovery of aluminum deposits and magnetic rocks on the lunar surface—had revolutionary implications for the uncertain future of manned space exploration and for accepted scientific theories about the origins and nature of the moon.

And like each of the previous lunar missions, Apollo 16 encountered a last-minute crisis that nearly turned its "anticlimactic" moon landing into an unmitigated disaster.

Fortunately, Apollo 16 boasted one of the coolest crews to date. Commander John Young, a forty-one-year-old former navy test pilot, had flown in space more times than anyone else in the NASA astronaut corps except Apollo 13's Jim Lovell. His three previous missions included the Gemini 3 flight in 1965, America's first two-man space shot; Gemini 10, a 1966 Earth orbital test of rendezvous techniques; and Apollo 10, the May 1969 "dress rehearsal" for man's first lunar landing mission. Short, tough-minded, and, in the words of one colleague, "super-trained, always training," he made his critical opinions clear in no uncertain terms. And according to what became known as Young's Law, he believed, "You're not really an astronaut until you've flown in space."

By that definition, Young's fellow crewmen were only would-be astronauts, but neither could be considered typical rookies. LMP Charley Duke, a six-foot-tall air force colonel born in Charlotte, North Carolina, had served as CAPCOM for the Apollo 11 landing, a role in which he demonstrated his easygoing grace under pressure. Command module pilot Thomas K. "Ken" Mattingly, a thirty-six-year-old navy captain from Chicago, had gone through almost an entire prime crew training regimen for Apollo 13, only to be bumped at the last minute after being exposed to German measles. Like Duke, he felt that the real Apollo 16 mission was something of a déjà vu.

The Apollo 16 astronauts, however, almost had to abort their moon landing mission because of a crisis in lunar orbit. The trouble began on the morning of April 20, 1972, just as Young and Duke completed the undocking maneuver at fifty thousand feet, and prepared to make their final descent to the surface in the lunar module *Orion.* Mattingly, who was attempting to pilot the command module *Casper* back into a seventy-mile-high circular orbit, suddenly reported that his engine control system had gone haywire.

"I don't know what's wrong with this thing," Mattingly reported. "It feels like it's going to shake the spacecraft apart."

Houston immediately ordered Young and Duke to delay their descent at least one more orbit until the critical malfunction in the command module could be diagnosed and corrected. Both feared for the worse.

"The way [Mattingly] described it," Duke recalls, "it sounded really bad to me. I thought, 'What we're going to do is end up joining back up again, and coming home.' And sure enough, the first response from Houston was, 'No, you're not going to land.' That was a real low point. Not because of any worry about not getting home because you could use the lunar module to burn out of orbit. But it was the fact that you'd trained all that time, and been up there [in lunar orbit] for a day. And you can see your landing spot every time you come around—there it is, fifty thousand feet below you, eight miles down—and they were going to say no."

But the men back at Mission Control were actually trying to find a way to save the landing attempt without violating the mission rules. They advised Mattingly to switch on the command module's backup guidance system. That immediately stopped the spacecraft's tailspin, and put her back on proper orbital trajectory. After a delay of six hours and three extra orbits, Houston gave the GO for landing.

At 8:32 P.M. Houston time on April 20, 1972, the *Orion* touched down at her designated landing site in the southeastern facial quadrant of the moon.

"That was probably the most exciting point in the mission," says Duke, "even more exciting than stepping out on the surface. While we'd trained for the landing, we were just coming out of this low point of the flight, psychologically anyway, and the thing had worked great—we recognized the landing spot, we were within a hundred meters of where we were supposed to land . . ."

The Apollo 16 astronauts stayed on the moon for a record seventy-one hours, nearly three full days. During that time, they made three excursions in their Moon Rover, reaching a record top speed of 10.5

m.p.h. while covering over twelve linear miles. They also bagged a record 210 pounds of lunar rock and soil samples.

Apollo 16's landing site was in part of the lunar highlands known as the Descartes region. Located below the equator about 250 miles south-southwest of the Apollo 11 landing site, it was believed to be an area of extensive ancient volcanism. But as Young and Duke quickly discovered, it was also a land full of startling surprises.

The most startling surprise was the astronauts' discovery of magnetic rocks around their landing site. Up to that time, scientists believed that the moon had virtually no magnetic field, which, in turn, implied that the moon had never had a molten liquid core like the earth's. But Young and Duke discovered rocks that were not only magnetic, but also exhibited "reverse polarity," i.e., some pointed up and others pointed down. That suggested that all the prevailing theories might be hogwash, that the moon really may have had a molten core and a magnetic field that "flip-flopped" from north to south.

While orbiting over the landing site, CMP Ken Mattingly made a surprising discovery of his own. Using sophisticated photo-mapping and sensor equipment, he took photographs and measurements that revealed the lunar highlands of the Descartes region to be rich in aluminum. In addition to reaffirming the vast differences between the geological composition of various lunar regions, Mattingly's discovery rekindled the possibility that man might one day mine the moon for precious ores and metals.

The Apollo 16 mission also featured a surprising stunt back down on the lunar surface during the astronauts' second EVA that almost proved fatàl. As Duke, the guilty party, recounts: "Dave Scott [of Apollo 16] had done that neat deal of dropping a hammer and a feather, and John and I were wondering, 'What are we going to do?' I said, 'Well, it's an Olympic year, so let's have a Moon Olympics.' John was gonna set the broad-jump record, and I was gonna set the high-jump record.

"Well, when the time came to try my high jump, I uncoiled and sprang straight up, but instead of going straight up along with me, the backpack started going over my back, and I started losing my balance. I don't know how high I jumped, but it was way off the ground, and I just knew when I hit the surface, I was going to split my suit open and I was dead. I actually had a moment of panic set in. 'Man, I'm in trouble!' I started scrambling, trying to get my balance, really never did, and ended up falling."

Luckily, Duke survived the fall without splitting open his suit or breaking any bones, but he admits that it was a humiliating experience.

"I was very embarrassed that I'd done such a stupid stunt. From the time we set down on the moon, it seemed like I always had something to say. But when I hit the surface and bounced up, I got very quiet. I wanted to just crawl back inside the LEM and forget about that little deal, but Houston TV was looking right at me. I got so quiet, John kept asking me, 'Are you okay? Are you okay?' "

Even so, Duke was still able to enjoy one final surprise on the lunar surface—a flash of déjà vu from his dream back on Earth. It happened on the morning of April 23, 1972, during the astronauts' third excursion in the Moon Rover when he took his turn at the rudder, and suddenly found himself rolling up a crystalline black and gray hill identical to one he had imagined in his sleep. As he recalls: "When we crossed the hill that was in my dream, it looked so familiar, I thought, 'Gosh, I've been here before!' "

Then, as if on some unconscious cue, the Moon Rover suddenly started to skid down the slope. Duke stopped daydreaming, and frantically wrestled the vehicle back under control.

"I almost turned the rover over," Duke apologized, adding, "I feel like I'm really sinking in."

"Charley, you really are," Young confirmed.

"You know, John, this black stuff is glass on these rocks," Duke observed as if to account for the Rover's accidental skid.

"Sure it is," Young returned.

Then the Apollo 16 commander spied a set of meandering parallel furrows up ahead, and exclaimed to his comrade in the driver's seat; "Charley, you said you were going to see some other tracks on the moon."

Duke would always delight at the way his dream had "come true" on the lunar surface, but he realized that the tracks had been made by their own Moon Rover, not by an alien vehicle. As he later assured; "We didn't see any other vehicles or any other astronauts up there—nothing like that."

APOLLO 17
DECEMBER 11–14, 1972

Here man completed his first explorations of the moon. May the spirit of peace in which we came be reflected in the lives of all mankind.

—plaque left on the moon by Apollo 17 commander EUGENE A. CERNAN

Although the Apollo 17 astronauts did not find life on the moon, they did achieve four historic "firsts." Theirs was the first mission to be launched at night. It was the first mission targeted for the easternmost sector of the lunar surface. It was the first mission to carry a cargo of live mice to the moon. And it was the first American mission to include a civilian scientist in the three-man crew.

But Apollo 17's chief historical distinction was being man's last lunar landing mission. In response to public, political, and budgetary pressures, NASA had canceled plans for the Apollo 18 and Apollo 19 flights. Having met the late John F. Kennedy's goal of putting a man on the moon in less than a decade, the U.S. manned space program was embarking on a new direction under the auspices of President Richard M. Nixon—the development of a reusable Earth-orbiting "space shuttle."

Not surprisingly, Apollo 17 sparked a new wave of nostalgic world interest. Over half a million people gathered at Cape Kennedy to witness the lift-off on December 7, 1972, the largest crowd since the Apollo 11 launch. The spectators included at least 200 members of Congress, 140 ambassadors and official dignitaries from foreign countries, and Apollo 11 commander Neil Armstrong, the first man to walk on the moon.

"Even as the world awaits this final launch, debate continues about the wisdom of the entire Apollo project," noted a *New York Times* editorial. "Some consider the moon landings the most brilliant scientific achievement in history; others regard the whole undertaking as a waste of resources needed for urgent requirements here on Earth. Yet one fact about the venture is beyond dispute: In the years 1969–1972 men landed on another celestial body for the first time and showed they could live and work in the bizarre and literally inhuman conditions on that foreign planet.

"Long after most other developments of the twentieth century are forgotten," *The Times* concluded, "future generations will recall this as the century in which men broke the bonds of terrestrial gravity, and began their cosmic destiny."

Besides being the last moon mission, Apollo 17 proved to be even longer, more ambitious, and, in the opinion of some scientists, even more productive than previous lunar expeditions. And that was just what the three proud astronauts selected for Apollo 17 had intended it to be.

"We may be the last on the moon," commander Eugene A. Cernan declared before the launch from Earth, "but we're the number one team."

Apollo 17 was Cernan's second trip to the vicinity of the moon. Like Apollo 16 commander John Young, he had also flown on the May 1969

Apollo 10 mission that orbited the moon but did not land. A thirty-eight-year-old navy captain from Chicago, Cernan was known for the "naughty boy" sense of humor he demonstrated at a presidential reception by sliding down a White House banister. He was also known for the spontaneous act of crying "son of a bitch!" when the Apollo 10 lunar module suddenly began to gyrate out of control in lunar orbit.

Although Cernan corrected the malfunction with the help of Mission Control, his pungent remarks were heard by radio and TV audiences all over the world. NASA was flooded with hundreds of letters both pro and con. Cernan coolly defended himself by issuing a disingenuously worded public apology. "To those of you who understood the significance of the moment," he proclaimed at a press conference following his return to Earth, "I say thank you."

CMP Ronald E. Evans, a thirty-nine-year-old navy commander from Kansas, did not have Cernan's salty tongue, but he was the most politically outspoken member of the astronaut corps. Evans accused the Nixon administration of supporting NASA "by talk but not by action," and dismissed opponents of the space program as "kooks who feel you ought to be spending the money on welfare." A veteran of more than a hundred combat missions in Vietnam, he prided himself in his recently acquired scientific expertise as well as in his battle scars.

"I consider myself a flying geologist," Evans told *The New York Times* prior to lift-off from the Cape. "I firmly believe that I'll be able to contribute quite a bit to science up there, and that extends man's capability beyond just flying the spacecraft up there and getting it back."

The most controversial member of the Apollo 17 crew, however, was LMP Harrison H. "Jack" Schmitt, a thirty-seven-year-old civilian geologist from Santa Rita, New Mexico, with a Harvard Ph.D., who had replaced astronaut Joe Engle after intensified political pressure to send a scientist to the moon. Certain insider and outsider critics objected that a lunar landing mission demanded the skills of a professional pilot. But as NASA officials pointed out, Schmitt had been an astronaut since 1965, and had spent more time in flight training during the preceding seven years than any other crewman.

Schmitt realized that the eyes of the world would be on him throughout the Apollo 17 mission.

"I suspect that people are expecting that when we fly a geologist to the moon that he is going to produce more from a scientific point of view by what he says, sees, and does than people who are not trained professionally," Schmitt confided to a reporter shortly before launch day. "I hope I'll be able to do that."

The five brown and white mice in the Apollo 17 cargo bay also had their own peculiar cross to bear. NASA scientists hoped that radiation-sensitive filmstrips implanted in the scalps of the mice would provide clues to the nature of the high-speed cosmic rays that had routinely penetrated and passed through the spacecraft and crewmen of previous Apollo missions. But the mice would have to make the ultimate sacrifice. Immediately after splashdown in the Pacific, they would be flown to a laboratory in the Samoan Islands to be killed and dissected.

Cernan and Schmitt touched down in the Taurus-Littrow region right on schedule at 2:55 P.M. on December 11, 1972. That in itself was something of a feat. Nestled in a remote and mysterious lunar shadowland roughly five-hundred miles northeast of the original Apollo 11 landing site, the Taurus-Littrow region was one of the least accessible areas on the lunar surface because of its proximity to the forbidding hemisphere twilight zone between the far and near sides of the moon.

The location of Apollo 17's landing site left the astronauts with precious little margin for error. A newly revised two-step descent procedure allowed them to make their final approach from an orbit of only forty thousand feet. But they still had to make their last and most crucial engine burn on the far side of the moon. Once they flew back into the light, they had less than ten minutes to reestablish radio contact with Houston and make their final course corrections before commencing their powered descent.

"The *Challenger* has landed!" Cernan exclaimed with unabashed joy when the LEM sunk its footpads into the lunar surface. "We is here, man! We is here!"

The Apollo 17 astronauts spent more time on the moon and more time out on the lunar surface than any of their predecessors. During their seventy-five-hour visit, they made three excursions on foot and in the Moon Rover, traveling a total distance of over twenty linear miles in over twenty-two hours of extravehicular activity. In the process, they had to display some old-fashioned "Yankee ingenuity" by repairing a damaged fender on their Moon Rover with some stiff-coated traverse maps and lamp clips. Their second excursion, which lasted seven hours and thirty-seven minutes, set a new EVA endurance record, and led to one of the most extraordinary geological finds in the history of the Apollo program.

NASA scientists believed that the narrow valley bounded by the

Taurus Mountains and the Littrow crater formation was formed by volcanic eruptions and massive molten lava flows at various stages of the moon's 4.5-billion-year history. They therefore predicted that the Taurus-Littrow region would yield some of the oldest and some of the youngest lunar rock specimens ever collected, and they were right.

But on the afternoon of December 12, 1972, Jack Schmitt made a surprising discovery in the Taurus-Littrow region that justified the controversial decision to send him to the moon. In the course of exploring the rim of the Shorty Crater, an otherwise negligible dimple named after a character in Richard Brautigan's best-selling novel *Trout Fishing in America,* Schmitt spied a patch of orange-colored soil.

"My first reaction was one of doubt," he remembers. "About thirty or forty minutes prior to that I had been briefly fooled by an orange reflection off the Moon Rover . . . I was too far from the Rover at that point for it to be a reflection . . . But the thought that was going through my mind was, 'Now, I don't want to be fooled again.' "

Cernan was also skeptical. "My first feeling was that Jack Schmitt had been on the moon too long. But sure enough, I went over there, and it was in fact orange."

Schmitt surmised that the orange soil was produced by a phenomenon known as a "fumarole," a venting of oxidized iron ore in gaseous form. In addition to debunking the theory that the moon was a colorless planet, Schmitt's discovery suggested that the Taurus-Littrow region really had been the site of extensive volcanism similar to the type that occurred during ancient epochs of Earth history, and that it might contain underlying layers of valuable metal deposits.

Schmitt says, however, that the highlight of his trip to the moon was not his discovery of the orange soil, but the sheer thrill of exploring the lunar surface on his own after years of watching previous moon missions on TV.

"Being there is the critical ingredient," he says. "That's what distinguishes a meaningful experience from one you've seen on television or in the movies or heard someone else talk about. I had tried to anticipate what it would be like for many years and particularly for the last fourteen months of training. But it was obvious that there was no way that one could have anticipated what it would be like to stand in the valley of Taurus-Littrow or in any spot on the moon, see this brilliantly illuminated landscape with a brighter sun than anyone had ever stood in before, with a blacker than black sky, and the mountains rising on either side to over

6,500 feet. And then to top the whole scene off in this blacker than black sky was a beautifully, brilliantly illuminated blue marble that we call Earth."

Although this was Cernan's second trip to the moon, it was his first landing mission, and he says that he felt a similar awe and wonder as he trekked across the lunar surface. "I remember looking back at the earth, and I could see day and I could see night at the same moment," he recalled in a recent interview. "It made me feel as if I was standing outside of time."

Cernan and Schmitt crawled back into the LEM for the third and final time at 12:38 P.M. on the morning of December 13, 1972. Before closing the hatch, they stopped to dedicate one of the multi-fragmented rocks they had collected to the children of all the nations of the earth. For the benefit of their viewing audience, Cernan also read the inscription on the plaque they would leave at their landing site:

"Here man completed his first explorations of the moon. May the spirit of peace in which we came be reflected in the lives of all mankind."

Then the Apollo 17 astronauts prepared for the next and most critical phase of that and every other lunar landing mission—their return to Earth.

Apollo 11 commander Neil Armstrong would always have the immortal distinction of having said man's first words on the moon. But it was Cernan who would have the privilege of saying the last words: "Okay, let's get this mother out of here!"

STAGE 3:

RETURN TO EARTH

Back to earth, the dear green earth.
WILLIAM WORDSWORTH
(1798)

14

THE SECRET TERROR OF SAYING GOOD-BYE

Return to Earth

Lunar Lift-off, Reentry, and Splashdown

It is never any good dwelling on good-byes. It is not the being together it prolongs, it is the parting.

ELIZABETH ASQUITH BIBESCO
The Fir and the Palm (1924)

All twenty-four men who flew to the moon knew they would have to face the same perils on the flight back to Earth, but only the twelve who landed on the moon knew the unique terror of facing the countdown to lunar lift-off. "Lift-off from the moon was probably the greatest anxiety of the whole flight," recalls Apollo 16's Charley Duke, "primarily because you really had very little to do during the last fifteen minutes of the countdown except sit there and think, 'What if it doesn't work? What am I going to do?' "

Although the on-board computer would execute most of the final countdown sequence, the lunar module had only one ascent engine. It was the one key component in the technological panoply of the spacecraft systems that lacked redundancy. There was no backup. No emergency escape booster. No second chance. The ascent engine had to light on time and keep on burning for just enough time to blast the LEM into lunar orbit—or else.

Apollo 11 astronauts Neil Armstrong and Buzz Aldrin deliberately refused to contemplate what might happen if the *Eagle*'s ascent engine failed on the moon. "That's an unpleasant thing to think about," Armstrong admitted at the crew's final prelaunch press conference back at the Cape. "We've chosen not to think about that at the present time. We don't think that's a likely situation. It's simply a possible one. At the present time, we're left with no recourse should that occur."

But the Apollo 11 commander would later admit that when he and Aldrin finally got ready to leave Tranquility Base, "We were not distracted by the question of whether the ascent engine would light, but we were surely thinking about it."

Command module pilot Mike Collins, who could only keep orbiting in the mothership while his comrades waited out the countdown to lunar lift-off, worried enough for all three of them. "I have been flying for seventeen years, by myself and with others," he would write in his post-mission memoir. "But I have never sweated out a flight like I am sweating out the LEM [lift-off] now. My secret terror for the last six months has been leaving them on the moon and returning to Earth alone; now I am within minutes of finding out the truth of the matter. If they fail to rise from the surface, or if they crash back into it, I am not going to commit suicide; I am coming home, forthwith, but I will be a marked man for life, and I know it. Almost better not to have the option I enjoy."

Despite the safe return of Apollo 11, the Apollo 12 astronauts felt considerable trepidation during the countdown to their lunar lift-off. Commander Pete Conrad tried to dispel their anxiety by recalling the equally critical engine burn he had to make to reenter the earth's atmosphere at the end of his Gemini 5 mission. But his anecdote failed to relieve the qualms of lunar module pilot Al Bean, who was then a rookie astronaut.

"Old Al had the fidgy-widgets," Conrad remembers, "and he kept going through his checklist looking at this and looking at that. I said, 'What's the matter, Al? You nervous? Worried about the engine not lighting?' And he said, 'As a matter of fact, I am.' So I went through my song and dance about Gemini 5, and I told him he might as well sit back and relax now. If the engine didn't fire, we would become the first permanent monument to the U.S. space program erected on the moon. Aside from that, there wasn't too much we could think about on it. I'm not sure that gave him the comforting, reassuring words he wanted."

Like Apollo 16's Duke, all the astronauts on the last four moon landing missions admittedly agonized in greater or lesser degree over the countdown to lunar lift-off, for they realized that the success of previous missions did not ensure the success of their own. But in contrast to Apollo 11's Collins, who declared that he was "very pleased to leave the moon" after his comrades rejoined him in the mothership, several also felt sharp pangs of regret as they prepared to depart the lunar surface.

"I was tired and grimy and ready to rest," recalls Apollo 14 lunar module pilot Ed Mitchell, "but I really regretted climbing up that ladder [to the LEM] for the last time because I knew I wouldn't have a chance to come back . . . The feeling before lift-off as we went through the checklist was, 'Take a good look. You're not going to see it again.' "

Even so, all twelve men who landed on the moon couldn't help reminding themselves that they would not see their home planet again either, unless the LEM's remarkably solitary power plant blasted them into lunar orbit. Designed and manufactured by Bell Aerosystems and the Rocketdyne division of North American Rockwell, the ascent engine consisted of a four-and-a-half-foot-high box filled with slightly less than five thousand pounds of a hypergolic fuel mixture of nitrogen tetroxide and hydrazine compounds. There were only two moving parts, a pair of ball valves that let the propellants flow into the combustion chamber. And there was no throttle or choke. The manually controlled master arm that ignited the engine at the climax of the automatic sequence had only two positions—ON and OFF.

Thanks to the one-sixth gravity of the moon, the ascent engine could perform its crucial function with remarkably little effort. Where the Saturn 5 engines had to generate 7.7 million pounds of thrust to launch the 7.5-million-pound booster from the Cape, it took only 3,500 pounds of thrust to lift the 4,800-pound upper half of the LEM off the lunar surface. And where the Saturn 5 typically needed nearly ten minutes to get from the launchpad to Earth orbit, the average elapsed time from lift-off to lunar orbit was about seven minutes.

The Apollo lunar modules, however, did not have any auxiliary or second-stage ascent engines to share the load. Although each LEM was equipped with sixteen reaction control system (RCS) rockets for rendezvous maneuvering, they could produce only four hundred pounds of total upward thrust. The astronauts had to depend solely on the ascent engine to get them from their stationary position on the lunar surface to the desired orbital altitude of sixty thousand feet by accelerating from zero velocity to over six thousand feet per second.

Houston always made sure to time the countdown to lunar lift-off so the LEM's ascent engine would ignite just as the command module completed another regularly scheduled pass across the front side of the moon. In theory, this precaution would make it easier for the CMP to affect an emergency rendezvous. But the mothership, which was in a sixty-mile-high orbit, could descend no lower than ten thousand feet above the surface. And unless the LEM managed to climb above thirty thousand feet, it would not be able to clear the peaks of the lunar mountains. Neither the CMP nor the men in the trench at Mission Control could do anything to save the astronauts if the LEM's ascent engine failed.

The transcripts of the Apollo 11 radio dialogue offer only a hint of the tension in Houston and on the moon during the final minutes of the countdown sequence.

CAPCOM (RON EVANS): "Tranquility Base . . . Just a reminder here. We want to make sure you leave the rendezvous radar circuit breakers pulled; however, we want the rendezvous radar mode switch in LGC just as it is on surface fifty-nine . . . Our guidance recommendation is PGNCS, and you're cleared for takeoff."

ALDRIN: "Roger, understand. We're Number One on the runway."

CAPCOM (EVANS): "Tranquility Base, little less than ten minutes here. Everything looks good, and we assume the steerable is in track mode AUTO . . ."

ALDRIN: "Stop, push-button reset, abort to abort stage reset."

ARMSTRONG: "Reset."

ALDRIN: "Deadband minimum, ATT control, mode control AUTO."

ARMSTRONG: "AUTO, AUTO."

Houston reported there was less than one minute remaining in the countdown.

ALDRIN: "Got your guidance steering in the AGS."

ARMSTRONG: "Okay, master arm ON."

ALDRIN: "Nine . . . eight . . . seven . . . six . . . five . . . Abort stage . . . Engine arm ascent . . . Proceed!"

A split second later, the ascent engine ignited as a set of explosive charges simultaneously blew away the nuts, bolts, wires, and water hoses connecting the upper and lower sections of the LEM. Then the astronauts felt no more than "maybe half a G or two-thirds of a G" worth of gravitational pressure as they began to rise from the lunar surface.

"That was beautiful!" Aldrin cried, dutifully informing Houston that the spacecraft was already climbing at a rate of over thirty-six feet per second.

"There was no time to sightsee," Aldrin would report in his autobiography. "I was concentrating intently on the computers, and Neil [Armstrong] was studying the attitude indicator, but I looked up long enough to see the flag fall over."

Unlike a Saturn 5 booster, the LEM did not lurch upward with a defeaning roar or belch out a terrifying holocaust of smoke and fire. But as the remote-controlled TV cameras set up by the astronauts on subsequent moon landing missions would document, a lunar lift-off was in many ways more spectacular than a lift-off from the Cape because of its almost inaudible gracefulness.

"It's like you're riding in a very fast elevator," remembers Apollo 12's Al Bean. "First of all, there's a big bang as you separate stages, but you don't hear anything after that. You never hear the main engine because of the vacuum of the lunar atmosphere. You hear the valves move, but you don't hear them firing. All you hear is a little thump-thump-thump kind of sound."

Bean says that the thumping nevertheless helped dissipate the intense anxiety he had felt throughout the countdown, and made him realize that the wisest thing to do was to sit back, relax, and try to observe the visual drama of lifting off from the moon.

"You don't have a lot of instrument displays telling you how the lunar module's engine is working because there's nothing you can do about it anyway," he says. "I wanted to see the lift-off, and I said to myself, 'I think I'll just look out the window and enjoy it.' And so I did. As we lifted off, I could see these sparkling things being blown off the insulation of the descent stage. It looked like the ripples you see when you drop a rock in a pond of water, this metallic insulation going out in concentric rings. It looked like those rings just went on forever. I can remember looking down and saying, 'I sure hope that insulation doesn't land on any of the experiments on the surface because it might change the thermal properties.' "

But like all his fellow Apollo astronauts, Bean admits that he was startled by the LEM's abrupt change in trajectory less than half a minute after lift-off. "When you get up about five hundred feet, you tip over at about a forty-five-degree angle, which is a much steeper angle than you pitch over into in a launch from the Cape when you're trying to get out of the earth's atmosphere. On the moon, you don't have any atmosphere to worry about, so you can tip over and get going into orbit a lot faster. I remember when the LEM tipped over, it seemed like we were almost horizontal—and then we just took off!"

The LEM would continue to gain altitude and velocity at exponential rates. Three minutes after lift-off, the spacecraft was 3,000 feet above the surface, still climbing at 185 feet per second but with a horizontal velocity in excess of 1,500 feet per second or over 1,000 m.p.h. Before another minute elapsed, altitude was 32,000 feet and forward speed nearly 2,500 feet per second. By the seven-minute mark, the LEM was nearly 60,000 feet high and traveling at over 6,000 feet per second.

A few moments later, the LEM entered a ten-mile-by-forty-five-mile elliptical lunar orbit, and the astronauts shut down the ascent engine, relieved that the most precarious phase of the mission was behind them. According to Bean; "It was a really great feeling to be going fast again. We still had the rendezvous ahead of us, but somehow we felt that would be an easy thing to do."

The command module pilot seldom agreed with that assessment, for he knew that the rendezvous phase was actually a complex and equally precarious game of tag that could take anywhere from three to five hours. And unlike his comrades in the lunar module, the CMP had to perform the multitudinous duties associated with maneuvering the mothership all by himself, including 850 separate computer key strokes.

"That's a helluva job for one guy in a spacecraft that's designed for three men to operate," notes Apollo 12 command module pilot Dick Gordon. "All the mechanics that you have to go through not only to calculate the rendezvous but to affect it, make those engine burns and all that by yourself is a pretty significant task."

The two astronauts in the lunar module also had quite a bit of work left to do. At this point, the LEM was several hundred linear miles behind the command module. But since the LEM was also at a much lower altitude (forty-five miles vs. sixty miles), it was orbiting "faster" than the mothership. The flight plan therefore called for the LEM to catch up with the CM by maintaining its lower orbit, then climb up to sixty miles and commence a tricky sequence of course correction prerequisite to the docking attempt.

The lunar module pilot, who usually handled most of the thruster firings along the way, had to take special care not to let the LEM tumble into a "whifferdill," the term for a dizzying orbital tailspin, which could deplete the remaining RCS rocket fuel. When the LEM finally came within fifty feet of the command module, the LMP would then have to turn the spacecraft 180 degrees around so that the lifesaver-shaped drogue was facing the needle-pointed probe on the nose of the mothership.

Docking was by definition a hit-or-miss proposition. The CMP initiated the maneuver with a final engine burn that caused the mothership to lurch toward the LEM. When the two spacecraft made contact, the CMP fired a nitrogen canister that started the retraction cycle. If the probe and drogue failed to lock properly, the astronauts inside the LEM could still get back to the command module by donning their space helmets and making an orbital EVA from hatch to hatch. But all three crewmen hoped such emergency "spacewalks" would not be necessary.

Apollo 11's Collins got what he later called "the surprise of my life" immediately after triggering the retraction cycle, for the LEM suddenly started spinning and rocking out of control. "Instead of a docile little LM, I suddenly find myself attached to a wildly veering critter that seems to be trying to escape. Specifically, the LEM is yawing around to my right, and we are misaligned by about fifteen degrees now. I work with my right hand to swing *Columbia* around, but there is nothing I can do to stop the automatic retraction cycle. All I can hope for is no damage to the equipment, so that if this retraction fails, I can release the LEM and try again."

Miraculously, Apollo 11's docking problem solved itself. *Eagle* and *Columbia* veered back into proper alignment as if guided by an invisible hand, then the probe and drogue locked together with a climactic bang.

Although the astronauts inside the LEM had risked their lives on the success of the rendezvous and docking maneuvers, the CMP was almost always the happiest member of the threesome when his comrades finally crawled back inside the mothership. That was certainly true of Apollo 12's Dick Gordon.

"I've never seen a happier guy in my life," recalls crewmate Al Bean. "He was more happy that we got back than we were happy to get back. He was just bananas. He wanted to get us a drink of water. He wanted to fix us some food. He couldn't do enough. It was like returning home to your mother after a twenty-year absence. It was really an incredible moment, and he was just great."

"I really was happier than hell that they had returned safely, and that two good friends were back," Gordon admits, "because we had talked about and gone across that bridge about coming home alone, and that was part of the job and part of the thought process that you go through. And if they couldn't have gotten off the lunar surface, that was exactly what would have had to have taken place. And you gotta do it [return to Earth alone] because there's nothing you can do for them if they don't make it back to the command module. That's kind of emotionally impacting when you think about it."

When Gordon noticed how dirty his "good friends" had gotten on the moon, however, he immediately ordered them to go back inside the LEM and clean up.

"It wasn't his decision, it was the ground's," says Bean. "We were really much dirtier than we imagined we'd be. The zippers on our pressure suits started clogging up and quit working. We hadn't taken precautions because we didn't know they were needed. But it was a big deal. In zero G, when you're dirty this stuff floats all around and gets in your eyes and gets in the environmental control system and clogs the filters. That's why we were concerned about letting the dirt get inside the command module—plus we had to go all the way home in it."

Conrad and Bean quickly realized the hopelessness of trying to dry clean their thoroughly soiled "moon cocoons," and ultimately decided to take them off and leave them behind. As Conrad recalls; "Al and I wound up climbing out of the LEM in nothing but our altogether. Somebody on the ground later said we were the first lunar 'streakers.' "

The astronauts' reunion in lunar orbit was always the day's most triumphant moment, for it marked the true beginning of the trip back to Earth. "Once we got joined back up," recalls Apollo 16's Charley Duke, "I sort of had the feeling, 'Well, my part's done. I can relax now.' "

But all three crewmen also realized that their newfound sense of security was an illusion, for they still had 240,000 miles to go and at least one more major hurdle to clear. Shortly after completing the docking maneuver, the CMP would trigger a set of explosive bolts that blasted the lunar module free of the command module. Then he fired the thrusters of the mothership for a few seconds to put a safe distance between the two spacecraft, and watched the LEM sail off into a low level lunar orbit of its own. The now empty LEM might remain aloft for several weeks or even several months, but it would eventually lose orbital momentum and crash to the surface.

On the first three lunar landing missions the departure of the LEM signaled the end of the astronauts' visit to the moon. But after Apollo 14 demonstrated that NASA had corrected the flaws that led to the Apollo 13 accident, space agency officials decided it was worthwhile for Apollo 15 and Apollo 17 to remain in lunar orbit for another two days to conduct various scientific experiments. (Apollo 16 was originally scheduled to follow a similar flight plan, but unanticipated command module guidance problems forced the astronauts to cancel their experiments and return home.)

The specific chores the Apollo 15 and Apollo 17 crewmen performed in lunar orbit fell into three main categories. First, they attempted to chart the topography of the lunar surface with a wide-angle panoramic camera, a narrow-lensed mapping camera, and a laser altimeter. The second item on the checklist called for them to extend a twenty-five-foot-long mechanical scoop from the nose of the command module in an effort to collect radioactive particulate samples. The third and final assignment was launching a 78.5-pound reconnaissance satellite designed to gather data on the mysterious concentrations of magnetic rock known as "mascons."

All the astronauts on all six successful Apollo lunar landing missions, however, eventually had to make the same crucial engine burn to get out of lunar orbit and start the final leg of their voyage. This was the "trans-Earth injection" or TEI, or as Apollo 11's Collins termed it, "the get-us-home burn, the save-our-ass burn, the we-don't-want-to-be-a-permanent-moon-satellite burn."

Due to the decision to delete the "Return to Earth" program from the already overloaded memory of the on-board computers, the astronauts had to rely on Houston to radio up all the key navigational and guidance data as calculated by the computers down at Mission Control. Then they had to fire the service module engine for exactly the right length of time while temporarily out of radio contact on the back side of the moon.

The TEI burn was never more than three minutes long from start to finish, but it was powerful enough to make the spacecraft accelerate from an orbital speed of about five thousand feet per second to the required "escape velocity" of over eight thousand feet per second or roughly 5,318 m.p.h. As the astronauts were slingshot around the front side of the moon, they were treated to one of the most thrilling experiences of the entire mission.

"Coming off the moon is something to behold," recalls Apollo 14's Stu Roosa. "You are just hauling out of there on the world's fastest elevator, and you have much more of a feeling of speed than you did when you left Earth orbit because back then you still had to worry about the transposition and docking maneuvers and you really didn't have time just to sit back and watch the earth get smaller. But coming out of lunar orbit after the TEI burn, you have your windows pointed at the moon so you can take pictures, and oh, boy, the way you're just hauling out of there is really something special."

Apollo 17's Gene Cernan says the sight of the rapidly shrinking moon inspired an unexpected nostalgia. "Having lived there for three days, the moon has a familiarity about it. You say, 'That's home . . . There's where we left the Rover . . . And there's where we explored . . . And there's the mountains and the valleys.' And in a sense, it has been a home, a life-sustaining home, for you. And when you see those things, you leave it with the same kind of feeling of awe that you left Earth with several days before."

The TEI burn was literally and figuratively the turning point for every moon landing mission. If the service module engine had failed to light or misfired or burned too long, the astronauts would have either remained in lunar orbit or been hurtled into space with no hope of being rescued before their oxygen supplies ran out.

"That was the last big hurdle," notes Apollo 16's Duke. "After that, to me, it was home free. We still had reentry to go and we had an EVA to do on the way back, but for some reason, you just knew you were gonna make it. You might not land right right on your splashdown site, but you knew you were gonna be all right."

The length of the homeward voyage varied from mission to mission because of the regularly changing distances between the moon and the earth, but it usually took about three full days, the first two of which seemed to pass the slowest. By the end of Day One, the astronauts had traveled only 5,000 miles. On Day Two, at about 35,000 miles out from the moon, they crossed the invisible boundary of the earth's "sphere of influence." Thanks to the increasing gravitational pull of their home planet, the spacecraft could cover the remaining 200,000 miles in only twenty-four more hours. But for the three crewmen inside, the return trip never went fast enough.

"It seemed like six times as long coming back as going out," says Apollo 12's Conrad. "Going out, you're all keyed up, you're ready to go, you've got this thing to do, you're going to land on the moon. But coming back, you've got only one more square to fill on the checklist, getting through reentry. And by that time, even if it's your first trip, everybody's an old hand. You've been in space for nine or ten days or whatever, and it's getting to be old hat."

Fellow crewman Al Bean retreated to his sleeping berth for most of the return trip. "Time seems to drag," he recalls. "The days are long because you really don't have a lot to do. You've done your job, you've got your [moon] rocks and everything else. You know you're going to land. You know you're going to reenter the earth at the right place because Mission Control calibrates that for you once they track you for a while, so that isn't even a big worry. For me, that was the bottom stage of the whole mission, the low point."

Apollo 14's Ed Mitchell was one of the few astronauts who claimed to enjoy the last leg of his mission almost as much as the first. Instead of feeling bored, he reports having "a profound personal experience" that could only have occurred on the allegedly tedious trip back to Earth.

"First of all, you're in space, in the void," Mitchell explains. "Then you see the earth, and you see yourself approaching this planet and we know that's home. The very profound shift in thinking that takes place as a result of that experience, changing from intellectually knowing that we live on a planet to the deep gut reaction of seeing that planet 240,000 miles away, is very visceral. In my case, it made me realize that science may have had the wrong view of things. It is not just a material universe. It has purpose and design and was created. You started getting very global in your perspective, and that's the perspective I've maintained. They say that travel is broadening, and that's about as broadening as you can get."

Mitchell adds that his existential musings continued over "a considerable period of time," and swung up and down various emotional peaks and valleys: "The peaks were the cognition that it is a harmonious, purposeful, creative universe. It was as though I could identify it, and feel it, and be it. The valleys came in recognizing that humanity wasn't behaving in accordance with that knowledge, that humanity was behaving, as we often say, like lemmings rushing toward the sea. The pathos, the deep sadness, came in simply contemplating the human condition—the lack of awareness of what we really are and what the universe really is."

Although most of the other Apollo astronauts say they shared Mitchell's basic perceptions, few chose to dwell on them as intensely. Instead, like Bean, they spent the majority of their spare time sleeping, or, in some cases, listening to music. The Apollo 11 crew's favorite tape was a recording of the song "Everyone's Gone to the Moon." Apollo 16's Charley Duke, a diehard country and western fan, kept replaying Bob Wills's classic "San Antone Rose." The only real "action" on the voyages back to Earth consisted of well-rehearsed TV performances like Apollo 11 lunar module pilot Buzz Aldrin's demonstration of how to put ham spread on a piece of bread in zero G, and the in-flight EVAs made by the command module pilots of the last three lunar landing missions.

The "spacewalks" always took place early on the second day of the return trip after the spacecraft had reentered the earth's sphere of influence. As the command module zoomed along at about 2,300 m.p.h., the CMP attached himself to a nylon tether, opened the hatch, and pulled himself along the handrail on the fuselage of the spacecraft. The official purpose of these exercises was to retrieve various film cassettes located at the end of the service module. Apollo 15's Al Worden got the job done in sixteen minutes. Apollo 16's Ken Mattingly, who also had to set up a microbial ecology evaluation device (MEED), stayed out about an hour. Apollo 17's Ron Evans was out for forty-seven minutes, but seemed to have the most fun.

"Hot diggety dog!" Evans cried as he floated alongside the spacecraft *America.* "Talk about being a spaceman—this is it!"

By the end of Day Two, the astronauts prepared to "fill the last square" on their checklist—reentry to Earth. That usually required a minor course correction burn to fine-tune the spacecraft's speed and trajectory. In the case of Apollo 11, for example, this burn consisted of an eleven-second rocket thrust that altered *Columbia*'s velocity by 4.8 feet per second out of a total speed of 4,075 feet per second. Similarly, the CMP had to make sure the spacecraft remained comfortably inside the forty-mile-wide "reentry corridor" specified by Mission Control so that it would slice into the atmosphere at exactly 6.5 degrees below the horizon. Notes Apollo 11's Collins: "Now that's precision, and I appreciate it. Too shallow, and we skip back out; too deep, and we burn up. Each sounds as grim as the other, and I don't want to edge .01 degree in either direction; 6.5 please!"

At the beginning of Day Three of the return voyage, the astronauts are within ten thousand miles of Earth, and are getting cabin fever.

"The right side of the equipment bay, wherein are located old launch day urine bags, discarded washcloths, and worse, is now a place to be avoided," Collins wrote in his memoir. "The drinking water is laced with hydrogen bubbles . . . These bubbles produce gross flatulence in the lower bowel, resulting in a not-so-subtle and pervasive aroma which reminds me of a mixture of wet dog and marsh gas . . . Things which were fun a couple of days ago, like shaving in weightlessness, now seem to be a nuisance. There is no sink in which to wash the hair, or even enough water to rinse the face . . . All these things are small potatoes, but they do produce—in me, at least—an overlay of impatience and irritation."

It was about this time that the crewmen of Apollo 12 reported sighting what could only be described as an unidentified flying object. But what they thought was "one of the great UFO stories of all time" turned into a uniquely symbolic embarrassment. As Apollo 12 command module pilot Dick Gordon recalls, "We were probably the dumb shits of the world because just before reentry we reported seeing a bright light in the middle of the Indian Ocean, and we couldn't figure out what the hell it was. We knew it wasn't clouds, and we knew it wasn't lightning or this type of thing. We were just amazed. Here was this big glow in the middle of the Indian Ocean, and we were trying to figure out what sort of phenomenon that was, then all of a sudden, Dong! we realized that it was the reflection of the moon."

Fortunately for the astronauts, the homeward journey was almost over. The CMP fired a set of explosive bolts that jettisoned the service module, reducing the total weight of the spacecraft, originally 7.5 million pounds including the Saturn 5, to just 11,000 pounds. Then he turned the mothership into the BEF (blunt end forward) position to protect himself and his fellow crewmen from the scorching heat of reentry.

The command module always hit the earth's atmosphere at over 36,000 feet per second. The astronauts temporarily lost radio contact with Mission Control as their spacecraft plunged through a sparkling layer of ionized particles. Then the parachutes opened, and the three crewmen were socked with 6.5 Gs. Approximately fourteen minutes later, the spacecraft would smack into the Pacific Ocean.

Within ten to fifteen minutes, the rescue boats arrived at the splashdown site. A school of frogmen dove into the water and threw a flotation collar around the bobbling spacecraft. Then a helicopter appeared, and prepared to ferry the astronauts to a nearby helicopter.

"That's where it all hits home," recalls Apollo 12's Dick Gordon. "When that damn command module splashes down in the Pacific Ocean, it's done and over. You're back, and you're safe, and you know damn good and well you did a damn good job, and it was a successful mission. There's a time between splashdown and the time you get back to the ship that's a good feeling—it's controlled jubilation, is what it is."

15

INSTANT CELEBRITIES

Life After Splashdown: Part One

1969–1972

I see thy glory like a shooting star
Fall to the base earth from the firmament.

WILLIAM SHAKESPEARE
King Richard II

All twenty-four men who flew to the moon were in for a shock upon their return to Earth. The timing and nature of that shock varied from mission to mission and crewman to crewman, but the tremors always began with the impact of splashdown. As the matter-of-fact voice of Mission Control warned the Apollo 11 astronauts just before their reentry on July 24, 1969: "I just want to remind you that the most difficult part of your mission is going to be after recovery."

That sage warning came from CAPCOM Jim Lovell, a veteran of the Apollo 8 lunar orbiting mission who would later command unlucky Apollo 13. Lovell was alluding to the fact that Apollo 11 crewmen Neil Armstrong, Buzz Aldrin, and Mike Collins were scheduled to spend no less than seventeen days in quarantine so that NASA doctors could make sure they had not been exposed to some unknown disease during their unprecedented excursion to the lunar surface.

"Keep the mice healthy," Collins replied, cheerfully reminding Mission Control to safeguard the army of contamination detection rodents stationed in the quarantine laboratory in Houston.

Immediately after splashing down in the Pacific Ocean, the Apollo 11 astronauts zipped on their BIGs, the suffocatingly hot "biological isolation garments" that covered them from head to toe. Then the rescue helicopter flew the BIG-baggaged threesome to the deck of the aircraft carrier *Hornet,* where they entered the mobile quarantine facility (MQF), a specially equipped silver-skinned house trailer Collins archly described as their "aluminum coffin."

Inside the MQF, two obliging human guinea pigs—flight surgeon Dr. William Carpenter and mechanical engineer John Hirasaki—helped the astronauts unzip their BIGs and get readjusted to the all but forgotten oppressiveness of terrestrial gravity. Then the three crewmen took turns in the shower, shaved, and put on clean blue NASA jumpsuits, still feeling stiff and heavy-legged.

A few short minutes later, Armstrong, Aldrin, and Collins were summoned to the rear window of the MQF where they stared through the parted curtains like caged animals as the ship's band played "Ruffles and Flourishes" and President Richard M. Nixon officially welcomed them back to Earth.

"Neil, Buzz, and Mike," Nixon began, "I want you to know that I think I'm the luckiest man in the world. And I say this not only because I have the honor to be the president of the United States, but particularly because I have the privilege of speaking for so many in welcoming you back to Earth. I could tell you about the messages we received in Washington. Over one hundred foreign governments, emperors and presidents and prime ministers and kings, have sent the most warm messages that we have ever received. They represent over two billion people on this earth, all of them who have had the opportunity through television to see what you have done . . .

"I called the three of, in my view, three of the greatest ladies and most courageous ladies in the whole world today, your wives. And from Jan and Joan and Pat, I bring their love and their congratulations. We think that it is just wonderful that they could have participated at least through television in this return; we're only sorry that they couldn't be here. And also, I've got to let you in on a little secret—I made a date with them."

A wave of polite laughter rippled across the deck of the *Hornet* as the president paused for effect.

"I invited them to dinner on the thirteenth of August, right after you come out of quarantine," Nixon announced. "It will be a state dinner held in Los Angeles. The governors of all the fifty states will be there, the ambassadors, others from around the world and in America. And they told me that you would come, too. And all I want to know—will you come? We want to honor you then."

"We'll do anything you say, Mr. President," Armstrong promised. "Just anything."

"One question I think all of us would like to ask," Nixon replied. "As we saw you bouncing around in the boat out there. I wonder if that wasn't the hardest part of the journey. Was that—did any of you get seasick?"

"No, we didn't," the Apollo 11 commander reported, adding, "And it was one of the harder parts, but it was one of the most pleasant, we can assure you."

There was a brief exchange of small talk about the recently rained-out All Star baseball game, followed by more reassurances from Armstrong that he and his fellow crewmen did indeed feel as good as they looked. Then Nixon closed off his speech with a dramatic declaration about the significance of the Apollo 11 mission:

"I was thinking as, you know, as you came down and we knew it was a success, and it had only been eight days, just a week, a long week. But this is the greatest week in the history of the world since the Creation. Because as a result of what happened in this week, the world is bigger infinitely, and also as I'm going to find on this trip around the world and as Secretary [of State William] Rogers will find as he covers the other countries and Asia, as a result of what you've done the world's never been closer together before. And we just thank you for that. And I only hope that all of us in government, all of us in America, that as a result of what you've done, we can do our job a little better. We can reach for the stars just as you have reached so far for the stars . . .

"We don't want to hold you any longer. Anybody have a last request? How about promotions, do you think we could arrange something?"

Another wave of polite laughter rippled across the deck.

"We're just pleased to be back and very honored that you were so kind as to come out here and welcome us back," Armstrong answered. "And we look forward to getting out of this quarantine—and talking without having glass between us."

After the ship's chaplain offered a presidentially requested prayer of thanksgiving, Nixon departed, and the astronauts repaired to the midsection of the MQF for their first home-cooked meal in over a week, whetting their already ravenous appetites with some long-awaited and very stiff cocktails. Then Aldrin crawled into his bunk to take a nap while Armstrong and Collins engaged in a vicious game of gin rummy. Later that afternoon, the *Hornet* steamed into Pearl Harbor, Hawaii, where cheering throngs of local servicemen and civilians watched a tractor trailer transfer the MQF to a C-141 cargo plane that promptly took off on a six-hour nonstop flight to Houston.

The Apollo 11 crewmen arrived at the Manned Spacecraft Center in the middle of the night, and dutifully crawled out of the MQF and into the more spacious but still claustrophobic confines of the lunar receiving laboratory (LRL). They proceeded to spend the next two and a half weeks undergoing comprehensive physical and psychological examinations, writing their official flight reports, and sheepishly blowing kisses to their wives through hermetically sealed window panes.

In keeping with previously established mission rules, the Apollo 12 and Apollo 14 crewmen had to endure similar quarantines before NASA doctors concluded that man did not risk any known form of contamination from exposure to the lunar surface. But even though the astronauts who made the last three lunar landings were spared the discomfort and indignity of the BIGs, the MQF, and the LRL, they, like their predecessors, also found that "the most difficult part" of their missions came after splashdown.

"You know before you go that it's going to change your life," says Apollo 12's Al Bean. But none of the twenty-four men who flew to the moon had any way of knowing just how abrupt the initial change would be—or that they would become the victims of their own success.

Virtually all of the astronauts dreaded the customary round of post-splashdown press conferences, speeches, and goodwill tours even more than being holed up in LRL quarantine. Most were by nature and profession nonverbal men with little inclination to analyze, describe, or reveal their innermost thoughts and feelings. Fixed in the public eye, they felt defensive and insecure, fell back on platitudes, clichés, and bureaucratic jargon. They looked silly, sounded sillier, and deeply resented what Apollo 15's Jim Irwin would aptly label "this instant celebrity status."

"We knew we had to become instant public relations experts to fit into this instant celebrity status," says Irwin, "and that was a very sobering thought. I dreaded public speaking. The night before I had to give a speech, I always had a fitful sleep and my stomach was in knots. And here I was coming back from the moon, and I knew that's all I'd be doing for at least six months. I did not have any joyful anticipation as to what was ahead."

The Apollo 11 astronauts naturally had to face the most intensive and extensive series of welcome-home ceremonies. Though hardly as dangerous as making man's first moon landing, this phase of the mission was, as commander Neil Armstrong would point out, "the part we're least prepared for." And as lunar module pilot Buzz Aldrin revealed afterward, it took a heavy psychological toll.

"It would take a couple of years for it to become clear to me, but that day on the U.S.S. *Hornet* [following splashdown] was actually the start of a trip into the unknown," Aldrin wrote in his post-mission autobiography *Return to Earth.* "I had known what to expect on the unknown moon more than I did on the familiar Earth."

The Apollo 11 crewmen got a startling preview of what was to come immediately upon entering LRL quarantine. Barely seventy-six hours after splashdown, three copies of *We Reach the Moon,* the first "instant history" of their mission, were deposited on their bedside tables. Published by *The New York Times* and written by space correspondent John Noble Wilford, the book featured five full chapters narrating man's first lunar landing.

The blizzard of press clippings that began piling up in the LRL attested to the world public's reaction to the events of July 20, 1969. In addition to running the largest front page headline in its history the following morning, *The New York Times* devoted the entire front section to related articles and commentary. "Not since the human race evolved has there been a comparable event, nor one so capable of lifting all mankind's horizons, dreams, and aspirations," declared a *Times* editorial. "What was fantasy to preceding generations is now accomplished fact. The achievement will be remembered so long as civilization survives."

While other major U.S. dailies also gushed with bold-faced bravos, most of the foreign press was equally if not more ecstatic. Headlines announcing man's first walk on the moon were bannered across the front pages of newspapers in England, France, Germany, and Jamaica. A Yugoslavian daily dubbed the Apollo 11 astronauts "Citizens of the Moon." "LA LUNA E NOSTRA," proclaimed Italy's *Carriere della Sport,* which translated, "The Moon Is Ours."

Even the Soviet Union's news organizations were obliged to report the news of Apollo 11's stunning success. Though the state-controlled TV network did not provide live coverage of man's first walk on the moon, taped footage of the feat generated such interest that Soviet leaders allowed their citizens to watch European television's live coverage of the splashdown. Following the recovery of the astronauts and their space capsule, Moscow radio informed its listeners, "Man's first flight to the moon has been completed. The glorious dream of visionaries and scientists has come true." Soviet president Nikolai V. Podgorny sent a congratulatory message to President Nixon in which he conveyed his personal best wishes to the "courageous space pilots." Russia's cosmonaut corps also offered their congratulations.

"We followed your flight closely and with great excitement," said the cosmonauts' message, which was broadcast over Moscow radio. "With all our heart we congratulate you on the completion of the remarkable trip to the moon and on your safe return to Earth."

Of course, there were also numerous critics of Apollo 11 both at home and abroad. Though the Communist Chinese all but ignored the mission, dismissing it as merely another capitalist propaganda ploy, liberal writers in the U.S. decried the irony of spending billions of dollars to land two men on the moon, insisting that such prodigious resources would be better spent on solving problems like air and water pollution. But some of the most visceral and vocal criticisms came from American blacks who claimed that millions of impoverished urban ghetto dwellers did not benefit in any tangible way from the space program.

"There ain't no brothers in the program where they can get into some of that big money," one patron of a Harlem bar told *The New York Times.* "The whole thing uses money that should be spent right here on Earth, and I don't like them [the news media] saying 'all good Americans are happy about it'—I damn sure ain't happy about it."

"It proves that white America will do whatever it is committed to doing," said Sylvia Drew, an attorney for the N.A.A.C.P. Legal Defense Fund. "If America fails to end discrimination, hunger, and malnutrition then we must conclude that America is not committed to ending discrimination, hunger, and malnutrition. Walking on the moon proves that we can do what we want to do as a nation."

There were also severe critics in the U.S. scientific community. "The moon rock samples, about which full columns of news are released by NASA, cannot even answer the few questions some geologists are interested in solving," charged Dr. S.E. Luria, an MIT biology professor and member of the National Academy of Sciences, in a September 1969 letter to *The New York Times.* "It is important that this be made clear because of the current discussion about big versus small Mars-landing programs—probably $3 billion a year for fifty years or $10 billion a year for fifteen years. This at a time when the Institute of General Medical Science of the National Institutes of Health has announced substantial cuts in new health-related research projects . . . It is time the American people were told frankly that the present space program is technically impressive, scientifically trivial, culturally misguided and socially preposterous."

Nevertheless, such howls of protest were all but drowned out by the even more thunderous and widespread official exclamations of joy and hope. Former president Lyndon B. Johnson declared that the Apollo 11 landing proved the U.S. "can do anything." President Nixon asserted that the feat would help revive the nation's "sense of purpose." In a speech to the International New Thought Alliance, the Reverend Dr. R.C. Barker opined that man could use the same powers of mind it took to land on the moon to create lasting world peace.

Armstrong, Aldrin, and Collins gamely tried to take all this hyperbole in stride, repeatedly reminding themselves and their families that they had to put both excessive praise and unfair pillories in the proper perspective. They knew there was no way to live up to their heroic public images or to fulfill the world's exaggerated expectations. But they were determined to represent their country with patriotic pride and graciousness—so long as the world and the NASA p.r. officers respected certain boundaries of personal protocol.

"I think it is possible," Armstrong had maintained at the crew's last prelaunch press conference, "to participate in an undertaking of this kind and still live a private life."

The Apollo 11 crewmen finally emerged from the LRL at 9 P.M. on Sunday August 10, 1969, to the rousing applause of some 250 attendant NASA personnel, and were immediately chauffered to their nearby suburban homes in three space agency staff cars. But because of the intrusive army of mass media staked out at every corner, they managed less than thirty-six hours of post-quarantine respite.

On Tuesday, August 12, the astronauts held their first postflight press conference at the Manned Spacecraft Center, a mutually discomforting hour-long session of superficial questions and answers. At 5 A.M. the next morning, the three crewmen and their families assembled on the tarmac at Ellington AFB, where they boarded Air Force II and took off for New York City, the first stop on the day's grueling transcontinental itinerary.

Over three million people cheered the Apollo 11 trio on their ticker tape parade through Manhattan, littering the streets with some three hundred tons of confetti and shredded paper. After accepting specially minted gold medals from Mayor Lindsay, the astronauts presented United Nations secretary general U Thant with a replica of the commemorative plaque they had left on the moon. Then they reboarded Air Force II and flew to Chicago, where they motorcaded through a crowd estimated at more than a million strong. The only damper on the occasion was a nonviolent march by five hundred black street gang members demanding that Mayor Richard Daley offer more construction jobs to minority youths.

Later that same afternoon, the Apollo 11 entourage arrived in Los

Angeles for a state dinner hosted by President and Mrs. Nixon. The guests included former Vice-President Hubert Humphrey, Reverend Billy Graham, Wernher von Braun, and diplomats from every major country in the Free World. Nixon awarded all three crewmen the Medal of Freedom, then pinned a fourth medal on Steve Bales, the Mission Control computer specialist who had given the GO for landing in the wake of the *Eagle*'s last-minute program overload alarms. Here the only sour note was the dutifully reported fact that the cost of the $43,000 dinner would ultimately be borne by the millions of uninvited American taxpayers.

On Saturday, August 16, Armstrong, Aldrin, and Collins were welcomed back to Houston with another downtown parade and an invitation-only gala in the Astrodome attended by over 55,000 well-connected local citizens. Frank Sinatra acted as master of ceremonies for an entertainment extravaganza starring singer Nancy Ames, comedian Flip Wilson, and impressionist Bill Dana, who had won national renown for his caricatures of the cowardly would-be astronaut José Jimenez. A few short weeks later, after an officially prescribed R & R period, the three crewmen were honored with hometown parades in Ohio, New Jersey, and New Orleans.*

In mid-September, the astronauts addressed a joint session of Congress in Washington, D.C., timorously taking their turns at describing the highlights of their mission according to a prerehearsed script. The next day, they listened to an even more anxiety-provoking State Department briefing in preparation for their upcoming "Giant Leap Tour" around the world.

Amid all the hoopla, the Apollo 11 crewmen also received some unsolicited advice from Frank Borman, who had become the president's most favored astronaut spokesman. According to Aldrin, the former Apollo 8 commander claimed, "If we played our cards right, we'd never have another worry."

"All things . . ." Borman reportedly informed them with a pregnant pause, "I mean, all material things . . . will be yours."

"Gee, thanks, Frank," Armstrong replied in a respectful tone as his comrades stared at Borman in silent puzzlement.

"A good bit later," Aldrin subsequently confided in his post-mission

*Collins, who was born in Rome, had decided to adopt New Orleans as his official American hometown because he liked the local food.

memoir, "I would howl with laughter at the absurdity of [Borman's] remark."

On September 29, 1969, the Apollo 11 astronauts, their wives, and a government-appointed support team began the "Giant Leap Tour" by flying to Mexico City, where President Diaz Ordaz paraded them through a crowd of fifty thousand spectators. The next day, they flew to Bogotá, Colombia, and rode in a similarly tumultuous parade to a reception at the presidential palace. Then, while Aldrin rushed back to Atlantic City, N.J., to address the AFL-CIO annual convention, Armstrong and Collins forged on to Buenos Aires, Brasilia, and Rio de Janeiro.

After reuniting for a two-day rest in the Canary Islands, the astronauts commenced a whirlwind European swing during which, as Joan Aldrin later reported, they met "three kings in two days." The first stop was Madrid for audiences with King Juan Carlos and his mentor Generalissimo Franco. Then it was wheels up for Paris to call on President Georges Pompidou. The following day, they paid their respects to Queen Juliana in Amsterdam and King Baudouin and Queen Fabiola in Brussels. Less than twenty-four hours later, they arrived in Oslo for a royal banquet hosted by King Olaf of Norway.

A previously scheduled visit to Sweden was canceled due to the country's policy of granting asylum to American draft dodgers unwilling to fight in the Vietnam war. Instead, the astronauts flew on to Bonn and then Berlin, where they kept a politically sensitive public speaking engagement on the friendly side of the Wall between East and West.

On October 14, Queen Elizabeth II honored the Apollo 11 crewmen and their wives with a royal reception at Buckingham Palace. Then the astronauts flew to Rome for an audience with Pope Paul VI, and a much-needed three-day rest period highlighted by an unscheduled private party at the villa of actress Gina Lollobrigida.

On October 18, they arrived in Yugoslavia, the only Communist country on their itinerary, and went duck hunting with the iconoclastic and surprisingly active septuagenarian President Tito. Two days later, they paraded through the jam-packed streets of Ankara, Turkey, then took off on an arduous overnight flight to Kinshasa, Zaire, where President Joseph Mobutu decorated them with the African nation's prestigious Order of the Leopard.

Nearly one million people cheered the Apollo 11 motorcade through Tehran on October 25, a day that climaxed with an invitation to the shah's birthday party. A throng of over 1.5 million greeted them in Bombay the following afternoon. The astronauts drew smaller but no less enthusiastic crowds in Dacca, Bangkok, Perth, Sydney, Guam, and Seoul. An estimated 120,000 hailed their motorcade through Tokyo, which eventually led them to the Imperial Palace and an audience with Emperor Hirohito.

The "Giant Leap Tour" finally ended with a White House dinner in Washington on November 6. President Nixon called it the most successful goodwill trip in U.S. history. Over the preceding thirty-eight days, Armstrong, Aldrin, and Collins had traveled 44,650 miles, visited twenty-three countries, met twenty heads of state, and been decorated nine separate times. But even though the government picked up the plane fare and lodging tab, the astronauts paid a heavy emotional price. Aldrin, who became mysteriously sullen and withdrawn early on in the tour, showed the most visible signs of stress.

"My prevailing memory of the trip" he wrote afterward, "is that there was liquor everywhere. There were almost always bottles of Scotch or gin in our hotel rooms. I marvel at how we all got through so much liquor with so few consequences."

Though Aldrin claims he got drunk only once on his trip around the world, he admits having frequent altercations with his wife, Joan. According to his memoir, "My mood was starting to swing from long, enthusiastic highs to long, immobilizing lows"—an early warning sign of the mental breakdowns to come.

Aldrin would also report that the usually unflappable Armstrong also displayed some surprising post-splashdown testiness, especially when the subject of "flicker-flashes," the mysterious light beams the crewmen had seen en route to the moon, was raised at the debriefing sessions in the LRL.

"Neil began to look doubtful and annoyed whenever the flashes were discussed," Aldrin claimed in his post-mission autobiography. "I had always envied his facility with words and his ability to give a speech, but now I sensed he was annoyed by my discussions of this phenomenon. I suspect it was an inevitable thing, but I'm certain it was nothing either of us had ever contemplated. We had been the backup crew on Apollo 8 and had become drinking buddies. Our friendship was casual and respectful and, as friends often have, we had something going for us: a total and enthusiastic immersion in our work. When the feeling [of annoyance] finally became apparent to me in the Lunar Receiving Lab, I quickly suppressed it. It struck me as an emotional and unscientific thing to feel, and human feelings weren't in keeping with the image we had as scientists and astronauts."

Part of the tension inside the LRL also grew from the fact that all three Apollo 11 astronauts now had to confront an agonizing midlife career crisis. At the time, NASA still planned nine more lunar landing missions culminating with Apollo 20. According to the rotation system, they could conceivably qualify for one of the last two flights, but in the interim, they would be required to endure yet another arduous stint on one of the backup crews. And as Aldrin later pointed out, it would be virtually impossible to top the feat he and Armstrong had already accomplished.

Ironically, Mike Collins was the first to realize that enough was enough. Although his role as command module pilot had necessarily denied him the exclusive privilege of setting foot on the lunar surface, he had let it be known well before the astronauts lifted off from Earth that Apollo 11 would probably be his last space flight. Having grown up as the worldwise son of a veteran foreign service officer, he was anxious to try his hand at international diplomacy, and said as much during his post-mission visits to the White House. In January of 1970, he resigned from NASA and the air force to become an assistant Secretary of State for Public Relations, a post he secured with the obliging help of President Nixon.

Collins's fellow crewmen decided to hedge their bets. That same month, Buzz Aldrin announced that he was taking an indefinite leave of absence from NASA—but retaining his commission as an air force colonel—to head up the USAF aerospace training school at Edwards Air Force Base in California. A few weeks later, commander Neil Armstrong, the laconic civilian test pilot who had won historical immortality with his "giant leap for mankind," unceremoniously retired from active flight status to become Deputy Assistant Administrator, Aeronautics, in the Office of Advanced Research and Technology at NASA headquarters in Washington.

The astronauts who followed in the footsteps of Apollo 11 found that the duration of their "instant celebrity" steadily declined in direct proportion to their network TV ratings. Apollo 12 crewmen Pete Conrad and Al Bean, who made man's second lunar landing late on the evening of November 18, 1969, drew only thirty million American viewers during their moon walks, less than one-third the audience for Armstrong's "giant leap." Likewise, the forty-day goodwill tour they began the following February attracted only a fraction of the crowds who had hailed their illustrious predecessors.

The welcome home ceremonies for Apollo 13 astronauts Jim Lovell,

Fred Haise, and Jack Swigert were embarrassingly meager. Although the astronauts' near fatal crisis in space had sparked worldwide concern, only ten thousand people turned out to cheer their motorcade through Chicago on May 1, 1970. Their parade through Manhattan was almost a nonevent. *The New York Times* blamed the "sparse" crowds on a lack of advance publicity, but the low attendance also signaled a sea change in public attitudes toward continued manned exploration of the moon.

Certain high-ranking NASA officials were also starting to have misgivings about Project Apollo for very different reasons. Manned Spacecraft Center boss Robert Gilruth reportedly became "very nervous" about risking more human lives after the Apollo 13 accident. Years later in an interview with author Joseph Trento, space agency administrator Dr. Thomas O. Paine recalled, "The lunar sites we were going to were turning out not to be all that different . . . We had a lot of single point failures and everything is never perfect . . . Bob Gilruth was very concerned . . . Bob thought, gee, Apollo 14 is plenty."

Although the astronauts and most of the NASA brass nevertheless wanted to attempt at least four more lunar landings after Apollo 14, it soon became clear that a political backlash posed an even greater threat to their plans than the potential safety hazards. Despite the myriad technological breakthroughs that had already flowed from the space program, once supportive middle American taxpayers did not feel that they were getting significant tangible returns on the $40 billion invested. They were also forking over many billions more each year to fight the ostensibly unwinnable war in Vietnam, and reeling from the much closer to home fallout from long-festering environmental and urban social problems. Space agency opponents on Capitol Hill joined forces with the Nixon administration to cut appropriations for lunar exploration. In early 1972, it was announced that Apollo 17 would be man's last flight to the moon for the foreseeable future.

In the meantime, despite considerable advance publicity, Apollo 14 crewmen Al Shepard, Ed Mitchell, and Stu Roosa managed to draw only scattered crowds for their ticker tape parade on May 7, 1971. Still more embarrassing as far as the Nixon administration and Mayor Lindsay were concerned, the motorcade was halted several times by bands of black demonstrators demanding that more funds be allocated for federal job assistance programs. A chanting mob of affiliated protestors also interrupted speeches at a United Nations ceremony later that afternoon.

Apollo 15 crewmen Dave Scott, Jim Irwin, and Al Worden, who had momentarily recaptured a worldwide TV audience thanks to Scott and Irwin's dramatic spins in the first Moon Rover, were the last to ride in a New York City motorcade. Scott claimed their parade on August 24, 1971, was "the most impressive thing I've seen since we left heaven." But instead of retracing the traditional route through the narrow canyons of Wall Street, the caravan cruised down Fifth Avenue, where the crowds were seldom more than three rows deep and local office workers did not have access to rolls of ticker tape.

Far more shocking and disheartening, the Apollo 15 crewmen, who were known as the straightest of All American boys, soon become embroiled in a "commercialization" scheme that eventually scandalized the entire astronaut corps. In June of 1972, the Washington *Star* reported that the trio had smuggled 398 unauthorized first-edition postage stamp envelope "covers" on their flight to the moon; following their return to Earth, a friendly NASA subcontractor had sold a hundred of the covers to a German dealer who had then resold them to collectors for over $150,000. An estimated $21,000 was later deposited in trust accounts reportedly set up to pay for the education of the astronauts' children.

By the time the scandal hit the front pages, all three crewmen had already decided to get themselves out of the stamp deal, and refused to accept any of the proceeds. (Irwin claimed that his "spiritual encounter with God on the moon" was what made him see the light.) But it was too late to undo previous commercialization deals or escape public disgrace. Irwin, who had announced his intention to retire before the Washington *Star* exposé, left NASA at the end of the summer. Worden, who had also taken himself off active flight status, transferred to NASA's Ames Research Center. Scott was summarily dropped from the astronaut corps and reassigned to a technical assistant's post.

"I think Dave took it the hardest," Irwin recalled in an interview years later. "Al and I had already decided to leave the program anyway. But Dave wanted to keep flying as long as he could, and he was hoping that he'd eventually get promoted to three-star general someday. He probably would have made it—he was certainly as well qualified as anybody in the air force—if it hadn't been for the stamp incident."

What's more, the Apollo 15 crewmen were by no means the first or the only ones to participate in unauthorized profiteering. According to a report NASA's own in-house investigators released in September of 1972, previous Apollo commercialization schemes involving stamps, watches, medals, and figurines had grossed more than $1 million. The report stated that in one such deal fifteen unnamed astronauts, nine of whom were still at NASA, received $37,800 (or roughly $2,500 each) for autographing 30,000 stamp covers later sold to private collectors in the U.S. and abroad. Five of the astronauts said they had donated their shares to charity; the rest presumably put theirs to personal use.

In still other commercialization schemes, various outside parties repeated all the ill-gotten gains without bothering to give the astronauts a cut. Apollo 15 was once again an unfortunate case in point. Before lifting off from the lunar surface, Scott and Irwin unveiled a three-and-a-half-inch-tall abstract human figure by Belgian sculptor Paul Van Hoeydonck dedicated to all the pilots who had lost their lives in the U.S. and Soviet space programs. After the crewmen returned to Earth, they learned that Van Hoeydonck had made 950 copies of the "first art on the moon," which a prominent New York gallery was marketing at $750 apiece. The statuette was appropriately titled, "The Fallen Astronaut."

NASA's deliberately sketchy disclosures apparently represented only the tip of the iceberg. Apollo 11's Buzz Aldrin, for example, later admitted in his post-mission autobiography that he and his fellow crewmen had stuffed their personal preference kits (PPKs) with "several hundred" first-edition stamp covers, adding that he still had over two hundred envelopes stashed away for some future rainy day. And according to Jim Irwin, the stamp covers confiscated in the wake of the Apollo 15 exposés were eventually returned so the crewmen could establish educational trust funds for their children as originally planned.

Although space agency officials vowed to impose strict limitations on the contents of PPKs carried aboard future space flights, the fallout from the Apollo commercialization scandal shattered the astronauts' heroic and squeaky clean public images. The average American taxpayer was becoming further disillusioned by the fact that unlucky Apollo 13 and the four successful lunar landing missions had cost over $400 million per shot without producing any obvious benefits that might alleviate the burdens of their daily lives, much less bring an end to the war in Vietnam and help create a lasting world peace.

Not surprisingly, the astronauts on man's last two moon shots experienced the most short-lived "instant celebrity" upon their return to Earth. The only Apollo 16 crewman to ride in a parade was lunar module pilot Charley Duke, who was honored with an affectionate little motorcade in his boyhood hometown of Lancaster, South Carolina, on May 26, 1972. Although he and fellow crewmen John Young and Ken Mattingly later received Distinguished Service Medals and a perfunctory round of presidential handshakes, they were not invited to address a joint session of Congress or dispatched on a global goodwill tour.

The Apollo 17 launch inspired a modest resurgence of nostalgic public interest, but MSC director Christopher C. Kraft, Jr., later revealed that NASA actually had to pay the major TV networks to provide live coverage of Gene Cernan and Jack Schmitt's excursions across the rugged valleys of Taurus-Littrow. After splashdown in the Pacific on December 19, 1972, Cernan, Schmitt, and command module pilot Ron Evans were welcomed aboard the aircraft carrier *Ticonderoga* by Senator Barry Goldwater (R-Arizona) and a contingent of likeminded politicians who urged renewed emphasis on space exploration for the collective benefit of all mankind, but they were merely crying in the wind.

President Nixon, still gloating about his landslide victory over Senator George McGovern in the 1972 campaign, had already charted an opportunistic new course for NASA in hopes of exploiting the economic and strategic potential of space. The process had actually begun early in his first term when Nixon was still rankling about having to carry on the legacy of the late JFK. So intense and deep-seated was Nixon's jealousy that he vetoed plans to use the aircraft carrier *John F. Kennedy* to recover the Apollo 11 astronauts after their splashdown on July 24, 1969, and insisted on sending the USS *Hornet* instead.

In the spring of 1970, Nixon appointed Vice-President Spiro T. Agnew to head a commission that would formulate plans for the future of the U.S. manned space program. But even as Nixon and Agnew publicly endorsed ambitious proposals to send men to Mars and other planets, newly installed budget director George Schultz began making sharp cuts in the funds earmarked for the remaining missions on the Project Apollo itinerary. The Nixon administration also vetoed the air force's proposed Manned Orbiting Laboratory (MOL) "spy platform," reluctantly preserving only a scaled-down civilian space station project dubbed "Skylab."

Over the next two years, NASA administrator Paine waged a bitter behind-the-scenes battle against the White House over alternative proposals for a reusable "Space Shuttle." The irrepressible Wernher von Braun, who still dreamed of sending at least two six-man missions to Mars, urged the development of a giant liquid-fueled spacecraft with an estimated cost of $10–12 billion. But Paine realized that the only way to win congressional approval was to design a "national vehicle" capable of hauling both commercial and military payloads into Earth orbit at a relatively inexpensive price per pound.

Paine also concluded that by accepting interim cuts in the NASA budget he might be able to obtain more money to invest in the space shuttle. But unlike his shrewd predecessor James Webb, who arbitrarily doubled the estimated cost of Project Apollo before making his bid for funding, Paine submitted what he thought was a very realistic $8 billion shuttle proposal, only to be ambushed by White House budget cutters just when he thought a final compromise had been struck.

In January of 1972, President Nixon formally asked Congress to approve construction of a Space Transportation System (STS), not for the $10–12 billion Von Braun had recommended or even the $8 billion requested by Paine, but for a far more politically palatable $5.1 billion. With the exception of stunned NASA officials and the disillusioned members of the astronaut corps, few Americans realized the truly ominous implications of this bare-bones space shuttle budget. Rather than incorporating the liquid fuel systems of Apollo-Saturn rockets, the vehicle would have to use cheaper solid fuel boosters whose thrust could not be cut back or shut down once the initial fire had been lit. Worse, the design lacked an operationally feasible emergency escape system for the astronauts.

For the first time in the history of the U.S. manned space program, safety factors were being sacrificed solely in the interest of saving money. Or so claimed critics at almost every level of the NASA hierarchy. Outraged by the obstinance and ignorance of the Nixon administration, Robert Gilruth refused to participate in the shuttle project, and decided the time had come for him to retire. The equally infuriated Wernher von Braun also handed in his resignation, and accepted a private sector job at the Fairchild Corporation.

Although most of the Apollo astronauts continued to maintain a resolute optimism and idealism, they were also practical-minded souls who could read the writing on the wall. The era of the pilot-astronaut was obviously giving way to the new age of the scientist-astronaut. There were no less than thirty-two trainees from the fourth, fifth, sixth, and seventh astronaut classes standing by. NASA would surely recruit scores of younger (and daresay, sharper?) men and women for the STS program. And no matter how rewarding future shuttle missions proved to be, they could never compare to leaving the earth and landing on another planet.

By the time the Apollo 17 crew returned to Earth on December 19, 1972, ten of the twenty-four men who had flown to the moon, including all three Apollo 11 astronauts, had either left NASA or resigned from active flight status. Of those who remained, just five got another chance to fly in space. Apollo 12's Pete Conrad and Al Bean commanded the Skylab 2 and Skylab 3 missions in the summer of 1973. Apollo 10's Tom Stafford participated in the Apollo-Soyuz joint mission with the Soviets in the summer of 1975. Apollo 16's John Young and Ken Mattingly were the only two who stayed on long enough to fly on the space shuttles.

By July 20, 1974, the fifth anniversary of Neil Armstrong's "giant leap," most voting-age Americans could still name the first man to walk on the moon. But as one commemorative editorial noted, "Many Americans would be hard-pressed to name all three participants" on the Apollo 11 mission. Precious few people outside NASA could name all eleven astronauts who followed Armstrong onto the lunar surface. Apollo 16's Charley Duke, the tenth man to walk on the moon, later told a group of high school students that by April 20, 1977, "only my wife and Momma remembered" that the date marked the fifth anniversary of his own lunar landing. Duke claimed that by the next year, only his mother seemed to remember, and that he "started pinning notes up around the house" to remind his wife and children.

For better or worse, the majority of Apollo astronauts discovered that their "instant celebrity status" was just that, a flicker of flame that burned out shortly after splashdown. And yet with the passage of time, each of the astronauts gradually gained a clearer and more profound understanding of how "the greatest adventure on which man has ever embarked" had transformed his own life and the lives of all mankind.

16

THE MELANCHOLY OF ALL THINGS DONE

Life After Splashdown: Part Two

1973–1988

Melancholy and remorse form the deep leaden keel which enables us to sail into the wind of reality.

CYRIL CONNOLLY
The Unquiet Grave (1945)

"Being an astronaut," Apollo 11 command module pilot Mike Collins observed in his post-mission autobiography *Carrying the Fire,* "is a tough act to follow." Collins wrote those words in 1974, just five years after his fellow crewmen Neil Armstrong and Buzz Aldrin made man's first lunar landing, but their truth rings ever more loudly with each passing year.

As if to prove once and for all that they were neither the mythic heroes nor the faceless robots most of the American public believed them to be, the twenty-four men who flew to the moon went off in twenty-four different directions following their return to Earth. One of the Apollo astronauts became a painter. One became president of a major airline. A third won election to the U.S. Senate. Others dabbled in international oil trading, real estate, beer distributing, farming, teaching, and parapsychology. Another became a full-time witness for Jesus. Still another suffered a series of mental breakdowns. One died of cancer. But only one continued to be a NASA space pilot.

In most cases, being an astronaut did not prove to be the springboard to great wealth or even guarantee financial security. While complete financial information is not available, only two of the astronauts—Borman and Shepard (at least until the Texas real estate crash)—appear to have become multimillionaires. But with the exception of civilians Armstrong and Schmitt, all could count on comfortable military pensions to supplement their incomes.

And yet despite the wide divergence in career choices, there was a remarkably consistent pattern to the astronauts' lives after splashdown. Apollo 11's Armstrong and Aldrin naturally experienced some unique aftereffects by virtue of being the first men to walk on the moon, while the crewmen of Apollo 13 were haunted by the fact that theirs was "the flight that failed." But even they were not exceptions to the general rule. As Apollo 12's Al Bean observes: "I think that the common denominator in what people did afterwards is that they became more like they really were before they left for the moon."

"Nobody that I know changed their religious beliefs as a result of going to the moon," Bean points out. "Take Jim Irwin [of Apollo 15], who had converted to Christianity at a revival before he left on his lunar mission. Being on the moon in a similarly exciting situation, he was converted to the belief that he had a responsibility to carry the Word of God. It provided, in a way, a continuity in his life, a furthering and a fulfilling. But if [Apollo 12's] Pete Conrad, the fellow I went with, who's a very good man but not a profoundly religious individual, had become religiously oriented like that, I would've been completely amazed."

In fact, as the born-again Irwin and other former moon voyagers note, the ways they and their fellow astronauts have changed closely correspond to their respective seat assignments on the spacecraft. The individual differences seem to depend not on when they flew to the moon or on the outcome of their mission, but on whether they served as commanders, lunar module pilots, or command module pilots. That, in turn, may have been a kind of self-fulfilling prophecy, for each man was originally selected and groomed for his respective role according to the NASA bureaucracy's assessment of his previous experience, piloting talent, and psychological profile.

The nine Apollo lunar mission commanders, six of whom actually touched down on the moon, were among the most intense, inscrutable, and least reflective members of the astronaut corps, the true zealots of space exploration. Like the Jesuit priests of the Renaissance, they had rational minds wedded to an absolute faith in their vocation. That made them relentless and willful; observant without being sensitive; decisive but never impulsive; responsive only to the directives of Mission Control and their own inner discipline. Even their highly motivated peers were awed by their dedication, and humbled by their self-assurance.

Not surprisingly, the Apollo commanders were outwardly unfazed, unchanged, and in some cases, seemed genuinely unimpressed by their trips to the moon. They took it all in stride, like the stoics they were, conscious every minute of their official responsibilities. Each mission had been planned and rehearsed down to the most minute detail with nothing left to the imagination, and it was the commander's duty to bring it off that way. They followed instructions and performed their assignments with skill and dispatch, always coolly, sometimes cleverly or jokingly, but never with much imagination. The most memorable wrinkle any of them added to their flight plan was Apollo 14 commander Alan Shepard's lunar golfing stunt. In a recent interview, Apollo 16's John Young flatly declared that "going to the moon was no big deal."

The commanders' style was clearly reflected in the leadership roles they assumed after leaving NASA. Apollo 8's Frank Borman, the commander of man's first lunar orbiting mission, who had always impressed his fellow astronauts as a corporate executive type, resigned from the space agency in 1970, and went on to become chairman of Eastern Airlines, a post he held for nearly a decade; shortly before the company's 1986 merger with Texas Air, he resigned to pursue private business interests. Apollo 10's Tom Stafford, who commanded the "dress rehearsal" for man's first lunar landing mission, later commanded the Apollo-Soyuz flight in July of 1975, then rose through the air force ranks to become a three-star general. In 1979, he resigned to become vice-chairman of Gibraltar Exploration Ltd., an Oklahoma oil firm.

Apollo 12's Pete Conrad, who commanded man's second lunar landing mission, later commanded Skylab 2, which orbited the earth for twenty-eight days in the spring of 1973; he resigned from NASA in December of that same year to become senior vice-president for marketing with Douglas Aircraft in Long Beach, California; he has since transferred to the company's St. Louis office. Jim Lovell, who had to fight for his life during the ill-fated Apollo 13 mission, also retired from NASA in 1973, and proceeded to battle Ma Bell as group vice-president of Centel, an independent telecommunications firm, in Chicago.

Alan Shepard, who sparked controversy within the astronaut corps by using his political clout to get assigned to command Apollo 14, capitalized on his fame as the first American in space to make himself a millionaire following his resignation from NASA in 1974. In addition to securing one of the coveted Coors beer distributorships awarded to well-connected Houston businessmen in the mid-seventies, he joined former mayor Louie

Welch in a suburban real estate development that created further controversy when home buyers alleged that the lots they had been sold were located in a flood plain.

Meanwhile, Apollo 15 commander Dave Scott, who had suffered embarrassment during the Apollo commercialization scandal in 1972, gave up his dreams of becoming a three-star general, and resigned from NASA and the air force in 1977. He is now president of Scott Science and Technology, an independent "technology transfer" firm based in Lancaster, California.

Apollo 17's Gene Cernan, who made man's last moon landing, left the space agency in 1976 to join Coral Petroleum, an international oil trading company that later collapsed in the wake of a federal price control violation scandal. Perhaps the most sensitive and articulate of the Apollo commanders, Cernan emerged from the scandal unscathed after investigators concluded he had nothing to do with the Coral schemes, and has since started his own independent energy firm. Recently remarried, he also works as a part-time space correspondent for ABC News.

Predictably enough, Apollo 11's Neil Armstrong, the first man to walk on the moon, has proved to be the most private and enigmatic of the former mission commanders. Many of Armstrong's peers had regarded him as the archetypal robotlike astronaut, but he had surprised them all by demonstrating uncommon grace and diplomatic skill during the Apollo 11 crew's post-splashdown goodwill tour, a fact that eventually came back to haunt him in a most unexpected way.

At a formal dinner hosted by British Prime Minister Harold Wilson, Armstrong earned thunderdous applause for a speech explaining how his ability to navigate a course to the moon was a direct result of the invention of a chronometer by an obscure eighteenth-century British scientist. Later, in Tokyo, when the Japanese ambassador to the U.S. put the Apollo 11 astronauts on the spot by noting that tradition called for the guest of honor to sing a song, Armstrong rose from his chair and trilled an old Ohio ditty about an Indian boy and an Indian maiden in a canoe on a moonlit night. According to the song, the two were so in love that their paddles became wings and flew them off to heaven.

By virtue of the international notoriety gleaned from his historic feat, Armstrong was in by far the best position to cash in on his name and fame, but he proved to be a reluctant hero. He left the astronaut corps shortly after completing the Apollo 11 goodwill tour, and moved to Washington to take a $35,500 a year job overseeing NASA's research into advanced aircraft. He accepted only one of the several dozen honorary public service posts he was offered, the chairmanship of the Peace Corps National Advisory Committee. In 1971, he resigned his NASA job to become a professor of aeronautical engineering at the University of Cincinnati.

Unlike former air force ace Chuck Yeager, who became a hot TV advertising celebrity after the publication of Tom Wolfe's book *The Right Stuff,* Armstrong seldom tried to use his hero's image for financial gain. He made a few television commercials for Lear Jet after joining the company's board of directors, and appeared in ads for the "new" Chrysler Corporation his friend Lee Iacocca was leading back to profitability. But for the most part, he avoided the public eye, declining virtually all of the hundreds of speech-making requests he received each month and refusing to grant in-depth retrospective interviews to the media.

Armstrong did find a way, however, to put his astronaut experience to work for the greater benefit of mankind through his role in academia. In the spring of 1976, he announced that he and a team of research associates were experimenting with equipment adapted from an Apollo life support systems backpack pump to improve blood pumps used in artificial hearts and kidneys. "The Apollo pump occurred to me because of its efficiency from the power and weight standpoint," Armstrong told a reporter. Though Armstrong's team confined their testing to animals, they reported that even without special protective linings, the Apollo pump proved gentler to blood cells than other pumps then in use.

In the wake of the January 28, 1986, Challenger tragedy, Armstrong agreed to serve as vice-chairman of the presidential commission appointed to investigate the causes of the space shuttle explosion. Now age fifty-eight, Armstrong has returned to private business as an executive of CTA Inc., a firm based in Lebanon, Ohio, that develops computer software for aviation systems.

Besides adding an air of intrigue to his deliberately bland public persona, Armstrong's reclusiveness has created controversy among his former colleagues at NASA, especially as the agency has faced Congressional budget cutters. Critics score Armstrong for being a poor salesman, an ingrate who has neglected his obligation to promote the manned space program of which he was perhaps the most favored individual beneficiary. Armstrong's defenders point out that he was hired to be an astronaut, not a tub thumper, and insist that he is really a sensitive and articulate man who simply prefers to lead by example.

The glimpses Armstrong has given of what he felt and thought after touching down in the Sea of Tranquility are not full of joyous hyperbole or heartfelt wonder. Rather, they are testimony to the fact that he was on edge most of the time, and that what he enjoyed most about landing on the moon was the profound relief of realizing that he and Aldrin were going to survive their "impossible" mission after all. As he remarked at a press conference in July of 1970: "I particularly remember the elation of finding out that we indeed weren't going to sink into the surface, and we could continue with all the other planned activities."

Those do not sound like the words of a man who was lusting for immortality or reveling in boyish delight over the fact that he was about to make a "giant leap for mankind."

In sharp contrast to Armstrong, who maintains that he would definitely fly back to the moon if he ever gets a second chance, former crewmate Buzz Aldrin candidly admits that he found man's first lunar landing mission to be a harrowing experience with devastating emotional repercussions. As he confided at the Apollo 11 astronauts' tenth anniversary press conference in the summer of 1979, "I'm not sure I would go again."

Aldrin is a uniquely troubled but archetypal representative of his breed. Although twelve Apollo astronauts touched down on the moon, the six who seemed to be the most deeply touched by it were the six lunar module pilots, the men assigned to play supporting roles for the mission commanders. The fact that the LMPs registered far more intense reactions—and later underwent more dramatic life-style changes—than the CDRs was a natural consequence of their specialized roles. As Apollo 15's Jim Irwin points out; "The people in my slot were sort of tourists on these flights. We monitored systems that were for the most part not associated with control of the spacecraft, so we had more time to look out the window, to register what we saw and felt, and absorb it."

Aldrin chronicled his own post-mission psychic turmoil in his 1973 autobiography *Return to Earth.* He reports that by the fall of 1970, after leaving NASA to head the USAF Aerospace Training School, he started a passionate extramarital affair with a woman in New York, and suffered embarrassing and debilitating bouts with depression. As he recalls, "The rule of my emotions was absolute and ruthless. In no way could I stop what I felt, but I hoped somehow not to feel anything at all. I yearned for a brightly lit oblivion—wept for it."

Aldrin adds that each time he decided to seek outside professional help, he "began to cry," for "the only help available was official air force treatment, and the matter would go on my record."

Finally, in the fall of 1971, Aldrin agreed to undergo psychiatric treatment at Brooks AFB in San Antonio, Texas. There, with the help of the drug thioridazine, he began to put the pieces of his life back together.

According to his book, he realized that the demands of an overbearing father, who still rankled that his son was "only" the second man on the moon, and the sudden uncertainty about the future of his preprogrammed career had catapulted him into despair.

"I had gone to the moon," he wrote later. "What to do next? What possible goal could I add now? There simply wasn't one, and without a goal I was like an inert ping-pong ball battered about by the whims and motivations of others. I was suffering from what poets have described as the melancholy of all things done."

By the end of the year, Aldrin and his wife Joan decided to divorce. The following spring, he retired from the air force and took a part-time job representing Volkswagen of America. But by the summer of 1973, Aldrin was back in psychiatric treatment under medication with the drug imipramine. He then determined to exorcise the demons in his soul by writing a confessional autobiography with Los Angeles journalist Wayne Warga.

In the decade and a half since, Aldrin has been a car salesman, a "free-wheeling rancher," a "consultant," and a science lecturer at the University of North Dakota. Having suffered through a second failed marriage, he has recently married for the third time. But at the age of fifty-eight, he still seems to be in search of a career goal worthy enough to snap him out of his melancholy. And as not always sympathetic former colleagues point out, Aldrin still wears his psychic scars on his chest, for he is the only Apollo astronaut who appears at public functions with his Distinguished Service Medal pinned to his civilian blazer, as if he feels the need to remind the world of his claim to fame. According to a close associate, his present occupation is "author and consultant in space architecture," but he "spends a great deal of time lobbying to ensure America's presence in space isn't put on the back burner."

Contrary to popular misconception, the other five Apollo lunar module pilots who landed on the moon did not "go crazy" following their return to Earth, much less become embroiled in all-out emotional battles similar to Aldrin's. But like their melancholic predecessor, each man did have to confront his own major midlife crisis when his astronaut days were over. And with one predictable exception, the LMPs wound up choosing what seemed like diametrically opposite new careers for what were actually very similar underlying motivations.

Apollo 12's Al Bean, the fourth man to set foot on the lunar surface, is living proof of his own point that most astronauts have tended to pursue interests that allow them to be "more like they really were before they left for the moon." In Bean's case, that was no accident. For as he recounted in a recent interview, he made a conscious decision to make his secret dreams come true just before he and Apollo 12 commander Pete Conrad attempted man's second touchdown on the moon.

"I remember thinking in lunar orbit before we went down to land, 'You know, when I get back from here—if I get back from here—I'm really going to try to live my life like I want to because it's a big risk to go to the moon,' " Bean recalls. "I came to grips with the thought that life is short and you want to make sure that you do what you have to do and need to do in your short lifetime. I had that philosophy before I went, but it was much stronger after I got back."

Bean, who also divorced his wife after splashdown, remained in the astronaut corps long enough to fly on the fifty-nine-day Skylab 3 mission in the summer of 1973, thereby compiling a total of over 1,671 hours in space, which still ranks him fifth among all American astronauts. He spent the rest of the decade serving in the NASA administrative ranks. But all the while, he quietly cultivated his suppressed desire to become a painter by training at various art schools.

In 1981, after enjoying some local commercial success, Bean resigned from NASA to become a full-time painter. His best paintings deal with a subject he had at first tried to avoid: his experiences on the lunar surface. Though he did not become a born-again Christian, Bean says that his trip to the moon did lead to a kind of spiritual awakening that continues to inspire his art.

"I think it's the awesomeness of it," he says, "the fact that you are far away from the earth and the earth is small yet you know it's big from being on it, and the fact that you can get to a place where it looks little even though you know it is big, brings out these kinds of feelings in people. Either that God allowed this to occur, or conversely, that this is just science and God had nothing to do with it."

Apollo 14's Ed Mitchell, who secretly conducted ESP experiments on his way to the moon, had a similar epiphany during his flight back to Earth. Like Bean, he suddenly came to the realization that "science may have had the wrong view of things. It is not just a material universe. It has purpose and design and was created." Mitchell, who also divorced shortly after splashdown, resigned from NASA in October 1972 to form an independent technology consulting firm and pursue his interest in parapsychology. The following spring, he founded the Institute of Noetic Sciences in Palo Alto, California, a nonprofit corporation "dedicated to research and education in the processes of human consciousness to help achieve a new understanding and expanded awareness among all people."

More recently, Mitchell has become active in raising public consciousness about the importance and potential benefits of future space exploration. In 1985, he helped found the Association of Space Explorers, a nonpartisan confederation of astronauts and cosmonauts from the U.S., the U.S.S.R., and twenty other countries. He has also published dozens of magazine articles on noetics, and edited a volume of essays in parapsychology entitled *Psychic Exploration.*

Not surprisingly, the LMP who turned most strongly to organized religion was Apollo 15's Jim Irwin. Having felt "the presence of the Lord on the moon," Irwin says he heard the call to "spread the Word" following his return to Earth. But as he admitted in a recent interview, answering the call was not easy, especially in the wake of the Apollo 15 commercialization scandal just before his resignation in the summer of 1972.

"I never prepared myself for this other aspect of the moon," Irwin confides. "We are trained in a very practical, scientific approach, with these other things being on the fringes of our experience. Even though I was a Christian before the flight, I didn't delve into it, didn't share, didn't talk about it really. I guess we are all conditioned by the experiences in our lives to where we are more open to things at a certain point in time than we have been previously, and maybe I was conditioned in my own life so that I was more receptive to [hearing the call to spread the Gospel] when I was on the moon."

In any event, Irwin claims that he was amazed to find he no longer dreaded public speaking—at least not when he spoke on behalf of Christianity. "When I came back, I realized that for the first time in my life, I had something important to say," recalls the eighth man to walk on the moon. "I found I could relax, I could be completely myself, and not some phony person . . . I was really just trying to communicate some feelings that I thought were very important . . . I thought they should become important to everybody on the earth if we were to survive, if we were to continue our quality of life . . . I also had a new love for people and for the earth, and I think that brings about a desire to communicate . . ."

Irwin, who did not divorce his wife, currently heads High Flight, an evangelical religious organization headquartered in Colorado. He has generally avoided public controversy since the Apollo commercialization scandal with the exception of one recent incident overseas. In the summer of 1986, Irwin and some fellow Americans were arrested in Turkey during what they claimed was an expedition to find Noah's Ark. Turkish officials jailed them on spy charges, then deported them back to the U.S.

Meanwhile, Apollo 16's Charley Duke has undergone a less-publicized but perhaps even more amazing religious conversion. Duke, who was Baptist before lifting off for the moon, did not feel the presence of the Lord on the lunar surface. But following his resignation from NASA in 1975, he rejected his old scientific beliefs in favor of a newfound faith in Christian fundamentalism, making what was in effect a kind of reverse odyssey from evolutionism to creationism. His moment of epiphany came in 1978 with the help of two Episcopal priests in San Antonio. "They laid their hands on me, and the next thing I knew I was back in space," Duke recently told an interviewer. "I started to cry—racking sobs—because I knew how much God loved me." Among other things, he says his faith helped save his marriage, and gave new purpose to his life and career. Now age fifty-three, he is the head of Duke Investments, a real estate firm based in New Braunfels, Texas.

Apollo 17's Jack Schmitt, the twelfth man on the moon, is the exception who proves the rule. Schmitt was the first scientist-astronaut in space. Unlike some fellow LMPs, he did not become more spiritual following his return to Earth, but he did make a fairly dramatic career change from geologist to politician. In 1976, he won election to the U.S. Senate from New Mexico on the Republican Party ticket, and immediately put his astronaut training to work as chairman of the subcommittee on space. However, Schmitt's opponents back home soon began to charge that he had his head in the stars and was neglecting the everyday needs of his constituents. ("What on earth has he ever done for us?" asked one anti-Schmitt political advertisement.) In 1984, he lost his bid for reelection.

Apollo 13's Fred Haise was the lone LMP who never got a chance to touch down on the lunar surface, but he was definitely touched by the life-or-death crisis he and his fellow crewmen survived following the explosion of their spacecraft's oxygen tanks. Besides bolstering his religious faith, the experience also confirmed his belief in the importance of space exploration. After resigning from NASA in 1979, he went to work for Grumman Aircraft in Bethpage, N.Y., the firm that built the Apollo lunar modules and is now a major contractor for the shuttle program.

The other eight men who flew to the moon but did not land were all command module pilots. As such, they were more akin in background and personality to the mission commanders than to the lunar module pilots. And like the CDRs, most CMPs made relatively predictable career changes into private business or space-related government jobs.

Apollo 8's Bill Anders initially considered becoming a physician in his home state of New Mexico, then decided to accept a job at an aerospace firm in Rhode Island. He is presently the general manager of the aircraft equipment division of General Electric in Dewitt, N.Y.

Apollo 11's Mike Collins did a two-year stint as deputy assistant Secretary of State for public affairs in Washington, then left the diplomatic corps to become curator of the National Air and Space Museum. In 1974, he published *Carrying the Fire,* by far the most entertaining and broadly informative astronaut autobiography. He is now a private consultant and free-lance writer based in suburban Virginia.

Apollo 12's Dick Gordon briefly managed the New Orleans Saints pro football team for former owner John W. Mecom, Jr., but wasn't able to turn the habitually losing franchise into a winner. He is now an executive with Astrosciences Corp., a California technology consulting firm. The only vestige of the "instant celebrity" he once enjoyed is the co-starring role he plays with former crewmate Al Bean and Apollo 7's Wally Schirra in a TV commercial for Actifed cold remedies.

Apollo 14 command module pilot Stu Roosa is a Coors beer distributor in Gulfport, Mississippi. Apollo 15's Al Worden is an aerospace consultant based in Palm Beach, Florida. Ron Evans of Apollo 17 operates a motivational sales training firm in Phoenix, Arizona.

Ken Mattingly of Apollo 16 is one of only two astronauts who stayed on long enough to fly in the shuttle program. In the summer of 1982, he helped open what many partisan critics decried as an ominous new era of space militarization by performing the first top-secret Defense Department experiments aboard the space shuttle in the summer of 1982. He also flew on the January 1985 shuttle mission, and conducted a second series of top-secret military tests. Mattingly resigned from NASA in June of 1985. He is now a rear admiral who directs the Navy's space sensors systems in Washington, D.C.

Jack Swigert of unlucky Apollo 13 also turned out to be the unluckiest of the CMPs. In the summer of 1978, just as he was resigning from the astronaut corps to pursue a political career, his father died. Swigert then ran unsuccessfully for a U.S. Senate seat in Colorado. Undaunted, Swigert later ran for and won election to the House of Representatives. But on December 27, 1982, just one week before he was scheduled to take his seat in Congress, he died of cancer at the age of fifty-one.

The twenty-three surviving men who flew to the moon comprise one of the world's most exclusive clubs. The twelve who actually walked on the lunar surface are an even more elite group. But in the two decades that have passed since man's first lunar mission, the Apollo astronauts have never had a full-fledged official reunion. According to Apollo 12's Al Bean, one reason is the prospect of facing the slings and arrows of "instant celebrity" in the inevitable confrontation with the media.

"I've often wondered if they asked the aviators who first flew above the clouds in airplanes the same sort of questions they asked us when we came back from the moon," Bean grumbled in a recent interview. "You know, questions like, 'Did you see God up there? Did you feel the urge to leave the earth forever? Does it make you feel superior to everybody on Earth because you're way up there and we're way down here below the clouds?' Nowadays, we don't expect anybody who takes an airplane trip to have all those crazy feelings. And I don't think that you have them when you go to the moon, I really don't."

Periodic efforts to stage unofficial astronaut reunions have also failed. Apollo 15's Irwin, for example, once hosted a retreat for the twelve moon walkers on a mountaintop in eastern Tennessee. But despite assurances that there would be no media representatives for miles around, only four other astronauts showed up. Apollo 11's Armstrong was one of those who declined Irwin's invitation, claiming that he did not want to put himself in a separate class from the twelve uninvited men who flew to the moon but did not land.

At the same time, the Apollo astronauts also realize that they are and forever will be uniquely linked—and fundamentally different from other people on Earth—by virtue of the extraordinary experience they have in common.

"You have to have a special bond," notes Apollo 17's Gene Cernan. "[Apollo 17 LMP] Jack Schmitt and I are the only two guys who shared those three days in the valley of Taurus-Littrow. We have done something together that is unique. That doesn't mean we'll go out drinking every day when we both happen to be in Washington or someplace like that at the same time. The bond we have is not necessarily a bond of friendship. But it's something very special—it's a bond of confidence."

Regardless of how much or how little their life-styles have changed, the Apollo astronauts also share at least one other common bond: a belief that "all things done" were not done in vain. They may be consummate realists, but they are also true idealists. As far as they are concerned, the liberals who carped that the Apollo budget would have been better spent on social programs are the cynics and myopics. For the history of the decade and a half since the last lunar landing attests that "the greatest adventure on which man has ever embarked" was well worth the unprecedented effort and expense—not only for the astronauts but also for all mankind.

17

THE $40 BILLION BARGAIN

Project Apollo's Benefits to Mankind 1973–1988

There never was a better bargain driven.
SIR PHILIP SIDNEY
The Bargain (1591)

"Going to the moon was a case where everybody on the planet won," declares Apollo 12's Alan Bean. "Those who took part in it won the most, which they deserved to do because they put more into it. But every citizen, whether they worked on the Apollo project or not, was a winner. Everybody benefited from it."

But unlike Bean and his fellow moon walkers, most of the earth's five billion average citizens still don't see exactly how they have benefited from Project Apollo. Neither do most of the two hundred million average U.S. taxpayers who financed the undertaking. As space shuttle astronaut Dr. Sally K. Ride noted in an August 1987 report to the NASA administrator entitled "Leadership and America's Future in Space," "there is considerable sentiment that Apollo was a dead-end venture, and we have little left to show for it."

Future historians may one day regard that sentiment as the grossest public misperception of the twentieth century. Like Columbus's discovery of America, man's conquest of the moon opened a new world that is anything but a dead end. Though few layman recognize or truly appreciate the cornucopia of benefits that have flowed from Project Apollo, virtually everybody on the planet, most especially the citizens of the U.S., already has plenty to show for it. And according to the astronauts and other informed observers, some of the greatest tangible and intangible benefits are yet to come.

One of the reasons so few laymen seem to recognize or appreciate this cornucopia is that many of the most important by-products evolved over a decade or longer until gadgets and gizmos once dismissed as "impossible dreams" became universally accepted facts of everyday life. As a result, the world's mass media, who were some of the primary beneficiaries of space-age innovations, have provided only sporadic coverage of the revolutionary technological and commercial repercussions from Project Apollo without giving headline credit where it is due.

NASA officials have also done a pitiful p.r. job. The agency annually publishes an encyclopedia of *Space Benefits* listing over five hundred specific "technology transfers" and their estimated dollar values, but it is hardly the kind of reading that appeals to the general public. And despite the mind-numbing compilation of specific inventions, *Space Benefits* does not mention what are already proving to be the most valuable benefits of all—the new insights, revolutionary ideas, and cosmic perspectives gleaned from man's first voyages to another planet.

Although bipartisan critics still grouse about the allegedly exorbitant expense of sending twenty-four men to the moon, it is now becoming indisputably clear that the venture would have been a bargain at twice the price. Project Apollo itself cost $24 billion in 1962 dollars. If the budgets of the Mercury and Gemini projects are added, the tally balloons to $40 billion, the equivalent of about $80 billion in 1988 dollars or roughly $3.2 billion per passenger. Yet, total expenditures on the space program between 1962 and 1972 amounted to only 0.3% (i.e., one-third of one percent) of the Gross National Product. The defense budget for fiscal year 1988 alone was over $290 billion, or more than three and a half times the inflation-adjusted total cost of going to the moon, accounting for six percent of GNP.

Although the $40 billion spent on the U.S. manned space program did not directly impact urban poverty, it did provide over 400,000 government jobs and funded more than 20,000 private contractors and subcontractors. And unlike the "employment" provided by most social programs, the vast majority of those jobs were not merely "make work" activities with no prospect of future returns to the employer or the employee. In the course of pioneering a new frontier, Project Apollo helped create whole new industries, processes, and products that had never before existed, which, in turn, helped revitalize the economies of the U.S. and the entire world.

The exact dollar value of Project Apollo's "spin-offs" is almost incalculable. The gross revenues generated by what NASA officials call "direct benefits"—i.e., items developed or invented expressly for the space program that were later transferred directly to the private sector—are well over $1 billion annually even under the narrowest definition of the term. But the economic impact of "indirect benefits" derived or adapted from the space program is many hundreds of billions even by the strictest accounting formula.

The birth of the home computer is a classic illustration of Project Apollo's extraordinary ripple effects. When the Soviets launched Sputnik in the fall of 1957, the state of the art IBM computer was bigger than half a block of two-bedroom homes, and there were only a handful of them in the entire U.S. The invention of the silicon chip in 1958—not President Kennedy's challenge in 1961 to land a man on the moon before the end of the decade—began the modern age of micro-electronics. But Project Apollo gave the new technologies of computerization and miniaturization a common and well-defined goal, a political urgency, and a multibillion-dollar economic incentive.

In less than ten years, the monstrous ancestors of the master computers at Mission Control were streamlined from block size to building size to floor size, gaining speed, accuracy, and power in quantum leaps with each reduction in bulk. The expertise acquired in downsizing the mainframes enabled NASA's computer specialists to design even more compact sibling models to fit inside the spacecraft. Hailed as miniature miracles in their own right, the seventeen-pound DSKY units aboard the command and lunar modules were a thousand times lighter than the IBM master computers in Houston, and boasted a 39,000 word vocabulary. Though their memory banks were still too small to accommodate the astronauts' seemingly indispensable "Return to Earth" program, the Apollo spacecraft's on-board computers augured the dawn of a new age for the whole planet.

In an epilogue to the Apollo 11 crew's official biography published in 1970, Arthur C. Clarke, the renowned science fiction novelist who also wrote equally accomplished books on scientific fact, envisioned the natural link between the cabin-size Apollo computer consoles and "pay TV" broadcast via satellite and underground cables. Clarke confidently predicted that: "By the 1980s, every home could have a display console on whose screen could be flashed instantly any picture or text stored in any library in the world. 'Orbital newspapers,' updated every hour, could be available on a global basis. Doctors, lawyers, engineers, scientists—in fact, all professional men—could have their own information channels which could keep them up-to-date in a way which today's journals and abstracting systems are hopelessly failing to do. If necessary, coding systems could ensure that only authorized viewers could tune in to these special services."

In 1977, just five years after man's last lunar landing, two California whiz kids named Steven Jobs and Steven Wozniak introduced the first home computer, the Apple I. Their amazing machine, which was made possible by Wozniak's invention of the floppy disk, spawned a whole new generation of consumer products and corporate spin-offs. By 1987, according to authoritative trade publication estimates, at least 15 million Americans owned home computers, and the industry's total annual sales were over $26 billion. What's more, the Apple Mac II, which retails for about $5,000, can execute commands even faster than the old IBM main frame computers at Mission Control, and can store two hundred times more information bits than the computers on board the command and lunar modules.

Could man have invented the home computer without Project Apollo? Yes, of course. But he probably would not have done so in this century. The young geniuses who founded Apple in 1977 did not have to start from scratch. The computer they built was really the offspring of over two decades of collective effort catalyzed by the space program. That effort included the national commitment to improve education in math, science, and engineering that was prerequisite to meeting JFK's challenge.

Though the home computer must properly be counted among the long-term indirect benefits of Project Apollo, it was preceded by a host of equally tangible direct benefits that came from the practical applications of space technology. The computer hardware and software developed to land a man on the moon was put to use by both public agencies and private corporations in developing down-to-earth solutions for some of the major social and environmental problems that had allegedly been neglected due to the government's expenditures on the space program.

For example, the Apollo computer simulations originally designed to take the astronauts on make-believe training missions later became key weapons in the fight to clean up the nation's air and water. TRW, the leading computer sim contractor, found that the same models could be adapted to measure emissions with nearly flawless precision. The data supplied by TRW's pollution-monitoring system was cited as the technical basis for environmental protection laws enacted in California, Alaska, Ohio, South Carolina, and Washington, D.C.

TRW also adapted the software developed for Project Apollo to protect the public from blackouts, fuel shortages, credit fraud, train wrecks, traffic jams, and overnight cash crises. Utilities used the company's spacecraft guidance program to regulate the flow of oil and gas pipelines and the output of electrical plants. Montgomery Ward used it for retail sales authorizations. Southern Pacific used it to dispatch railroad cars. The city of Los Angeles used it to control traffic lights, saving motorists an estimated nine percent in waiting time while noticeably reducing the number of left-hand-turn accidents. Banks used it to operate twenty-four-hour automatic tellers.

All of the above-mentioned TRW innovations were part of a much larger trend epitomized and fostered by the space program—the top-to-bottom computerization of twentieth-century society.

In addition, Project Apollo yielded an array of noncomputerized benefits, many of which had medical applications. Ultralightweight composites invented for spacecraft assemblies were used in making leg braces. Hospitals installed the Moon Rover astronauts' remote biomedical trackers on mobile carts to give more freedom of movement to partially disabled patients. Apollo telemetry systems were modified to transmit EKG data from ambulances to emergency rooms. Mass spectrometers designed to monitor crew respiration were adapted for surgical uses.

In a speech dedicating the new Manned Spacecraft Center in Houston back in the fall of 1962, President Kennedy had declared that man would fly to the moon on the shoulders of "a giant rocket more than three hundred feet tall made of new metal alloys, some of which have not yet been invented." One of those metals was a tungsten alloy with high ductibility that turned out to be an unexpected godsend for radiologists. Hairpin-shaped tungsten X-ray filaments last longer than other types and do not have "whiskers." As doctors and medical researchers attest, these otherwise esoteric improvements have greatly enhanced the diagnosis and treatment of cancer.

Space technology has made even greater contributions to the fight against world hunger through direct and indirect benefits to agriculture. Satellites have long since become key tools in weather and crop forecasting, as well as in large-scale drought and soil erosion photo-surveys. Solar energy collection panels based on Apollo prototypes are widely used in the midwestern wheat belt for more cost-efficient grain drying. Likewise, farmers have increased harvests and conserved precious water supplies by installing center-pivot irrigation systems that use special lubrication techniques developed for aerospace engines to reduce wear and heat stress.

By far the most famous space benefits are the astonishingly diverse and practical new consumer goods spawned by Project Apollo. Contrary to popular misconception, two of the most famous products often associated with the space program—Velcro and Tang—cannot properly be included in this illustrious list. The concept for the Velcro fastener, which consists of interlocking "male" and "female" fiber patches, was developed back in the late 1940s after Swiss researcher George de Maestral noticed the adhesive properties of some cockleburs he was examining through his microscope. Tang, the instant breakfast mix General Foods advertised as the favorite space beverage of the astronauts, made its commercial debut back in 1957; two years later, it was available in supermarkets all across the country. In 1965, Tang was listed as one of the optional drink selections for the Gemini 4 astronauts. But as Buzz Aldrin reports in his post-mission memoir, it was not one of the drinks chosen by him and the other members of the Apollo 11 crew.

Teflon, the synthetic coating featured on nonstick cookware, belongs in the same category as Velcro and Tang. Although it is widely believed to be a spin-off from the space program, Teflon's orgins trace back to 1938, when a DuPont researcher named Roy J. Plunkett noticed that certain combinations of fluorocarbons used in his refrigerant gas experiments produced a strange, inert waxlike substance. His discovery remained a military secret until the end of World War II as scientists assigned to the Manhattan Project tried to adapt the compound to the manufacture of atomic bombs. In 1956, a French company sold the first Teflon-coated frying pans to the public through a prestigious store on the Riviera. By the time President Kennedy issued his historic challenge to land a man on the moon in his May 25, 1961, address to Congress, Teflon-coated appliances were already becoming the rage in U.S. households.

However, Project Apollo did "invent" new uses for both Velcro and Teflon that their originators could scarcely have imagined. Confronted with the problem of securing various loose objects the astronauts would need to use during their voyages through the weightlessness of space, it was NASA engineers who realized that the solution lay in decorating the spacecraft walls with dozens of Velcro squares that could "mate" with Velcro squares glued onto the hand tools in the storage compartments. Synthetic garment fasteners thus became essential equipment holders. Likewise, the contract to manufacture nonflammable pressure suits for the Apollo crewmen, which went to the Owens-Corning Fiberglas company, specifically called for the Beta cloth outer shell to be coated with Teflon. The same miracle substance that made it easy for housewives to clean their frying pans helped make it easier for the astronauts to clean off the dust and debris they picked up during their EVAs on the moon.

And while Tang itself was an independently developed commercial product, the unique problems posed by Project Apollo led to a revolutionary breakthrough on the culinary front. Given the spacecraft's limited storage capacity and the special vicissitudes of zero G, it was obvious from the outset that ordinary foodstuffs would be too bulky to carry on an extended mission, as well as too difficult to prepare. The search for a solution to the dilemma of dining in space led the U.S. Army's Natick Laboratory to invent freeze-dried foods that have, in turn, had a very down-to-earth impact on the war against hunger. The basic concentrate is now sold to leading food processors like Nestlé and Lipton, who add extra ingredients to produce brand name soup and casserole mixes. But freeze-dried meals, which are cheaper and simpler to fix than traditional dishes, have also become the principal menu items in food programs for the blind and the elderly.

Over the past two decades, numerous other household and recreational goods with direct lineage from Project Apollo have transformed the life of the average consumer. Advances in traction engineering made in the course of developing the Moon Rover vehicles enabled Goodyear to introduce the first studless snow tire, a milestone in the improvement of winter auto safety. Insulation fabrics created to protect the astronauts from the intense heat and cold on the lunar surface were later woven into so-called space blankets that offered unprecedented comfort to military and civilian campers. Quartz crystal timing elements originally designed to guarantee the accuracy of the master clock at Mission Control were miniaturized and mass produced in the form of affordable wristwatches guaranteed not to lose or gain more than one minute per year.

Project Apollo inspired the invention of ingestible toothpaste, the nontippable life raft, the five-year flashlight, and the lightweight graphite composites now used in the manufacture of golf clubs, tennis rackets, and jet fighter fuselages. Black & Decker adapted the technology acquired in designing the battery-powered drills the astronauts used to bore core samples out of the lunar surface to create a whole new line of cordless power hand tools; the company's most popular and practical products include such appliances as the high-speed screwdriver and the Dustbuster vacuum cleaner. Optometry firms adapted the technology developed in inventing scratch-proof lenses for the Apollo space cameras to the manufacture of scratch-proof safety and athletic goggles like those worn by pro basketball star Kareem Abdul-Jabbar of the Los Angeles Lakers.

Project Apollo also pioneered a new frontier that did not reside in outer space: the management of the world's first large-scale multidisciplinary organization not dedicated to war. Just as the armies of past wars were reinforced by all the civilian "troops" on the home front, the effort to land a man on the moon drew on the resources of an entire nation. It required the coordinated labors of astrophysicists and arc welders, ship captains and seamstresses, uncommon geniuses and ordinary office clerks. Management techniques developed to cope with the most ambitious technological undertaking of all time have since been applied to institutions and enterprises ranging from cancer hospitals and drug processing labs to automobile factories and nuclear power plants.

Space Benefits, however, fails to assess what may well prove to be the source of Project Apollo's most long-lasting and truly life-changing benefits to mankind—its contributions to science and human knowledge about the cosmos.

Apollo 11's Mike Collins noted just before his historic lift-off from Earth on July 16, 1969, that he and his fellow astronauts believed that the moon would be "a Rosetta stone of life." The craters, canyons, rocks, rilles, mountains, and maria on the lunar surface promised to provide a 4.5-billion-year-old birth certificate and geological development record of man's nearest celestial neighbor. NASA scientists hoped that by deciphering the true history of the moon, they could fill in the missing link in the evolutionary history of the earth, whose oldest rock in scientific captivity is only 3.5 billion years old, and that by so doing, they would then be able to date and document the true origins of the human race.

Unfortunately, the task of reading the writing on the lunar Rosetta stones has proved far more difficult than anticipated. Part of the problem can be blamed on insufficient federal funding. The ALSEP experiments the Apollo astronauts deployed on the moon, which transmitted raw data on such phenomena as "moonquakes," "solar winds," and cosmic radiation, had to be shut down in 1978 due to congressional budget cuts. As a result, scientists have lost their electronic "touch" with the earth's closest celestial neighbor. Instead of being able to monitor the moon on a continuing up-to-the-minute basis, they are forced to extrapolate their theoretical and empirical studies from old and incomplete ALSEP transmissions.

As the fate of the infamous moon rocks illustrates, part of the problem can also be blamed on politics and bureaucratic bungling. According to official records, the Apollo astronauts brought back 842 pounds of lunar rock, soil, and core samples. The majority of that booty, some 615.34 pounds, remains in nitrogen sealed "pristine sample" containers stored at the Lunar Receiving Lab in Houston. The LRL cache also includes 55.44 pounds of opened samples that have been used in research projects. A so-called insurance sample weighing 117.48 pounds is kept at Brooks Air Force Base in San Antonio. An estimated 17.38 pounds is currently on loan to various scientists around the world. Another 19.36 pounds has been donated to various museum displays or presented as gifts to heads of state around the world.

But according to a NASA audit released in the fall of 1979, "substantial quantities" of moon rocks are "unaccounted for or missing." The exact size and weight of these samples is unknown. Curators of the LRL collection claim that 14.74 pounds have been "consumed" in scientific tests and sample preparation procedures and/or inexplicably "lost." But even that ostensibly candid estimate fails to account for the whereabouts of at least 2.26 pounds of the 842 pounds space agency officials say were unloaded from the Apollo spacecraft. It is possible, as previous journalistic investigations have suggested, that some of the moon rocks were stolen. Given the widespread public misperception that the moon rocks were, despite their extraterrestial origins, as valueless as zircon, it is even more likely that some of them were simply misplaced or thrown out with the trash. In any case, man has already squandered a "substantial" portion of the lunar material that he paid $40 billion to collect.

What's more, the Apollo astronauts discovered that the moon did not intend to reveal her secrets quickly or easily. If man's first lunar landings destroyed most of the traditional myths and romantic notions about the "silvery orb" that circles the earth, they also raised a host of provocative new questions for each old wives' tale they debunked. The moon turned out to be far more complex, more colorful, more vibrant—and in many ways, more mysterious—than primitive man or his modern-day descendants ever imagined. But even though the twenty-four men who flew to the moon did not return with all the answers, their pioneering venture in planetary exploration was an unqualified success from a scientific point of view.

First, the Apollo 11 astronauts demonstrated that man could land on and walk across the lunar surface, something that was by no means a guaranteed fait accompli before Neil Armstrong's "giant leap." In the fall of 1961, just after JFK issued his challenge to land a man on the moon before the end of the decade, Arthur C. Clarke published a foreboding best-seller entitled *A Fall of Moondust.* Basing his sci-fi narrative on a widely accepted scientific theory that the moon was merely a ball of loose particulates, Clarke described a lunar landing attempt in which the astronauts are enveloped by a cloud of moondust.

Armstrong and Aldrin proved that the real lunar surface had the hard-packed consistency of wet sand, a finding that increased rather than diminished its mystery. The astronauts found that their boots left shallow footprints, but when they attempted to penetrate the topsoil to collect core samples, they encountered surprising resistance. Just as confoundingly, after struggling to plant the American flag at a sufficient depth, they found that the pole kept tipping over, as if the horizontal resistance of the subsurface was inversely proportional to the vertical resistance. If they dug straight down, they hit rock; but if they dug sideways, they sliced into a soft muck.

Although the Apollo astronauts never encountered intelligent nonhuman life on the lunar surface, they did prove that the moon is not merely a "dead" planet. The rock samples returned from the Sea of Tranquility by the first two men on the moon were largely composed of basalt scarred with an acne of gas boils called vesicles; these volcanic characteristics supported the theory that the flatland maria were formed by ancient lava flows. And if so, that meant that the moon once had an active molten core, and may still. Adding to the intrigue was the fact that geological heat-flow experiments conducted by the crews of subsequent lunar landing missions all produced positive results, as did seismic tests for "moonquakes."

Likewise, the perpetually shuddering moon, whose axial "librations" had already been documented by astronomers, turned out to have a very active surface. Their soil samples showed that the loose particulates or "regolith" beneath their boots had been pulverized by eons of micrometeorite bombardment. Scientists estimated that these micrometeorite showers rained down at the rate of nearly ninety thousand pounds per twenty-four-hour day, and completely turned over the topsoil of the entire lunar surface every forty million years. Damage found on the undersides of various rock samples indicated that the lunar surface was simultaneously being transformed by storms of high-speed cosmic rays that "eroded" the terrain much like winds and rain on Earth.

The astronauts also proved that the moon is not merely a colorless heap of stone. The Apollo 15 crew, who explored the central mountains of Hadley-Apennine, returned with bags full of tiny green marblelike crystals, inevitably prompting jokes about the old folktale that "the moone is made of a greene cheese." The Apollo 17 crew, who explored the valley of Taurus-Littrow, bagged a headline-making chunk of orange soil. NASA analysts later concluded that both the green- and orange-colored samples were fragments of "fountain fired" volcanic glass.

Although the Apollo lunar booty did not settle the long-running debate about the origins and evolution of the moon, it burst the ballooning authority of all the leading orthodoxies. Rocks found in both the mountains and the mare suggested that the lunar surface had been sculpted by lava flows. But various mineral specimens found in those same areas indicated that meteorite impact may have also played a major role. Similarly, there was evidence in support of such competing lunar origin theories as "fission" (the moon broke off from the earth), "capture" (the moon was an errant body caught in the earth's gravitational pull), and "coaccretion" (the moon formed independently from the earth).

What has emerged is a new conceptual synthesis about the moon's creation known as the planetesimal "impact" theory. According to this hybrid hypothesis, one or more large planetesimals struck the earth during the early days of the solar system; the residue from this collision was hurled into orbit where it eventually formed the moon. Assuming the planetesimal or some piece of it simply bounced off the earth, this notion would be only a modified version of the "capture" theory. If the residue of the collision was actually a fragment torn from the earth, that would confirm the fission theory. But the beauty of the impact theory is that it accommodates both capture and fission, thereby accounting for the diversity of rocks found on the lunar surface. It also suggests that the moon may be more geologically similar to Earth than previously suspected—and have more to offer mankind.

As if to remind the world of the potential commercial rewards from scientific inquiry, the Apollo astronauts gathered conclusive evidence that the moon boasts bountiful deposits of minerals and precious metals. The rocks excavated from the highland and mountainous regions proved to be rich in magnesium, aluminum, and calcium. Other rock samples contained large amounts of titanium. (The Apollo 11 crew even discovered two new minerals: Tranquillityte, named after their landing site, and Armalcolite, an acronym formed from the letters of their last names.) While most data suggested the moon's core holds a relatively small amount of iron, the samples also indicated that the lunar crust contains larger and more various concentrations of iron ore compounds than the earth's crust. The bottom line on all this lunar geology is that the moon promises to be a very lucrative site for future ventures in extraterrestrial mining.

Against all odds and previous unmanned reconnaissance photos, many of the astronauts hoped to find water on the moon, and were deeply disappointed when they did not. But as they hasten to point out, they did not prove that there is not any water on the moon because the six successful lunar landing missions only managed to explore a very small portion of the visible surface. In fact, the positive results of the heat-flow experiments even raised the possibility that the moon might contain a large subterranean layer of ice. According to Apollo 16's John Young, "There is still a good chance that there is water on the North Pole of the moon."

In any event, the astronauts did return with plenty of evidence that the moon might be able to support life with a little outside help, a remarkable discovery that never received appropriate headline coverage in the mass media. Analysis of the moon rocks at the LRL showed that they, like most rocks found on Earth, contained fifty percent oxygen. If man could find a practical way to extract the oxygen, he could mine his own breathing supply from the lunar surface. Using the same basic process, he could also produce liquid oxygen to fuel his spacecraft and water to drink. And by perfecting the art of lunar oxygen mining, man would have the ability to establish a permanent colony on the moon and occupy it as long as he pleased.

No wonder the astronauts believe that Project Apollo's greatest benefits to mankind are yet to come. "The Moon has become man's first outpost in space," notes Apollo 15's Jim Irwin. "It gave us our first footstep on another planet."

Apollo 11's Neil Armstrong noted at his post-mission press conference back in August of 1969 that the success of man's first lunar landing ensured that some type of space program "will continue to exist for the rest of man's existence . . . We don't have the option any longer of saying yes or no to it. We only have the option of saying when."

"The path of evolution is in space as much as it is on Earth," maintains Apollo 17's Jack Schmitt, the former Republican senator from New Mexico, adding with admittedly partisan conviction, "I am absolutely convinced that within the next thirty, forty, or fifty years, there will be human colonies on other planets. The moon is a candidate for such settlements; so is Mars. Whether it will be our democratic civilization or the Soviet Union or the People's Republic of China, I don't know. That's something the citizens of the Free World are going to have to decide. But I do know which way I think the decision ought to be made. Because if we do not compete adequately in space, we will almost certainly cease to compete here on Earth."

STAGE 4:

2001 AND BEYOND

The earth is the cradle of humanity, but mankind will not stay in the cradle forever.

K. E. TSIOLKOVSKY

Outside the Earth (1903)

18

STAR WARS OR PEACE?

Mankind's Future in Space 1988–2001 and Beyond

The challenge is to tame and harness the space frontier—to go beyond Apollo . . ."

"Leadership and America's Future in Space"
DR. SALLY K. RIDE, August 1987

John Young was the only Apollo veteran still on active flight status at NASA on January 28, 1986. Then age fifty-five, the outspoken chief astronaut had flown on more space voyages than any other American. His flight log included the first and eighth Gemini missions, two trips to the moon (Apollo 10 and Apollo 16), and two key shuttle flights, the first STS mission in April of 1981 and the first mission carrying foreign guest astronauts in the fall of 1983. In 1981, he had received the Congressional Space Medal of Honor for his "demonstrated leadership, courage, and skill" on missions "pivotal to human progress in space." And though his dark brown hair had long since turned silver, Young still had a trim physique, a steely eyed glare, and a fire in his belly.

In early January of 1986, a few weeks before the *Challenger* launch, Young wrote a highly controversial internal memo citing a long list of safety flaws in the shuttle's design. Among the most serious flaws, he claimed, were the O-ring seals between the joints of the shuttle's solid rocket booster. Young charged that unless these problems were corrected, there was "a significantly higher probability of costing NASA orbiters and killing flight crews."

Like the cries of other whistle-blowers inside and outside NASA, Young's warnings were ignored, and the *Challenger* got the GO for launch on the bitterly cold morning of January 28, 1986. Moments after lift-off, the booster's O-ring seals failed, and the shuttle exploded, killing all seven crew members, one of whom was a civilian grade school teacher.

The *Challenger* tragedy shocked the nation and the entire world. As former astronaut Dr. Sally K. Ride, the first American woman in space, wrote in her August 1987 report entitled "Leadership and America's Future in Space," "For nearly a quarter of a century, the U.S. space program enjoyed what can appropriately be termed a 'golden age.' The United States was clearly and unquestionably the leader in space exploration, and the nation reaped all the benefits of pride, international prestige, scientific advancement, and technological progress that such leadership provides.

"However, in the aftermath of the *Challenger* accident," Ride continues, "reviews of our space program made its shortcomings starkly apparent. The United States' role as the leader of spacefaring nations came into serious question. The capabilities, the direction, and the future of the space program became subjects of public discussion and professional debate . . . The U.S. civilian space program is now at a crossroads . . . It is widely agreed that we are living off the interest of the Apollo-era technology investment, and that it is time to replenish our technology reservoir in order to enhance our range of technical options."

Like the U.S., mankind is also at a crossroads. It is time for the citizens of the twentieth century to decide upon their future in space and that of the twenty-first-century generation. Will there be continued superpower competition or a new era of international cooperation? Will the emphasis be on "star wars" or on the forging of a lasting cosmic peace?

Already, there are signs that a new and potentially far more destructive space race is well under way. What distinguishes this contest from the "old" space race that began with Sputnik in 1957 are the size and nature of the stakes and the number of players involved. The object is no longer putting the first man in Earth orbit or landing the first man on the moon. Now the name of the game seems to be technological, commercial, and military control of outer space. And the race is no longer a two-sided battle between the leading superpowers but a global contest involving no less than five factions of space-faring nations.

Although the Russians did not land cosmonauts on the moon, they have retaken the lead in many if not the most important aspects of modern space exploration. The U.S. has launched no space station since Skylab in 1974, and will not launch another until at least 1995. The U.S.S.R. has launched eight space stations since the mid-seventies. They are presently the "sole long-term inhabitants of Earth orbit" thanks to *Mir,* a space station capable of accommodating up to twelve cosmonauts at a time. In January of 1988, cosmonaut Yuri Romanenko set a new space endurance mark of 326 days, nearly four times the American record.

The Soviets are also forging ahead in interplanetary exploration with their sights set on Mars. They have announced plans to send three separate robotic spacecraft to Mars before 1995, and hope to launch an unmanned Martian soil sample return mission in the late nineties. Although the U.S. sent a series of Viking probes to Mars, the last one stopped transmitting photographs in 1982.

The Soviet Union, however, is by no means America's only space competitor, a fact that became, in Ride's phrase, "starkly apparent" in the aftermath of the *Challenger* explosion. Under the leadership of France and West Germany, the European Space Agency (ESA) has already produced sophisticated launch vehicles of its own, and is offering to carry U.S. satellites and commercial payloads into Earth orbit while NASA struggles to resurrect the shuttle program. The Japanese are formulating separate programs under the aegis of their National Space Development Agency. And the People's Republic of China is reportedly making slow but steady progress in developing a giant booster called "Long March."

The U.S., on the other hand, is still trying to recover from the *Challenger* disaster, a process that has prominently involved two former Apollo astronauts. The first step came immediately after the accident when President Reagan appointed a special investigative commission. Reagan originally asked Apollo 11 commander Neil Armstrong to head the task force, but Armstrong demurred, insisting instead on playing the lower-key role of vice-chairman while former Secretary of State William Rogers served as chairman. Critics later charged that Armstrong conspired with Rogers to soften the commission's findings, but the final report nevertheless scored NASA and prime contractors like solid rocket manufacturer Morton Thiokol for bureaucratic and technical malfeasance at almost every step before the *Challenger* launch. Chief astronaut John Young also sounded off following the *Challenger* accident, blaming NASA for blind adherence to an overly ambitious mission agenda. "There is only one driving reason that such a potentially dangerous system would ever be allowed to fly," Young charged in yet another controversial internal memo, "launch-schedule pressure."

The Rogers Commission reports has led to major shake-ups at NASA and in the ranks of private contractors, though the true scope and ramifications of those changes are still subject to bitter debate. In the spring of 1986, President Reagan appointed Dr. James Fletcher to replace James Beggs as the space agency's top administrator. But critics pointed out that Fletcher, who held the same post from 1971 to 1977, is a Mormon from Utah, the home of shuttle contractor Morton Thiokol's solid rocket motor facility. Adding to the controversy is the fact that NASA chose not to fire Morton Thiokol after the *Challenger* accident, claiming that finding a new company to do the job would cause at least two years of additional delay in getting the next shuttle launched.

Although Morton Thiokol has since embarked on a $400 million cost-only solid rocket motor redesign, the effort has been marred by repeated test failures, new accusations of corporate and technical malfeasance, and further delays. As of this writing, the shuttle *Discovery* is slated to be launched in late August of 1988. But according to Young, new problems keep cropping up with the shuttle parts that have been shipped to the Cape, and it will be "a miracle" if the *Discovery* is launched on schedule. And even if the *Discovery* does launch on time, it will have taken NASA over two and a half years to get the program back on track; by comparison, the delay following the Apollo 1 fire was only one and a half years.

Meanwhile, the U.S. is still foundering in the equally if not more important task of setting the nation's long-term direction in space. Even before the *Challenger* accident, the shuttle, which had been conceived by the Nixon administration as a "national space vehicle" that could serve the needs of both civilian and military users, had proven to be an economic bust. Originally budgeted at $5.6 billion in 1972 dollars, the spacecraft has so far cost over $30 billion in 1972 dollars; instead of flying thirty to sixty missions per year as projected, the shuttle fleet flew only nine missions in its 1985 peak year. With costs per mission averaging $120 million and cargo bay rentals only $71 million, each pre-*Challenger* flight racked up an average loss of nearly $50 million.

And unbeknownst to most of the American public, military spending on space has steadily outgrown civilian spending in recent years. According to a General Accounting Office report, the Pentagon space budget increased by 164 percent to $12.7 billion between 1981, the year of the first shuttle flight, and 1985. During that same period, the NASA space budget increased only thirty-nine percent to $6.9 billion. Though some of the military increase was attributable to President Reagan's Strategic Defense Initiative (SDI), popularly known as "Star Wars," the lion's share was spent on more traditional reconnaissance and communications satellite projects. But in the course of this spending spree, the Pentagon became almost totally dependent on the shuttle for its space transportation needs. As a result, the grounding of the shuttle fleet after the *Challenger* accident virtually crippled the military's ability to maintain, replenish, and renovate some of the nation's most vital high-tech defense systems.

In the fall of 1986, NASA administrator James Fletcher assigned a special task force headed by astronaut Sally Ride to come up with "a blueprint to guide the U.S. to a position of leadership among the space-faring nations of the earth." Many of the recommendations contained in her report were later incorporated into a January 1988 presidential directive. Unlike the late President Kennedy's challenge to land a man on the moon by the end of the decade, Reagan's directive was a masterpiece of political double talk, but it gave the U.S. what it had heretofore lacked—a national space policy.

Reagan's national space policy attempts simultaneously to serve three masters—the civilian sector, the commercial sector, and the military sector. The policy's stated long-range goal is "expanding human presence and activity beyond Earth orbit into the solar system." The near-term goal is to provide "assured access to space" through a "mix of vehicles" that include the shuttle and alternative space transfer systems. One of its basic principles states: "The United States is committed to the exploration and use of outer space by all nations for peaceful purposes and for the benefit of all mankind." But another adds: "The United States will pursue activities in space in support of its inherent right of self-defense and its defense commitments to its allies."

Translated to specifics, that means the U.S. intends to pursue the development of SDI, establish a permanent American presence in space via the space station, encourage private enterprise ventures in rocketry and space technology, and commence a systematic exploration of the cosmos—a tall and still only vaguely defined order.

A key part of the new national space policy is "Project Pathfinder," a research and technology program designed to set the direction for future space missions. Pathfinder's objectives include the development of space vehicles, planetary rovers, cryogenic fuel depots, in-space assembly and construction plants, and closed-loop (i.e., self-sustaining) life support systems. But the most intriguing aspects of the project are the specific space exploration initiatives under consideration, all four of which were examined in detail in the Ride report. They are: (1) Mission to Planet Earth;

(2) Exploration of the Solar System; (3) Humans to Mars; and (4) Outpost on the Moon.

The main feature of the proposed "Mission to Planet Earth" is the establishment of orbital observation systems constructed and maintained with international cooperation. Step one calls for the launching of four sun-synchronous polar platforms (two by the U.S., one by the European Space Agency, and one by the Japanese) beginning in 1994. Step two calls for five geostationary platforms (three U.S., one ESA, and one Japanese) to be launched between 1996 and 2000.

The second initiative, "Exploration of the Solar System," "would build on NASA's long-standing tradition of solar system exploration and would continue the quest to understand our planetary system, its origin, and its evolution." The proposal includes a "comet rendezvous asteroid fly-by" mission intended to gather data on how the solar system began, and a "Casini mission" to Saturn in 1998, which would look for pre-biotic chemical links between the earth and the ringed planet. But the "centerpiece" of this initiative would be the unmanned exploration of Mars in three robotic missions, the first of which "would bring a handful of Mars back to Earth before the year 2000."

Humans to Mars is by far the most glamorous—and the most controversial—of the three proposed initiatives. The project would begin with a series of robotic missions in the 1990s while using by then available Space Station facilities to do further research on the biomedical effects of long-duration space flight. Assuming all goes well, NASA astronauts would then make three "fast piloted" round-trip missions to Mars, each lasting about one year from lift-off to splashdown, to investigate the feasibility of establishing an outpost on the Martian surface by 2010.

Among the proponents of the Mars venture is astronomer Carl Sagan, who calls the moon "a pretty boring place" and insists on the need for a space project that could capture the national imagination. But as Ride points out, the Mars mission "would require a concentrated massive national commitment . . . to a goal and its supporting science, technology, and infrastructure for many decades." She also notes that such an ambitious and spectacular undertaking could have an adverse affect on Soviet-American relations: "There is a very real danger that if the U.S. announces a human Mars initiative at this time, it could escalate into another space race."

Ride's fourth proposed initiative—"Outpost on the Moon"—has received the least favorable public reaction even though it is in many ways the most feasible and promises the greatest long-term rewards.

Phase I: "Search for a Site" calls for a series of unmanned robotic lunar missions in the 1990s that would among other things attempt to find water in the still-uncharted polar regions of the moon. Phase II: "Return to the Moon" would feature the first manned lunar landing missions since Project Apollo. But instead of flying directly from the Cape, the astronauts would stop off at the Space Station to board a "lunar transfer vehicle." Each crew would stay on the moon for up to two weeks at a time to assemble the equipment needed to build a temporary shelter. They would also attempt to mine oxygen from the lunar soil.

By the year 2001, the moon base would include a habitation area, a research facility, a rover, soil-moving machinery, and an oxygen mining pilot plant. Each three-member crew would be able to stay for an entire lunar night, the equivalent of fourteen Earth days. By 2005, up to five astronauts would occupy the base for several weeks at a time.

Phase III: "At Home on the Moon" would begin upon the completion of a permanent lunar outpost on or about the year 2010. Up to thirty people would occupy the facility for periods ranging from thirty days to several months. Such extended visits would necessarily depend on successful construction of a "closed loop life support system" keyed to an "operational" lunar oxygen plant, which would produce a breathing supply for the astronauts and liquid oxygen to fuel their spacecraft. Their main duties would consist of "frontline scientific research and technology development," as well as the mining of precious metals like iron, aluminum, and titanium, drug processing in the uncontaminated low gravitational lunar environment, and satellite launching, which would require only one-twentieth the energy needed to launch from Earth.

The reasons for going back to the moon are multifarious, encompassing both the cosmic and the mundane. One is basic science. The moon very likely holds the key to understanding the origins and history of the earth. But because the Apollo missions landed in only six different spots on the lunar surface and explored them for extremely brief periods of time, man's closest celestial neighbor still remains a mystery. Did the moon and the earth evolve separately or in tandem? Was the moon once capable of supporting life, and if so, what kind? Is there, as astronaut Young and many others believe, still a good chance of finding water on or around the North Pole of the moon? Does the moon abide a host of exotic and/or valuable ores and minerals not discovered on previous manned and unmanned missions?

The moon is a natural laboratory for scientific research. Its far side is the ideal place for a giant telescope (possibly constructed out of glass blown from lunar sands) that could afford vast new glimpses into deep space astronomy. The moon might also prove an ideal place for experiments in superconductivity, and for studying the long-term effects of micro-gravity on human cardiovascular systems with an eye to the prevention and cure of heart disease. Last but by no means least, the moon could be an ideal place for humans to adapt the skills necessary for exploring and colonizing other planets.

By far the most dramatic attraction of Project Pathfinder is the prospect of discovering life on other planets, especially intelligent nonhuman life. Given the known dimensions of the universe, the odds in favor of such a discovery seem far greater than the odds against. Our own solar system, which contains nine planets, thirty-two moons, and the sun, is itself part of the Milky Way Galaxy, which is 100,000 light years in diameter (each light year is equivalent to over 6 trillion miles) and contains over 100 billion stars. And the Milky Way is but one of some 10 billion galaxies, each of which also contains over 100 billion stars.

That means that the chances of finding a solar system similar to ours must be tabulated from a figure of no less than 10 billion times 100 billion. And assuming that these solar systems contain untold hundreds of billions of planets, the chances of finding some form of life—planet or humanoid—are manyfold greater. Not surprisingly, when asked in a recent interview if he believed in the possibility of finding other life in deep space, the anything but starry-eyed John Young replied, "I'm not through looking for life in our solar system."

Unfortunately for Young, he probably won't be able to conduct his search in person. In the spring of 1986, following his controversial post-*Challenger* memos, he was reassigned from his post as chief astronaut to become a special assistant to the director of the Johnson Spacecraft Center. At the time, Young still had the assignment of commanding the scheduled June 1989 shuttle mission to launch the Hubble Space Telescope. But in late March of 1988, Daniel C. Brandenstein, his successor as chief astronaut, scratched him from the flight roster and replaced him with a forty-three-year-old rookie commander. Though Young officially remains eligible for reassignment, he is now fifty-seven years old, and the next opening on a shuttle mission will not be until the early 1990s when he will be over age sixty.

If, as many expect, Young resigns from NASA before the twentieth anniversary of man's first lunar landing in July of 1989, the astronaut corps will lose its last living link to the glory years of Project Apollo. But like most of his former Apollo colleagues, Young clings to the hope that no matter what course the U.S. space program takes in the future, it will model its undertaking after the "the most hazardous and dangerous and greatest adventure on which man has ever embarked." As he told an interviewer shortly after being grounded: "You have to do it right and make the right decisions and get the right advice and do it. That never changes. You can't let the cost and the schedule drive you to your knees like we did with *Challenger.* You've got to hang in there like we did in the old days, during the Apollo days . . ."

EPILOGUE: THE MESSAGE OF APOLLO

We shall not cease from exploration and the end of all our exploring will be to arrive where we started and know the place for the first time.

T. S. ELIOT
Four Quartets (1943)

Although Project Apollo was one of the most extensively documented undertakings in human history, many of the earth's five billion inhabitants still refuse to believe that twelve astronauts really did set foot on the moon. Exactly how many people cling to this preposterous heresy is unknown because there has never been a worldwide opinion poll on the subject. But just as the members of the Flat Earth Society in London continue to dispute evidence that the world is round, untold numbers of serious and not-so-serious disbelievers continue to insist that man's first lunar landings were actually a series of government-sponsored Hollywood hoaxes.

Shortly after the Apollo 11 and Apollo 12 missions, NASA public affairs officer Julian Scheer mischievously fueled the flames of doubt at the tenth annual meeting of a drinking fraternity known as the Man Will Never Fly Memorial Society. Scheer delighted some two hundred admittedly inebriated members of the society by narrating a film of astronaut training exercises at a terrestrial "moonscape" in Michigan that bore an indistinguishable resemblance to the real lunar surface.

"The purpose of this film is to indicate that you can really fake things on the ground—almost to the point of deception," Scheer informed his audience, devilishly inviting them to "come to your own decision about whether or not man actually did walk on the moon."

Apollo 11 commander Neil Armstrong was also challenged with the hoax theory at his first post-splashdown press conference in August of

1969. But instead of dismissing the notion as a joke, Armstrong refuted it with a wry and most convincing observation: "I think that one would find that to perpetrate such a hoax accurately and without a few leaks around the [space] agency, would be very much more difficult than actually going to the moon . . ."

Man had also left plenty of evidence of his visit to the moon for anyone—or anything—who cared to look. The complete list of artifacts the Apollo astronauts discarded on the lunar surface has long since been lost in the bowels of the NASA archives, but it consisted of no less than ninety tons of equipment and refuse. Among the most prominent items on the inventory were: six LEM descent stages, three Moon Rovers, six ALSEP packages, six Hasselblad cameras, one timer, twelve Portable Life Support System (PLSS) backpacks, twelve pairs of lunar boots, six American flags, assorted film canisters, armrests, garbage bags, and commemorative plaques, and two golf balls. "We've turned the moon into a used car lot," Apollo 15's Jim Irwin notes wryly. "But most of the things we left on the surface, including the rovers, reflect light. So I like to think that the moon is a little bit lighter and brighter place thanks to man's visit."

However, on the first anniversary of man's first lunar landing in July of 1970, Armstrong admitted that he had been so far disappointed to find that most of the world did not seem to grasp what he called "the message of Apollo." According to Armstrong, that message is "that in the spirit of Apollo, a free and open spirit, you can attack a very difficult goal and achieve it if you can all agree on what that goal is . . . and that you will work together to achieve it.

"I had really hoped that the impact would be more far-reaching than it has," Armstrong told the assembled media. "Not that we haven't had a great appreciation for the lunar landing around the world, and especially in other countries . . . But I had hoped that it might take our minds away from some of the more mundane and temporal problems that we as a society face, which are very similar to the problems that society of every age has faced, and look a little farther into the future with an aim toward solving problems before they become problems."

In the fall of 1987, Walter Cronkite, the former CBS television news anchor widely revered as "America's most trusted man," tried to put the Apollo 11 astronauts' feat in proper perspective during an interview for the twentieth anniversary issue of *Rolling Stone* magazine. "I would think that the most important event of our [past] twenty years has got to be the walk on the moon," Cronkite opined. "In future history books, this will be comparable to Columbus's discovery of America, and our political and economic concerns will fade into memory and will scarcely be an asterisk."

Nevertheless, Armstrong and his fellow moon voyagers now find themselves in much the same kind of historical twilight zone that swallowed up Christopher Columbus following his return to Spain. Queen Isabella had expected Columbus to find an alternative route to India; she had no idea that he had discovered a "New World" and neither did he. In fact, on the homeward leg of his second trans-Atlantic voyage, Columbus made his crewmen swear that they had landed on mainland China, though they had actually reached the island now known as Cuba. The Apollo astronauts are the prisoners of a myopic twentieth-century society with little or no appreciation for their role in opening up the new world of human space exploration and planetary colonization.

"It may be that the problems we have on Earth can't be solved unless we go into space," noted science author Isaac Asimov observed in a recent interview with CBS newsman Dan Rather. "We go out there not just for adventure, not just to satisfy a few scientists, but to solve our problems here on Earth."

Ironically, by making the "impossible dream" come true, the twenty-four men who flew to the moon inadvertently fulfilled the prophecies of ancient myth and modern science fiction. Back in 165 A.D., Lucian of Samosata claimed that he and his fictional crew of moon voyagers had discovered "an enormous mirror suspended over a rather shallow well" during their visit to the lunar surface. According to Lucian, the mirrored well enabled them to "see each city and nation" back on Earth "as clearly as if you were standing over it." The Apollo astronauts did not find Lucian's mythical well, but for the first time in human history, they did see their home planet in its entirety from an outside perspective.

"Space travel has given us a new appreciation for the earth," declares Apollo 15's Irwin. "We realize that the earth is special. We've seen it from afar, we've seen it from the distance of the moon. We realize that the earth is the only natural home for man that we know of, and that we had better protect it."

"You get the feeling that the earth is one unit," observes Apollo 11's Mike Collins. "You don't have the feeling that it's a fragmented place, that it's divided up into countries. You can't see any people, you can't hear any noise. It just seems like a very small, fragile, serene little sphere. It makes you wish that it really were as serene as it appears. It makes you wonder what any of us might do to make it a happier and more peaceful place."

Future historians may very well conclude that one of Project Apollo's most important legacies was its seemingly anticlimactic role in promoting international cooperation in space. The Apollo-Soyuz mission in 1975, which featured the rendezvous and docking of an American spacecraft and a Soviet spacecraft in Earth orbit, was the first extraterrestrial joint venture between the superpowers. Though subsequent proposals for more cooperative missions have so far failed to be acted upon, various political leaders are now advocating the idea of joining forces for the conquest of Mars. In the interim, both sides have been broadening the demographic base of their space programs by actively encouraging the participation of their respective allies.

As of this writing, 204 men and women from eighteen different countries have flown in space on American or Soviet missions. In addition to citizens of the U.S. and the U.S.S.R., the crew rosters include representatives of such ideologically divergent nations as West Germany, East Germany, France, Cuba, Mexico, North Vietnam, Saudi Arabia, and Poland. But despite differences in race, sex, and political persuasion, most of the astronauts and cosmonauts who have flown in space returned to Earth with remarkably similar epiphanies about their home planet.

"I saw that the earth is very small and very fragile," reports Soviet cosmonaut Vladimir A. Solovyov, who orbited the globe for 237 days in the space station Mir. "I did not see any borders or boundaries between countries."

"The earth looked so blue and so round, and so small, so delicate," recalls cosmonaut Alexei Leonov, who became the first man to "walk" in space back in 1965, adding that what he saw on his then-unprecedented excursion outside his spacecraft made him realize that his home planet was "a house of all people that must be defended from harm."

"Before I flew I was already aware of how small and vulnerable our planet is," declared East German cosmonaut Sigmund Jahn after returning from a Soviet orbital mission. "But only when I saw it from space, in all its ineffable beauty and fragility, did I realize that humankind's most urgent task is to cherish and preserve it for future generations."

"Only a child in its innocence could appreciate the purity and splendor of this vision," declares French astronaut Patrick Baudry, who flew on the eighteenth American space shuttle mission in the summer of 1985.

Reinhard Furrer, a West German astronaut who flew on the twenty-second shuttle mission in the fall of 1985, waxes with similarly poetic exuberance in describing his experiences in space: "I would have wished that after my return people had asked me how it was out there. How I coped with the glistening blackness of the world and how I felt being a star that circled the earth."

In the course of pioneering a new frontier, the twenty-four men who flew to the moon also pioneered this new cosmic perspective on the human condition. Some skeptics insist that man will have to make peace on Earth before he can hope to make peace in space; the Reagan administration's Star Wars initiative is but the latest evidence supporting that rather pessimistic view. Others claim that only an outside threat in the form of alien invaders or life-threatening meteoroids could unify the divisive nations of the world. But according to the Apollo astronauts and fellow space travelers from all over the globe, the transformative experience of viewing our home planet from space certainly hasn't hurt the prospects for forging a lasting harmony on Earth.

"I think the missions to the moon unified the spirit of man," says Apollo 15's Irwin. "We had to be unified in our country to make them a success. But yet in their success, they unified the spirit of the whole world. We all rejoiced, not necessarily because Americans did it, but because another human being finally reached out and touched the face of the moon."

Just as the importance of Columbus's voyages to the New World was not recognized during his own lifetime, the true significance of man's first lunar landings will probably not be fully appreciated in this century or even the next. Where Columbus merely helped chart the boundaries of his home planet, the Apollo astronauts literally opened the door to infinity.

"We've shown that the curve of evolution has been bent," declares Apollo 17's Jack Schmitt. "We've shown that mankind is willing to commit himself to living in different environments than those in which the species evolved. Now we know that if we really want to, we can go settle anywhere in the solar system where it's reasonably compatible to do so."

By reaching out and touching the face of another planet, the twenty-four men who flew to the moon helped the human race take the first small steps from infancy to adulthood. Thanks to their heroic voyages outside the earth, they and the citizens of their home planet suddenly became citizens of the universe. And that was truly "a giant leap" for all mankind.

GLOSSARY OF ACRONYMS

ALSEP Apollo Lunar Scientific Experiments Package
AOS Acquisition of Signal
BEF Blunt End Forward
BIG Biological Isolation Garment
CAPCOM Capsule Communicator
CDR Commander
CM Command Module
CMP Command Module Pilot
CSM Command and Service Module
DOI Descent Orbit Insertion
DSKY Display and Keyboard (computer)
EMU Extravehicular Mobility Unit
EVA Extravehicular Activity
GET Ground Elapsed Time
LEC Lunar Equipment Conveyor
LEM Lunar Excursion Module
LET Launch Escape Tower
LLRV Lunar Landing Research Vehicle
LLTV Lunar Landing Training Vehicle
LM Lunar Module
LMP Lunar Module Pilot
LOI Lunar Orbit Insertion
LOS Loss of Signal
LRL Lunar Receiving Laboratory
LRV Lunar Roving Vehicle
MCC Mission Control Center
MEED Microbial Ecology Evaluation Device
MESA Modular Equipment Stowage Assembly
MET Modular Equipment Transporter
MQF Mobile Quarantine Facility
MSC Manned Spacecraft Center

MSOB Manned Spacecraft Operations Building
OPS Oxygen Purge System
PAO Public Affairs Officer
PDI Power Descent Initiative
PLSS Portable Life Support System
PPK Personal Preference Kit
RCS Reaction Control System
REFSMMAT Reference to Stable Member Matrix
RTCC Real Time Computer Center
SM Service Module
S-IC Saturn 5 first stage
S-II Saturn 5 second stage
S-IVB Saturn 5 third stage
TEI Trans-Earth Injection
TLI Trans-Lunar Injection
VAB Vehicle Assembly Building

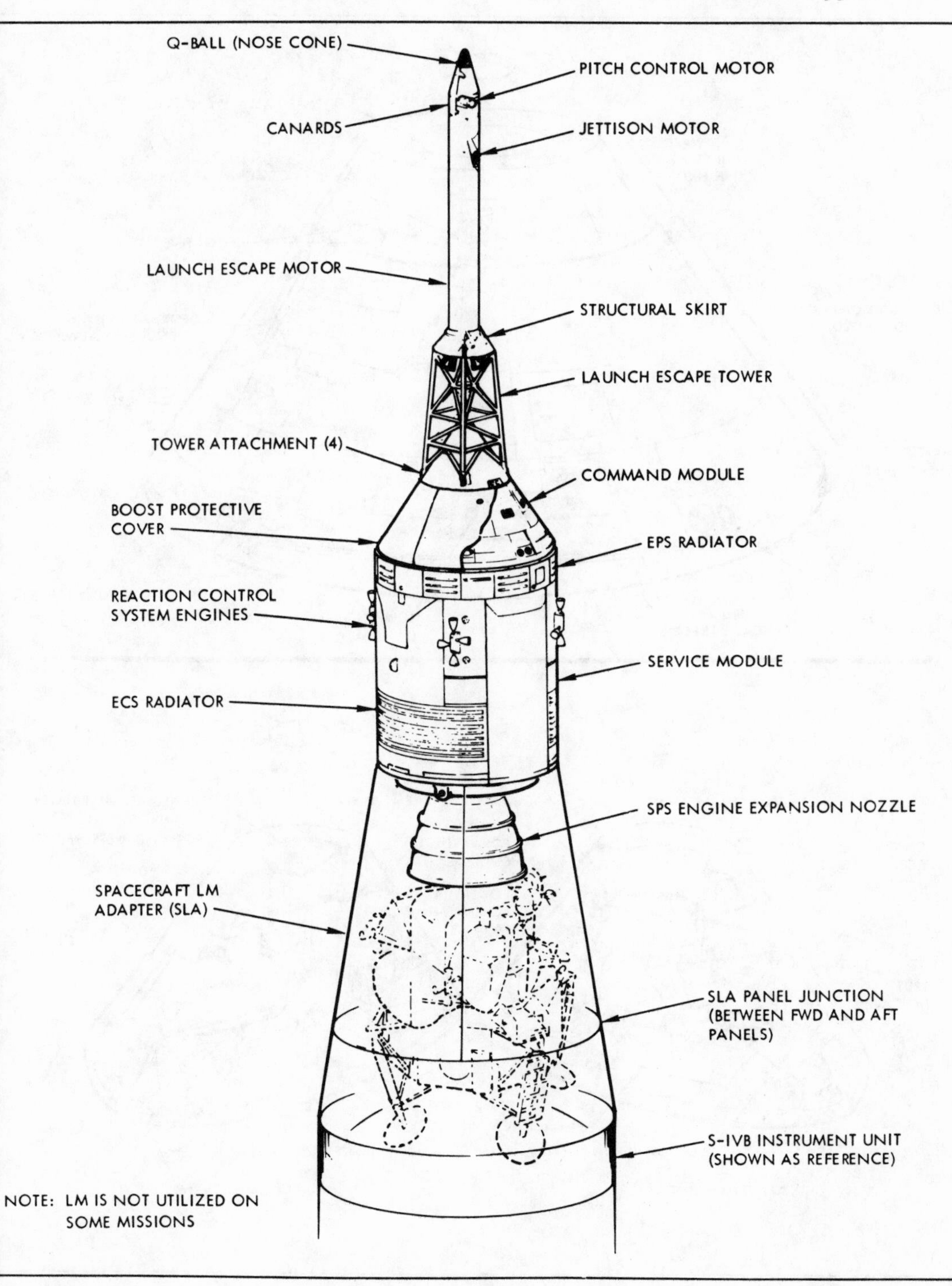

Spacecraft (NASA)

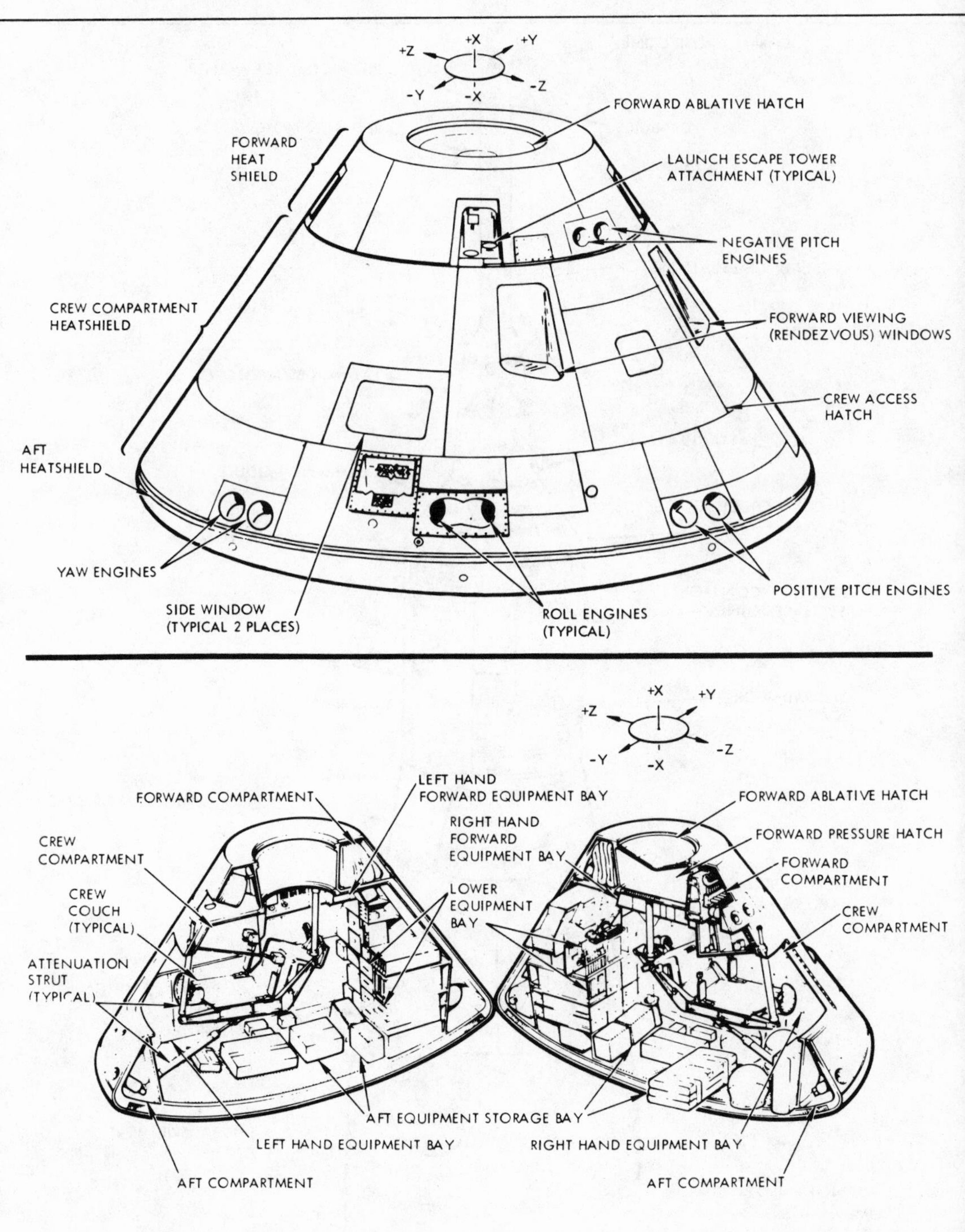

Command module (NASA)

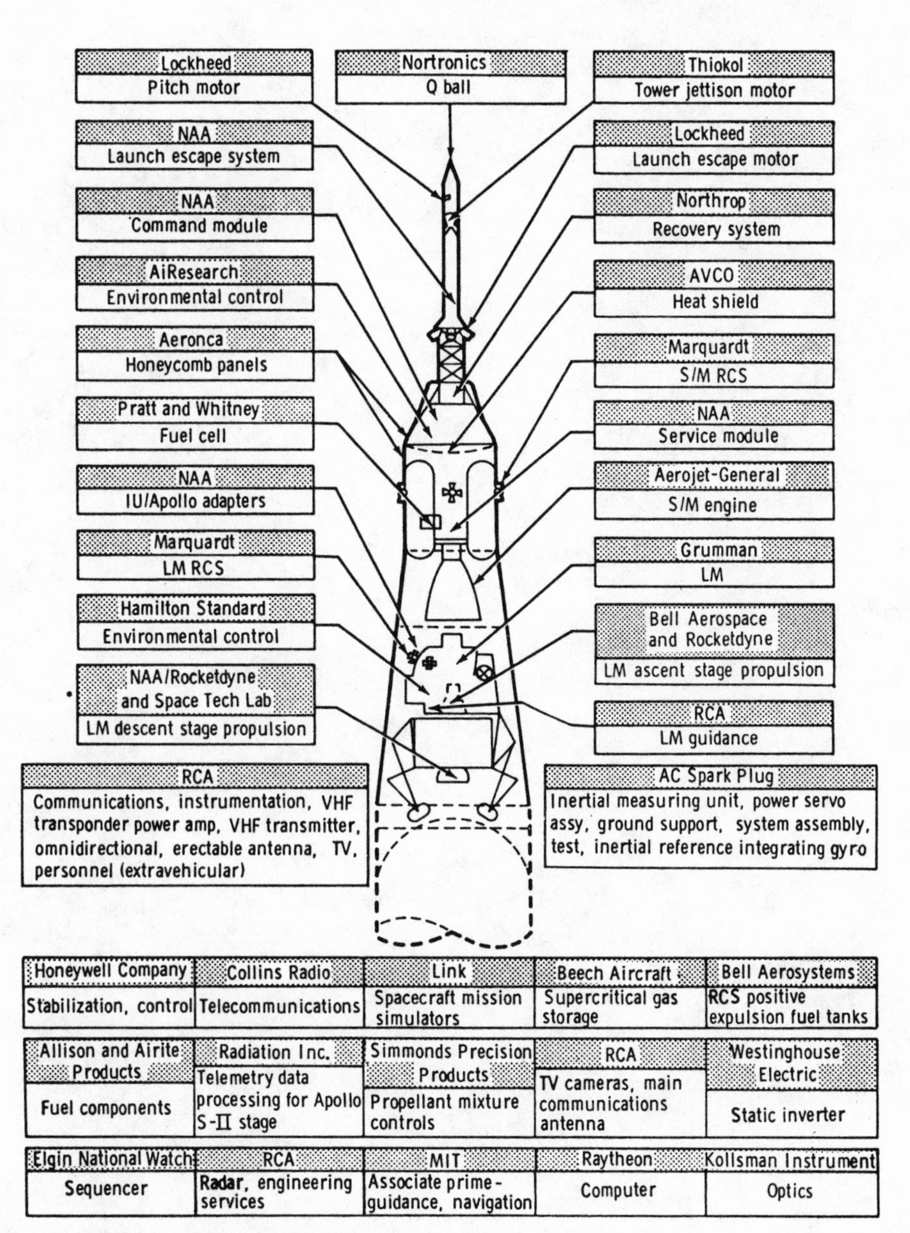

*STL named sole contractor January 1965.

Major spacecraft component manufacturers

PROJECT APOLLO
Flight Log

MISSION	DATE	CREW (CDR-CMP-LMP)*	OUTCOME
Apollo 1	January 27, 1967	Virgil I. Grissom Edward H. White II Roger B. Chaffee	All three astronauts died in launchpad fire during countdown demonstration exercise at Cape Kennedy, Florida.
Apollo 7	October 11–22, 1968	Walter M. Schirra, Jr. Donn F. Eisele Walter Cunningham	Helped get U.S. manned space program back on track by making first successful test flight of command and service module combination in wake of Apollo 1 fire.
Apollo 8	December 21–27, 1968	Frank Borman James A. Lovell, Jr. William A. Anders	Made man's first voyage to the moon. Guided spacecraft into lunar orbit, but did not attempt to land on the surface because they did not have lunar module. Astronauts read excerpts from the Book of Genesis while circling moon on Christmas Eve.

*CDR-mission commander
CMP-command module pilot
LMP-lunar module pilot

PROJECT APOLLO
Flight Log *(Continued)*

MISSION	DATE	CREW (CDR-CMP-LMP)*	OUTCOME
Apollo 9 (CM—*Gumdrop*) (LM—*Spider*)	March 3–13, 1969	James A. McDivitt David R. Scott Russell L. Schweickart	Made manned test of command module/lunar module combination in Earth orbit. Confirmed the feasibility of lunar orbital rendezvous and docking maneuvers crucial to success of first moon landing mission.
Apollo 10 (CM—*Charlie Brown*) (LM—*Snoopy*)	May 18–26, 1969	Thomas P. Stafford John W. Young Eugene A. Cernan	Made man's second round-trip voyage to the moon in "dress rehearsal" of first lunar landing mission. Stafford and Cernan flew lunar module down to 50,000-foot orbit around lunar surface, then reunited with Young in command module and returned to Earth.
Apollo 11 (CM—*Columbia*) (LM—*Eagle*)	July 16–24, 1969	Neil A. Armstrong Michael Collins Edwin E. Aldrin, Jr.	Made man's first moon landing in the Sea of Tranquility. Armstrong became first man to walk on the moon with his "giant leap" off the ladder of the lunar module *Eagle.* Aldrin became second man to walk on the moon, and helped plant American flag and plaque: "We came in peace for all mankind."

*CDR-mission commander
CMP-command module pilot
LMP-lunar module pilot

PROJECT APOLLO
Flight Log *(Continued)*

MISSION	DATE	CREW (CDR-CMP-LMP)*	OUTCOME
Apollo 12 (CM—*Yankee Clipper*) (LM—*Intrepid*)	November 14–24, 1969	Charles Conrad, Jr. Richard F. Gordon, Jr. Alan L. Bean	Made the first pinpoint lunar landing in the Ocean of Storms. Conrad and Bean became third and fourth men to walk on the moon.
Apollo 13 (CM—*Odyssey*) (LM—*Aquarius*)	April 11–17, 1970	James A. Lovell, Jr. John L. Swigert, Jr. Fred W. Haise, Jr.	Suffered the first accident in space when spacecraft's oxygen tanks exploded during translunar voyage. Heroic grace under pressure enabled them to circle the moon and return safely to Earth.
Apollo 14 (CM—*Kitty Hawk*) (LM—*Antares*)	January 31–February 9, 1971	Alan B. Shepard, Jr. Stuart A. Roosa Edgar D. Mitchell	Helped get Apollo program back on track by making man's third successful lunar landing in the Fra Mauro region. Mitchell conducted secret ESP experiments on the way to the moon. Shepard hit golf balls on the lunar surface.

*CDR-mission commander
CMP-command module pilot
LMP-lunar module pilot

PROJECT APOLLO
Flight Log *(Continued)*

MISSION	DATE	CREW (CDR-CMP-LMP)*	OUTCOME
Apollo 15 (CM—*Endeavor*) (LM—*Falcon*)	July 26–August 7, 1971	David R. Scott Alfred Worden James B. Irwin	Made first landing in the lunar mountains at site in the Hadley-Apennine region. Scott and Irwin drove first four-wheeled vehicle (the Moon Rover) across the lunar surface. They also found 4.5-billion-year-old "Genesis rock." Irwin claimed he "felt the presence of the Lord."
Apollo 16 (CM—*Casper*) (LM—*Orion*)	April 16–27, 1972	John W. Young T. Kenneth Mattingly, II Charles M. Duke, Jr.	Landed in the Descartes region to gather evidence of ancient volcanic activity on the moon. Young and Duke drove the second Moon Rover vehicle across the lunar surface.
Apollo 17 (CM—*America*) (LM—*Challenger*)	December 7–19, 1972	Eugene A. Cernan Ronald E. Evans Harrison H. Schmitt	Made sixth successful Apollo lunar landing in Taurus-Littrow region. Schmitt, a geologist, was first scientist to fly in space. Third and final Moon Rover mission. Left plaque: "Here man completed his first explorations of the Moon. May the spirit of peace in which we came be reflected in the lives of all mankind."

*CDR-mission commander
CMP-command module pilot
LMP-lunar module pilot

PROJECT APOLLO
Flight Log *(Continued)*

MISSION	DATE	CREW (CDR-CMP-LMP)*	OUTCOME
Apollo-Soyuz	July 15–24, 1975	Thomas P. Stafford Donald K. Slayton Vance D. Brand Aleksei A. Leonov Valery N. Kubasov	First joint space mission for U.S. and U.S.S.R. Made successful rendezvous and docking with Soyuz spacecraft carrying cosmonauts in Earth orbit.

*CDR-mission commander
CMP-command module pilot
LMP-lunar module pilot

CHAPTER NOTES

The purpose of these notes is to cite primary sources and to give due credit for quotations drawn from secondary sources. In instances where appropriate citations are made in the text of the narrative, they are not repeated below.

Prologue

xi Quote from President John F. Kennedy speech dedicating the Manned Spacecraft Center, made at Rice University in Houston, Texas, September 12, 1962.

xi Quote from Neil A. Armstrong, Apollo 11 mission commentary transcript, July 20, 1969.

xi Estimate of television audience for Armstrong's walk on the moon made by European Broadcasting Union, reported by *The New York Times,* July 22, 1969.

xii "We came in peace . . ." plaque left at Sea of Tranquility by Apollo 11 astronauts, July 21, 1969.

STAGE ONE: GO FOR THE MOON

1. This Time Is for Real

3 All quotations of Apollo 14 astronaut Edgar Mitchell from co-author Al Reinert's interview, West Palm Beach, Florida, June 1981, unless otherwise noted.

4 Quotations of Apollo 15 astronaut James B. Irwin on these pages and all pages below are from co-author Al Reinert's interview, Colorado Springs, Colorado, April 1981, unless otherwise noted.

4 Quotations of Apollo 12 astronaut Alan Bean on these pages and all pages below are from author's interviews in Houston, Texas, April 1987, unless otherwise noted.

6 "It's one of the phases . . ." Apollo 11 press conference, July 5, 1969.

7 "I'm sick of steak . . ." *First on the Moon: A Voyage with Neil Armstrong, Michael Collins, and Edwin E. Aldrin, Jr.,* Gene Farmer and Dora Jane Hamblin (Boston: Little, Brown, 1970), p. 41.

7 "had an air of studied casualness . . ." *Carrying the Fire: An Astronaut's Story,* Michael Collins (New York: Random House, 1974), p. 360.

7 "Dr. Thomas Paine . . . told us that concern . . ." *Return to Earth,* Col. Edwin E. "Buzz" Aldrin, Jr., with Wayne Warga (New York: Random House, 1973), p. 217.

7 Aldrin quotes from *Return to Earth,* pp. 214–18.

7 "I had a hard time . . ." and all other quotations from Apollo 16 astronaut Charles Duke from author's interview, New Braunfels, Texas, June 1987, unless otherwise noted.

12 Quotations of Apollo 14 astronaut Stuart Roosa on this page and all pages below are from co-author Al Reinert's interview, Austin, Texas, September 1980, unless otherwise noted.

12 Quotations of Apollo 12 astronaut Richard Gordon on this page and all pages are from co-author Al Reinert's interview, Houston, Texas, November 1979, unless otherwise noted.

13 Quotations from Apollo 17 astronaut Eugene Cernan on this page and all pages below are from co-author Al Reinert's interview, Houston, Texas, July 1980, unless otherwise noted.

15 Norman Mailer, *Of a Fire on the Moon* (New York: New American Library, 1970), p. 17.

17 "There's frost on the sides . . ." and all other quotations from Apollo 16 astronaut T. Kenneth Mattingly are from co-author Al Reinert's interview, Houston, Texas, October 1978, unless otherwise noted.

2. Man and the Moon

21 All quotations from Jules Verne's *From the Earth to the Moon,* from *Works of Jules Verne,* edited by Charles F. Horne (New York: F. Tyler Daniels Co., 1911).

21 All quotations from *A True History* by Lucian of Samosata taken from edition translated by Paul Turner (Bloomington, Indiana: Indiana University Press, 1958).

24 "the man with lanthorn . . ." *A Midsummer Night's Dream* by William Shakespeare, Act V, Scene 1.

24 "The moone is made of a greene cheese." First cited in *A Pistle to the Christian Reader* by John Firth (1529), according to *The Home Book of*

Proverbs, Maxims, and Familiar Phrases, selected and arranged by Burton Stevenson (New York: Macmillan, 1948).

25 K. E. Tsiolkovsky quotation cited in article by Dr. Wernher von Braun, *The New York Times,* July 17, 1969.

27 "the first real space travel movie . . ." *The Film Enclyclopedia: Science Fiction,* edited by Phil Hardy (New York: William Morrow, 1984), p. 78.

3. It Feels Just Like It Sounds

29 "This ride up the elevator . . ." *Carrying the Fire,* p. 363.

30 "You can face due south . . ." and all other quotations of Apollo 11 astronaut Michael Collins from co-author Al Reinert's interview, Washington, D.C., May 1977, unless otherwise noted.

31 Quotations of Apollo 12 astronaut Charles Conrad on this page and all pages below are from co-author Al Reinert's interview, Long Beach, California, August 1981, unless otherwise noted.

42 Public Affairs Officer (PAO) quotations on this page and all pages below are excerpted from mission commentary transcripts of Apollo 11–17 missions.

47 Quotations of Apollo 17 astronaut Harrison H. Schmitt on this page and all pages below are from co-author Al Reinert's interview, Albuquerque, New Mexico, May 1981, unless otherwise noted.

4. The Space Race

51 Account of T. Keith Glennan and early days of NASA based in part on *Prescription for Disaster* by Joseph Trento (New York: Crown Publishers, 1987).

52 Memo by President John F. Kennedy cited in "Space and Power Politics" by John M. Logsdon, *The New York Times,* July 16, 1969. Account of decision to go to the moon based on Logsdon article and on "Kennedy Listened, Raised Questions, Deliberated, Then Acted," by Theodore C. Sorenson, *The New York Times,* July 16, 1969.

52 "When I heard President Kennedy . . ." from "Experts Were Stunned by Scope of Mission," by Dr. Robert R. Gilruth, *The New York Times,* July 16, 1969.

5. Slipping the Surly Bonds

59 " 'High Flight' is without a doubt the best known . . ." *Carrying the Fire,* p. 99.

60 "All three of us are very quiet . . ." *Ibid.,* p. 369.

64 "The third stage is crisp and rattly . . ." *Ibid.,* p. 370.

67 "The first few minutes in orbit . . ." *Ibid.,* p. 371.

71 "It doesn't make a hell of a lot of sense . . ." *Ibid.,* p. 375.

72 "just a tiny bit rattly," Apollo 11 mission commentary transcript, July 16, 1969.

75 "thrashing around like three great white whales . . ." *Carrying the Fire,* p. 382.

75 "It requires a very precise . . ." *Ibid.,* pp. 384–85.

6. Chariots of Fire

79 Radio dialogue of Virgil I. Grissom from Apollo 1 countdown demonstration transcript, January 27, 1967.

80 "If you have a problem . . ." attributed to Walter Schirra in *The All American Boys,* Walter Cunningham (New York: Macmillan, 1977).

81 "Fire! We've got a fire . . ." Apollo 1 countdown demonstration transcript, January 27, 1967.

82 "The country hadn't recruited us . . ." *The All American Boys,* p. 265.

82 "That may sound like . . ." *Ibid.,* p. 4.

83 "The truth of this is not easily . . ." *Ibid.,* p. 5.

84 Donald K. Slayton quotations from *The All American Boys,* p. 200.

85 "How we were picked . . ." and all below quotations by Apollo 16 astronaut Charles M. Duke, Jr., are from author/co-author interviews in New Braunfels, Texas, June 1987.

85 "There isn't any magic . . ." *First on the Moon,* p. 104.

89 "There is something extremely unsettling . . ." *Carrying the Fire,* p. 123.

90 "Running on the beach is grand . . ." *Ibid.,* p. 150.

91 "We got high . . ." *The All American Boys,* p. 70.

91 Account of astronauts' sex lives and quotations by Apollo 13 astronaut John L. Swigert and Apollo 7 astronaut Walter Cunningham from *The All American Boys,* pp. 184–98.

93 "The monkey I had bought for Joan . . ." *Return to Earth,* p. 174.

94 "ten thousand dollars invested in worthless stock . . ." *The All American Boys,* p. 173.

95 Account of problems encountered by Apollo 4, 5, 6, and 7 based in part on *Of a Fire on the Moon,* pp. 187–88.

96 "spectacular but superfluous . . ." *The All American Boys,* p. 200.

97 "Please be informed . . ." and all other radio dialogue in this chapter by Apollo 8 astronauts from Apollo 8 mission commentary transcript, December 24, 1968.

97 "I think the Russians would have . . ." *Prescription for Disaster,* p. 82.

97 "You're it." *Return to Earth,* p. 201.

98 "The LM hadn't even flown men yet . . ." *Carrying the Fire,* p. 317.

99 "not truly a flight into the unknown . . ." Apollo 11 press conference, July 5, 1969.

102 "Neil used to come home . . ." *First on the Moon,* p. 31.

103 "I think we're going to the moon . . ." Apollo 11 press conference, Houston, Texas, July 5, 1969.

7. Life in Zero G

104 Aldrin quotes in this chapter from *Return to Earth,* pp. 222–24.

109 Collins quote, *Carrying the Fire,* p. 384.

115 Collins quote, *Ibid.,* p. 386.

8. Hello, Moon

126 Collins quote, *Carrying the Fire,* p. 392.

128 Aldrin quotes, *Return to Earth,* pp. 226–27.

128 Collins quotes, *Carrying the Fire,* pp. 398–99.

134 Aldrin quote, *Return to Earth,* p. 227.

138 Collins quote, *Carrying the Fire,* p. 400.

STAGE TWO: MEN WALK ON MOON

9. A Giant Leap

147 "Apollo 11, Apollo 11 . . . Good morning from the Black Team . . ." and all other radio dialogue in this chapter from Apollo 11 mission commentary transcript, July 20–21, 1969.

148 "By now we had worked out . . ." *Return to Earth,* p. 227.

149 Gilruth quote from *Prescription for Disaster,* p. 64.

150 "I reserve my God-given right . . ." *Return to Earth,* p. 203.

150 "I could by holding my breath . . ." *First on the Moon,* p. 19.

151 "I remembered one science fiction story . . ." *Return to Earth,* p. 306.

152 "We kept this secret . . ." *Ibid.,* p. 200.

152 "Neil equivocated . . ." *Ibid.,* p. 206.

153 "In the eleven weeks . . ." *Carrying the Fire,* p. 7.

154 "The most dangerous items . . ." Apollo 11 press conference, July 5, 1969.

155 "I got as far as my liquid . . ." *Return to Earth,* p. 228.

156 "We're supposed to know how . . ." *First on the Moon,* p. 234.

160 "The computer isn't going to dodge boulders . . ." Apollo 11 press conference, July 5, 1969.

164 "It was not a serious program alarm . . ." Apollo 11 press conference, August 12, 1969.

164 "In the simulator we have . . ." *Ibid.*

165 "I was surprised by the size . . ." *First on the Moon,* p. 241.

166 "I was tempted to land . . ." *Ibid.,* p. 242.

166 "I was absolutely adamant . . ." *Ibid.,* p. 242.

167 "I changed my mind a couple of times . . ." *Ibid.,* p. 242.

167 "In this zone . . ." *Return to Earth,* pp. 230–31.

167 "I felt no apprehension . . ." *Ibid.,* p. 231.

168 "The landing was so smooth . . ." *Ibid.,* p. 231.

168 "Boy, Charley, I thought . . ." *First on the Moon,* p. 243.

169 "We gave in to our excitement . . ." *Return to Earth,* p. 231.

169 "If there was an emotional high point . . ." *The New York Times,* July 21, 1979.

172 "I would like to have observed . . ." Apollo 11 press conference, August 12, 1969.

173 "The sky is black . . ." *First on the Moon,* p. 256.

174 "That's part of the exercise . . ." Apollo 11 press conference, August 12, 1969.

175 "There was, in fact, widespread discussion . . ." *Return to Earth,* p. 234.

175 "I had thought about what I was going to say . . ." Apollo 11 press conference, August 12, 1969.

176 "I thought it [the article 'a'] had been included . . ." *Chariots for Apollo: A History of Manned Lunar Spacecraft,* Courtney G. Brooks, James H. Grimwood, and Loyd S. Swenson, Jr. (Washington, D.C.: NASA, 1979), p.346.

179 Aldrin quotes from *Return to Earth* pp. 234–47.

182 "There's one sight I'll never forget . . ." *The New York Times,* July 21, 1979.

182 Aldrin quotes, *Return to Earth,* p. 235.

183 "there was a distinct smell . . ." *First on the Moon,* p. 304.

184 "There is a problem . . ." *Ibid.,* p. 297.

185 "The thing which really kept us awake . . ." *Ibid.,* p. 304.

185 "It's a beginning . . ." Apollo 11 press conference, August 12, 1969.

10. Snoopy and the Surveyor

189 "If you were going to make a movie . . ." *The All American Boys,* p. 221.

189 "a raw sex appeal . . ." *Ibid.,* p. 222.

190 "When you think you have the door . . ." *Ibid.,* p. 222.

192 All radio dialogue in this chapter from Apollo 12 mission commentary transcripts.

11. Unlucky Thirteen

201 "It may turn out . . ." and other quotations of Walter Sullivan in this chapter from *The New York Times,* April 12, 1970.

205 All radio dialogue in this chapter from Apollo 13 mission commentary transcripts.

207 "When we arrived . . ." *The All American Boys,* p. 229.

209 "We know human suffering . . ." *The New York Times,* April 14, 1970.

12. The Moon Shot Never to Be Forgot

218 "If you've seen one . . ." *The New York Times,* January 31, 1971.

218 "We feel they've gotten there . . ." *Ibid.*

224 "It's a beautiful day . . ." and all other radio dialogue in this chapter from Apollo 14 mission commentary transcripts.

225 "For the first time, men have made . . ." *The New York Times,* February 1, 1971.

13. The Moon Rovers

232 "As I stand out here . . ." and all other radio dialogue in this chapter from Apollo 15 mission commentary transcripts.

235 "I had to go through a lot of psychiatric help . . ." *The New York Times,* July 27, 1971.

235 "I would be volunteering . . ." *Ibid.*

243 "Charley, you said you were going to see . . ." and all other radio dialogue in this chapter from Apollo 16 mission commentary transcripts.

244 Sullivan quote from *The New York Times,* April 24, 1972.

248 "Even as the world waits . . ." *The New York Times,* December 7, 1972.

249 Schmitt quotes, *Ibid.*

250 "The *Challenger* has landed!" and all other radio dialogue through the end of this chapter from Apollo 17 mission commentary transcripts.

STAGE THREE: RETURN TO EARTH

14. The Secret Terror of Saying Good-bye

256 "We were not distracted . . ." Apollo 11 press conference, August 12, 1969.

256 "I have been flying . . ." *Carrying the Fire,* p. 420.

258 "There was no time to sightsee . . ." *Return to Earth,* p. 239.

261 "Instead of a docile little LM . . ." *Carrying the Fire,* p. 423.

263 "the get-us-home burn . . ." *Ibid.,* p. 245.

266 "Hot diggety dog!" Apollo 17 mission commentary transcripts.

266 "The right side of the equipment bay . . ." *Carrying the Fire,* p. 444.

15. Instant Celebrities

268 "I just want to remind you . . ." Apollo 11 mission commentary transcripts.

268 "Keep the mice healthy . . ." *Ibid.*

271 "the part . . ." Apollo 11 press conference, August 12, 1969.

271 "It would take a couple of years . . ." *Return to Earth,* p. 12.

272 Soviet and other foreign and domestic reactions to Apollo 11 mission excerpted from *The New York Times,* July 24–29, 1969.

273 "The moon rock samples . . ." letter to *The New York Times,* September 26, 1969.

275 Borman conversation reported in *Return to Earth,* p. 55.

277 "My prevailing memory . . ." *Ibid.,* p. 74.

277 "Neil began to look doubtful . . ." *Ibid.,* p. 22.

279 "The lunar sites . . ." *Prescription for Disaster,* p. 90.

284 "Only my wife and momma . . ." *The New York Times,* June 15, 1979.

16. The Melancholy of All Things Done

285 "Being an astronaut . . ." *Carrying the Fire,* p. 463.

289 "The Apollo pump . . ." *The New York Times,* March 28, 1976.

290 "The rule of my emotions . . ." *Return to Earth,* p. 288.

291 "I had gone to the moon . . ."*Return to Earth,* p. 300.

294 "They laid their hands . . ." *Texas Monthly,* February 1988.

17. The $40 Billion Bargain

299 "By the 1980s, every home could have . . ." *First on the Moon,* p. 390.

STAGE FOUR: 2001 AND BEYOND

18. Star Wars or Peace?

320 "I'm not through looking . . ." author's interview with Apollo 16 astronaut John Young, Houston, Texas, July 1987.

321 "You have to do it right . . ." *The Houston Chronicle,* March 27, 1988.

Epilogue

323 "We shall not cease . . ." *Four Quartets,* T. S. Eliot, 1943.

323 "The purpose of this film . . ." *The New York Times,* December 18, 1969.

324 "I would think that the most important event . . ." *Rolling Stone,* November 5, 1987.

326 "I saw that the earth is very small . . ." author's interview with cosmonaut V. A. Solovyov, Mexico City, October 1987.

326 "The earth looked so blue and round . . ." *The Home Planet,* Kevin W. Kelly for the Association of Space Explorers (Boston: Addison-Wesley, 1988).

326 "Before I flew I was already aware . . ." *Ibid.*

326 "Only a child in its innocence . . ." *Ibid.*

327 "I would have wished that . . ." *Ibid.*

ACKNOWLEDGMENTS

This book is the literary sibling of a full-length feature film of the same name produced and directed by Al Reinert, who conducted the principal interviews with the Apollo astronauts. The book is not the shooting script for the film, nor is the film a celluloid condensation of the book. Just as each represents a distinct art form, each is a self-contained work in its own right, but the two may be best enjoyed in tandem as complementary answers to the same question: what was it like for man to leave the earth and land on another planet?

The author owes an immeasurable debt to co-author Reinert for his inspiration and general editorial guidance, as well as for his fine interviews and research over the course of eight years. In addition to being a working colleague, Al has been and is the greatest friend any person could have.

The author would like to give special thanks to Betsy Broyles Breier, who worked cheerfully and tirelessly on behalf of both the book and the film, and to Marc E. Grossberg, who did likewise. Other members of the Broyles family who deserve special thanks are Mr. and Mrs. William D. Broyles, Sr., and William D. Broyles, Jr., who made some invaluable editorial suggestions.

This book could not have been completed without the contributions at various stages of five dedicated researchers: Andrew Crispin, Harrell Hays Perkins, David Haley, Travis Brandenberg, and Alex Dunne. Many valuable suggestions were made by five diligent manuscript readers: Peter Graves, Lucy Butler, Michael Zilkha, Christina Zilkha, and Bill Sadler, who also provided much needed sustenance along the way.

In addition to thanking all the Apollo astronauts interviewed for the book and film, the author would like to acknowledge the particularly patient and generous help of Charley Duke and Al Bean, who granted several follow-up interviews in person and over the telephone. Among those at NASA who were especially helpful were Terry White and Mike Gentry.

The author is similarly grateful for the support of several colleagues at *Newsweek* magazine: Dominique Browning, Robert Rivard, Michael Reese, Linda Buckley, Jennifer Foote, Janet Huck, Michael Lerner, Lester Sloan, and Pat Sotelo.

Other friends and associates who played important supporting roles include: Jeannetta Arnette, Victor Ayad, Peter Bean, Rosalie Barker, Franci Beck, John and Mary Bransford, John Broders, John Burke, Lisa Hart Carroll, Joseph Carruth, Harry Chittick, Sara Colleton, Gregory Curtis, Zan Eisley, Gary Fisketjon, Geneva Glover, James P. S. Griffith, Jr., Cornelia Guest, Larry and Maj Hagman, Charles E. Hardy, J. R. Hays, William and Dana Hurt, Christopher and Nima Isham, Mary Jameson, Angela Janklow, William H. Lane, Jr., Nicholas Lemann, Mark Lindquist, Kathleen Marcus, Peter McAlevey, Terry and Joan McDonnell, Jay McInerney, Jack McKeown, Don Messinger, Bill Morris, Anton Mueller, Valerie Nickerson, Tina Nides, Maria O'Brien, Maria Oharenko, Anne Dale Owen, J. D. Page, Thomas H. Pierce, Bart Richardson, John F. Ryan, Jim Signorelli, August Spier, Dean Smith, Dorrance Smith, Philip Susman, Richard Wells, Firooz Zahedi, the residents of the Regency House in Houston, Texas, and last but by no means least, the author's parents, Harry and Magee Hurt.

The author also counts himself fortunate to have had the advice and counsel of literary agent Amanda Urban, who transformed the original book proposal into a contract.

Finally, the author is blessed to have had the finest editor and publisher in the business. Morgan Entrekin provided the kind of financial and moral support most authors only dream about. He was a true and faithful friend at every step of the way. This book simply could not have been written without him.